Fundamental QSARs for Metal Ions

Fundamental QSARs for Metal Ions

John D. Walker
Michael C. Newman
Monica Enache

CRC Press
Taylor & Francis Group
Boca Raton London New York

CRC Press is an imprint of the
Taylor & Francis Group, an **informa** business

CRC Press
Taylor & Francis Group
6000 Broken Sound Parkway NW, Suite 300
Boca Raton, FL 33487-2742

First issued in paperback 2019

© 2013 by Taylor & Francis Group, LLC
CRC Press is an imprint of Taylor & Francis Group, an Informa business

No claim to original U.S. Government works

ISBN-13: 978-1-4200-8433-7 (hbk)
ISBN-13: 978-0-367-38052-6 (pbk)

Typeset by Exeter Premedia Services Pvt Ltd., Chennai, India

Visit the Taylor & Francis Web site at
http://www.taylorandfrancis.com

and the CRC Press Web site at
http://www.crcpress.com

Contents

Contents

Preface

The idea for writing *Fundamental QSARs for Metal Ions* was inspired by: (1) reading many of the publications from Michael Newman's laboratory on quantitative ion character-activity relationships (QICARs), and (2) co-authoring a 2003 review with Monica Enache on quantitative cation-activity relationships (QCARs) for metal ions. We use the term *quantitative structure-activity relationship* (QSAR) in the title because it's the more universally recognized acronym. However, the terms QCARs, QSARs, and QICARs are used interchangeably throughout the book, depending upon whose publication we're discussing.

When Monica, Michael, and I discussed writing the book, we knew it would be a challenging task to write the first book on QSARs for metal ions. At Monica's suggestion, we solicited the contributions of one chapter coauthor and two chapter authors who provided chemical properties of metal ions and descriptors to predict metal–ligand binding. Monica coauthored Chapter 2 with Maria Pele[*] and translated parts of Chapters 3 and 4 from Romanian to English. Chapter 3 was written by Valentina Uivarosi[†] and Chapter 4 was written by Laszlo Tarko.[‡] Michael wrote Chapters 1 and 8 and provided extensive constructive comments on Chapters 2–7. I wrote Chapters 5, 6, and 7 and reviewed Chapters 1–4 and 8.

Fundamental QSARs for Metal Ions was designed to provide guidance and information so that the regulatory and regulated communities could develop QSARs for metal ions as they do now for organic chemicals. Chapter 1 provides a historical perspective and introduction to developing QSARs for metal ions. Chapter 2 explains the electronic structures and atomic parameters of metals essential to understanding differences in toxicity. Chapter 3 describes the chemical properties of metals that have been and can be used to develop QSARs for metal ions. Chapter 4 illustrates the descriptors needed to develop metal ion–ligand binding QSARs. Chapter 5 discusses the 97 QSARs for metal ions developed from 1972 to 2003 and the 183 QSARs for metal ions that were developed from 2004 to 2011, since our 2003 QCARs review. Chapter 6 explains the differences between QSARs for metal ions and biotic ligand models. Chapter 7 lists the regulatory limits of metals and provides examples of regulatory applications. Chapter 8 illustrates how to construct QSARs for metal ions.

---

[*] Maria Pele, Ph.D., Professor, The University of Agricultural Sciences and Veterinary Medicine of Bucharest, Bd. Marasti 59, Sector 1, Bucharest, Romania.

[†] Valentina Uivarosi, Ph.D., Faculty of Pharmacy, Department of Inorganic Chemistry, University of Medicine and Pharmacy "Carol Davila," Bucharest, Romania.

[‡] Laszlo Tarko, Ph.D., Romanian Academy Center of Organic Chemistry "Costin D. Nenitzescu," Sector 6, Splaiul Independentei 202 B, Bucharest, Romania.

About the Authors

John D. Walker has 44 years of professional academic, industrial and government experience in environmental toxicology and chemistry. Since 1989 he has been Director of the Toxic Substances Control Act Interagency Testing Committee (ITC), The ITC is an independent advisory committee to the U.S. Environmental Protection Agency (U.S. EPA) Administrator with representatives from 14 U.S. Government organizations. The ITC's statutory responsibility is to identify industrial chemicals that are likely to harm humans or the environment and recommend them for testing to develop data needed to assess their risks. Prior to being selected as the ITC's Director John was a senior scientist with the U.S. EPA, a regulatory toxicologist with the U.S. Food and Drug Administration, an environmental scientist with Lockheed Martin and a Research Associate at the University of Maryland.

John received his bachelor's degree in chemistry and biology magna cum laude from Kent State University where he was elected to Phi Beta Kappa. He studied microbiology at the University of Dayton where he was elected to Sigma Xi and studied aquatic toxicology at the Ohio State University's Franz T. Stone Laboratory while earning his Ph.D. His M.P.H. is from the Johns Hopkins University School of Public Health.

John is a Charter Member of the Society of Environmental Toxicology and Chemistry (SETAC), an Emeritus Member of the American Society for Microbiology and the American Academy of Microbiology and a former Editor of SETAC's International Journal, *Environmental Toxicology and Chemistry*.

John co-authored the *Laboratory Manual for Marine Microbiology* and edited the book, *QSARs for Pollution Prevention, Toxicity Screening, Risk Assessment and Web Applications*. He has authored or co-authored 160 peer-reviewed publications and has written 140 abstracts of presentations for national and international professional society meetings. John was the first recipient of the American Fisheries Society/U.S. EPA Science Achievement Award in Biology/Ecology and has been awarded 5 U.S. EPA Bronze medals. He was awarded the U.S. EPA's Unsung Hero Award for his work with Special Olympics. John is married with 4 children and 2 grandchildren.

Michael C. Newman is currently the A. Marshall Acuff Jr. Professor of Marine Science at the College of William and Mary's Virginia Institute of Marine Science, where he also served as Dean of Graduate Studies for the School of Marine Sciences from 1999 to 2002. Previously, he was a faculty member at the University of Georgia's Savannah River Ecology Laboratory. His research interests include quantitative ecotoxicology, environmental statistics, risk assessment, population effects of contaminants, metal chemistry and effects, and bioaccumulation and biomagnification modeling. In addition to more than 125 articles, he has authored 5 books and edited another 6 on these topics. The English edition and Mandarin and Turkish translations of *Fundamentals of Ecotoxicology* have been adopted widely as the textbook for introductory ecotoxicology courses. He has taught at universities throughout the world, including the College of William and Mary, University of California–San Diego, University of Georgia, University of South Carolina, Jagiellonian University (Poland), University of Antwerp (Belgium), University of Hong Kong, University of Joensuu (Finland), University of Koblenz–Landau (Germany), University of Technology–Sydney (Australia), Royal Holloway University of London (UK), Central China Normal University, and Xiamen University (China). He has served numerous international, national, and regional organizations, including the OECD, U.S. EPA Science Advisory Board, Hong Kong Areas of Excellence Committee, and the U.S. National Academy of Science NRC. In 2004, the Society of Environmental Toxicology and Chemistry awarded him its Founder's Award, "the highest SETAC award, given to a person with an outstanding career who has made a clearly identifiable contribution in the environmental sciences."

Monica Enache has 13 years of professional academic experience in Biology. Since 1999 she has been working at the Faculty of Biotechnology of the University of Agricultural Sciences and Veterinary Medicine of Bucharest, Romania. Prior to joining the faculty she carried out her PhD degree in Liverpool John Moores University (UK), and received her Bachelor of Science degree from the Faculty of Biology of the University of Bucharest (Romania). She is currently a member of the National Society of Cell Biology in Romania, and has authored or co-authored 10 publications and 14 abstracts for presentations at national or international scientific meetings.

1 Introduction

1.1 THE CONCEPT OF STRUCTURE–ACTIVITY RELATIONSHIPS (SARS)

SAR resides at the intersection of biology, chemistry, and statistics.

McKinney et al. (2000, p. 9)

McKinney et al. (2000) identified the essential steps in generating SARs. It is critical at the beginning to identify the mechanism underpinning the bioactivity of interest. Once the underpinnings are defined, the relevant toxicants and their relevant qualities can be identified with the intention of using them to understand and predict trends within the toxicant class. Next, a qualitative or quantitative approach that relates toxicant qualities to bioactivity is formulated. Approaches range from simple dichotomous categorizations to complex quantitative models generated with a variety of statistical techniques. Such SARs or quantitative SARs (QSARs) are relevant to the specified toxicant class and bioactivity. Additional SARs might be needed to address other classes or activities.

The QSAR approach for organic compounds is well established in contrast to the nascent approach for inorganic chemicals such as metal ions. As a late nineteenth-century QSAR example, the Meyer-Overton rule related anesthetic potency to its oil–water or oil–air partition coefficient. This theme of relating organic compound bioactivity or accumulation to lipophilicity still dominates much of QSAR literature about nonpolar organic contaminants. There are numerous cases where additional qualities based on other molecular structures or properties of organic compounds are included to develop QSARs for different organic contaminant classes. These qualities are often quantified in metrics of nucleophilicity, electrophilicity, molecular topology, and steric qualities (Newman and Clements 2008). QSARs based on lipophilicity, nucleophilicity, electrophilicity, molecular topology, or steric qualities have been developed for pollution prevention, toxicity screening, risk assessment, and web applications (Walker 2003).

In contrast, qualitative rules such as the d-orbital electron-based Irving-Williams series (Brezonik et al. 1991) are well established for ordering the relative bioactivities of subsets of metals, but quantitative relationships for metal ions have remained inexplicably underdeveloped in toxicology and risk assessment (Newman et al. 1998; Walker and Hickey 2000). Fortunately, this underdevelopment is now recognized as such and is steadily being resolved, as illustrated by the studies described in Chapter 5.

1.2 METALS IN THE MOLECULAR ENVIRONMENT

Metals can be classified based on an array of qualities. Some are more useful than others for quantitatively predicting intermetal differences in bioactivity. Following the lead of numerous authors, most notably Nieboer and Richardson (1980), this treatment will focus on classification schemes that link biological mode of action to coordination chemistry.

There are several candidate metal classification schemes to employ for SAR and QSAR generation (Duffus 2002). The easiest to eliminate at the onset is classification based on whether the metal is an unstable or stable nuclide. This classification is irrelevant because our intent is not prediction of effects arising from different types of ionizing radiations. It is prediction of adverse effects from chemical interaction between metal and organism. Classification based on natural abundances such as *bulk*, *abundant*, or *trace* elements is unhelpful because we wish to make predictions for toxicological effects at unnatural, as well as natural, concentrations. However, there are cases in which natural abundance or natural occurrence information can provide valuable insight, as exemplified by the studies of Fisher (1986) and Walker et al. (2007), respectively. Another general classification of metals is the dichotomous division of metals as either being heavy or light metals. The general cutoff between these two groupings (circa 4 g cm^{-3}) has been applied loosely to highlight the toxicity of many heavier metals. Obviously, a dichotomous schema has minimal utility here, especially for creating QSARs. At a slightly finer scale, Blake (1884) did note more than a century ago a correlation between atomic number and metal toxicity. Conforming to the Irving-Williams series, toxicity to mice increased progressively with atomic numbers from manganese (atomic number 25, density 7.43) to copper (atomic number 29, density 8.96) (Jones and Vaughn 1978), but this increase also corresponded with the progressive addition of d-orbital electrons from $[Ar]3d^54s^2$ to $[Ar]3d^{10}4s^1$. Such a scheme based on density or atomic number does not incorporate important periodicities influencing metal toxicity. A schema framed around the periodic table seems more amenable because metal binding to critical biochemicals can easily be related to the classic periodicities therein. Certainly, trends in the nature and occupation of the outer valence shell can be discussed starting from this classic vantage, e.g., qualities of d- versus s- and p-block elements (Barrett 2002; Walker et al. 2003). However, this approach requires extension to generate related quantitative metrics of binding tendencies. For example, zinc ($[Ar]3d^{10}4s^2$) was less toxic in the above progression (Jones and Vaughn 1978) than might have been anticipated based on atomic number, density, or the number of d-orbital elections alone. With the maturation of coordination chemistry as a predictive science, relevant quantitative metrics have emerged that combine several metal ion properties into directly useful metrics. Continuing the example, the empirical softness index (σ_p) described later conveniently resolves the inconsistency just noted for zinc toxicity. These schemes framed on classic periodicity-related binding tendencies are favored here.

The primary purpose of classifying [metal ions] in (a), or hard, and (b) or soft, is to correlate a large mass of experimental facts. All the criteria used for the classification are thus purely empirical; they simply express the very different chemical behavior of various [metal ions].

Ahrland (1968, p. 118)

TABLE 1.1

Classification of Metal Ions According to Nieboer and Richardson (1980)

Metal Ion Class	Metal Ions
b	Au^+, Ag^+, Cu^+, Tl^+
	Hg^{2+}, Pd^{2+}, Pt^{2+}
	Bi^{3+}, Tl^{3+}
Intermediate or borderline	Cd^{2+}, Co^{2+}, Cr^{2+}, Cu^{2+}, Fe^{2+}, Mn^{2+}, Ni^{2+}, Pb^{2+}, Sn^{2+}, Ti^{2+}, V^{2+}, Zn^{2+}
	Fe^{3+}, Ga^{3+}, In^{3+}
a	Cs^+, K^+, Li^+, Na^+
	Ba^{2+}, Be^{2+}, Ca^{2+}, Mg^{2+}, Sr^{2+}
	Al^{3+}, Gd^{3+}, La^{3+}, Lu^{3+}, Sc^{3+}, Y^{3+}

Note: The actinides and lanthanides are class (a) metals. Although placed in this table as an intermediate metal ion, Pb^{2+} tends toward class (b) more than most intermediate metal divalent ions in that part of this table. Cd^{2+} also is classified as being along the line between class (b) and intermediate metal ions.

Pearson (1963) and Ahrland (1968) developed the hard and soft acids and bases (HSAB) concept that fulfills many of the practical requirements for metal ion SARs and QSARs. Their approach was to quantify differences in metal ion bond stability during complexation with different ligand atoms. The electrophilic metal ion was envisioned as a Lewis acid and the nucleophilic donor atom of the ligand as a Lewis base.* Class (a) and (b) metals were designated hard and soft Lewis acids, respectively. The soft/hard facet of HSAB theory refers to how readily the outer valence shell deforms during interaction between the metal ion and ligand donor atom. This quality of metal ions generally corresponds to nonpolarizable (hard, class [a]) and polarizable[†] (soft, class [b]) during interaction with donor atoms of ligands.

The class (a) and (b) metals are clustered predictably in the periodic table, with intermediate (borderline) metals being found between these clusters. The exact borders for these classes of metals vary in the published literature because the tendencies used to separate the metals are continuous and a discrete classification is partially arbitrary.

The widely applied Nieboer and Richardson (1980) tabulation of these metal ions classes is summarized in Table 1.1. The general trend in bond stability of the class (a) metal ions with various ligand donor atoms is $O > N > S$ and that for class (b) metal ions is $S > N > O$ (Nieboer and Richardson 1980). Borderline metal ions are more complex, having binding tendencies intermediate between class (a) and (b) metals. Interactions between the hard class (a) metal ions and ligands tend to be ionic in

---

* Recollect that a Lewis acid is a species that can accept an electron pair and a Lewis base is one that can donate an electron pair.
† A polar bond is one in which a dipole is formed along the bond axis. Polarizability in this treatment generally corresponds with the readiness of the valance shell to deform during metal-ligand interaction.

nature and those for class (b) tend to be covalent. Those of intermediate metal ions vary in degrees in the covalent nature to their bonds with ligands.

The coordination chemistry-based approach for qualitatively predicting trends in metal ion bioactivities has been applied successfully for several decades. In the early 1960s, Shaw (1961) drew from the field of coordination chemistry to relate metal toxicity to metal–ligand bond stabilities. Using the then-maturing HSAB theory, Jones and Vaughn (1978) related toxicity directly to a continuous metric of metal ion softness, σ_p. Williams and Turner (1981) extended this approach by adding more toxicity data and considering mono-, di- and trivalent metal ions. This general approach continues to be expanded and refined to generate metal ion QSARs.

1.3 METALS IN AND EFFECT ON WHOLE ORGANISMS

Coordination chemistry directly influences metal–biological interactions, although metal essentiality can introduce additional features (Fraústo de Silva and Williams 1993). Relevant interactions include adsorption to biological surfaces, bioaccumulation, and toxicological effect. This chapter broadly describes these biological phenomena and, through examples, relates them to metal coordination chemistry. Such relationships between metal ion characteristics and bioactivity were referred to as *ion character-activity relationships* (ICARs) by Newman and coworkers (e.g., Ownby and Newman 2003). The quantitative rendering of these relationships has been called, alternatively, *quantitative ICARs* (QICARS) by Newman et al. (1998) and *quantitative cationic-activity relationships* (QCARs) by Walker et al. (2003). ICARs will be discussed in the remainder of this chapter, while detailed discussions of QCARs and QICARs are presented in later chapters.

> Many attempts have been made to correlate the physiological action of the elements with their physical or chemical properties, but with only partial success.
>
> **Mathews (1904, p. 290)**

> [T]he degree of toxicity of ions is largely determined by their affinity for their electrical charges, this affinity determining the readiness with which they tend to abandon the ionic state to enter into chemical combination with protoplasmic compounds.
>
> **Erichsen Jones (1940, p. 435)**

> [T]he fungostatic action of metal cations is related to the strength of covalent binding to surface ionogenic groups on the cell ...
>
> **Somers (1961, p. 246)**

> The results of this investigation ... establish a toxicity sequence ... that is of very general significance in aquatic biology and one that is also firmly based on the principles of co-ordination chemistry.
>
> **Shaw (1961, p. 755)**

Quantitative ion character-activity relationships can be developed for a range of effects based on metal–ligand binding theory.

Newman et al. (1998, p. 1423)

Developing and validating Quantitative Cationic Activity Relationships or (Q)CARs to predict the toxicity [of] metals is challenging because of issues associated with metal speciation, complexation and interactions within biological systems and the media used to study these interactions.

Walker et al. (2003, p. 1916)

As reflected in these quotes, the idea that metal ion biological activity is relatable to coordination chemistry is more than a century old. What is new is our emerging capability to quantitatively predict metal–biological activity with coordination chemistry-based metrics. Our understanding of coordination chemistry has advanced substantially, bringing along with it an assortment of convenient metrics for quantifying differences in metal chemistries. Although the pioneering work of Alfred Werner that began the field of coordination chemistry took place a century ago, the HSAB concepts that permeate discussions here and in other chapters came together only in the last half of the twentieth century (e.g., Pearson 1963, 1966). Hard and soft acids and bases theory now has evolved to such an extent that it is applied to develop both organic and inorganic QSARs (Carlson 1990). An array of potential physicochemical metrics has emerged with more refinements made every year. They are actively being assessed for their relative advantages in facilitating quantitative prediction of metal bioactivity (e.g., Kaiser 1980). Reviews by Newman et al. (1998), Ownby and Newman (2003), and Walker et al. (2003) reconfirm the viability of predicting metal activity with metal ion coordination chemistry metrics. Studies such as those of Wolterbeek and Verberg (2001), Kinraide and Yermiyahu (2007), and Kinraide (2009) enhance their potential each year by comparing and refining metrics. Complementing this growth in physicochemical metrics is the increasingly comprehensive and sound effects database available for use in quantitative models. Enough progress had been made as we enter the new millennium that general metal selection approaches for developing these relationships are beginning to emerge (e.g., Wolterbeek and Verburg 2001). It is the explicit goal of this book to synthesize this recent work, and in so doing, facilitate further advancement toward establishing powerful QSARs for metals.

1.3.1 ACCUMULATION IN THE ORGANISM

A metal ion must interact with a biological surface before being taken up and having an effect. Such interactions can be conveniently modeled with the Langmuir model.

$$n = \frac{KCM}{1 + KC} \tag{1.1}$$

where n is the measured amount of metal adsorbed per unit of adsorbent mass, C is the measured equilibrium dissolved metal concentration, M is the estimated adsorption maximum for the adsorbent, and K is the estimated affinity parameter for the adsorbent that reflects the bond strength involved in adsorption (Newman 1995). Numerous examples show that adsorption differences can be related to coordination chemistry metrics. The K values for a series of class (a) metals adsorbing to *Spirogyra* sp. (Crist et al. 1988) were correlated with polarizing power, $Z^2 r^{-1}$, of the metals where Z is the ion charge and r is the ion radius (Figure 1.1 top panel). The $Z^2 r^{-1}$ metric quantifies the energy of the ion during electrostatic interaction with a ligand. The general conclusion from Crist et al.'s (1988) metal–algal sorption study was that the affinity parameter (K) was dictated by charge density if adsorption involved strong metal–ligand bonds, but by energy of hydration if weak metal–ligand bonds were involved. Similarly, maxima for adsorption of a series of class (a), intermediate, and class (b) metals to *Saccharomyces cerevisiae* (Chen and Wang 2007) increased as the $\chi^2 r$ of the metal ion increased (Figure 1.1 bottom panel). (χ is electronegativity and r is the ion radius.) The $\chi^2 r$ metric reflects the relative degree of covalent versus electrostatic nature of bonds during metal interaction with a ligand (Newman et al. 1998). Similar studies include those exploring simple or competitive metal adsorption to other microalgae (Crist et al. 1992; Xue et al. 1988), brown macroalgae (Raize et al. 2004), and bacteria (Can and Jianlong 2007; Zamil et al. 2009).

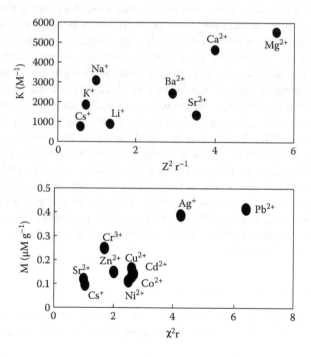

FIGURE 1.1 Trends in adsorption metrics as a function of coordination chemistry for alkali and alkali earth (class [a]) metals and the alga, *Spirogyra* sp. (top panel), and a series of class (a), intermediate, and class (b) metals and the bacterium, *Saccharomyces cerevisiae* (bottom panel). The units of K in the top panel are M^{-1} from an algal suspension with 0.5×10^{-3} to 0.1×10^{-1} M of dissolved metal.

Adsorption models such as that of Crist et al. (1992) can be linked to the biotic ligand model (BLM) with which aquatic toxicologists infer dissolved metal bioactivity (Di Toro et al. 2001). The BLM estimates equilibrium concentrations or activities of dissolved metals species that can compete for and adsorb to ligands on biological surfaces. The dissolved metal bioavailability and potential effect are assumed to result from its interaction with a ligand such as those of a gill sodium or calcium channel protein (Di Toro et al. 2001). For instance, cadmium diffusion through crustacean gill is facilitated by its interaction with a calcium-binding membrane protein (Rainbow and Black 2005). The interaction might simply lead to metal entry into the cell or production of an adverse effect such as ionoregulatory disruption. The relative amounts and binding qualities of calcium and cadmium competing for passage through the membrane via this calcium-binding protein would influence the rate of cadmium introduction into gill cells. QSARs and BLMs are discussed in Chapter 6.

Adsorption involved in other exposure routes, such as gut uptake, is also related to coordination chemistry. Movement across the gut can involve straightforward mechanisms such as those just described or more involved mechanisms. Two class (b) metals provide examples in which crucial complexation processes began in solution before any direct interactions occurred with the cell membrane. Methylmercury forms a stable bond with the sulfur atom of cysteine ($HO_2CCH(NH_2)CH_2SH$) and can enter cells as a complex via an essential amino acid uptake mechanism (Hudson and Shade 2004). Cadmium also strongly binds to compounds with thiol groups such as those of the tripeptide, glutathione (glutamic acid-cysteine-glycine), and cysteine. These cadmium complexes can gain entry to the gut epithelial cells by a specific organic anion transport mechanism (Pigman et al. 1997).

At the other extreme from this thiol-linked transport of class (b) metals, membrane transport of class (a) metals such as potassium or sodium does not involve covalent bonding. Large electronegativity differences between the metal ion and biochemical ligand donor atoms result in bonds of a predominantly ionic nature:* class (a) metals are present primarily as the free aquated ion and bind only weakly to biochemical groups (Fraústo de Silva and Williams 1993). They move into cells by passive diffusion or active pumping. Passive diffusion involves carrier (ionophores) or channel proteins (Ovchinnikov 1979). Cation permeability and ionophore/channel selectivity are determined by the qualities of associated ionic bonding, Van der Waals forces, and hydrogen bonds. Features such as charge density (that is, cation radius and charge) and related hydration sphere characteristics influence ionic interactions. As an example, the electrostatic interactions in the K^+ channel cavity strongly influence selectivity, greatly favoring K^+ over the smaller Na^+ (Bichet et al. 2006). Cation interactions during Na^+/K^+ ATPase pumping also involve noncovalent interactions with RCO_2^- groups that cause conformational shifts in the protein helices to facilitate transmembrane movement (Fraústo de Silva and Williams 1993).

---

* Fajan's rules dictate that the partial covalent nature of a metal's ionic bond with an anion will increase with high charge of either of the interacting ions, small cation radius, or large anion radius. The covalent nature of the bond also is higher with a noninert gas electron configuration of the cation, e.g., the more covalent nature of Cu^{2+} ($\{Ar\}3d^9$) bonds, in contrast to the ionic bonds formed by Na^{+1} ($\{Ne\}$) (Barrett 2002).

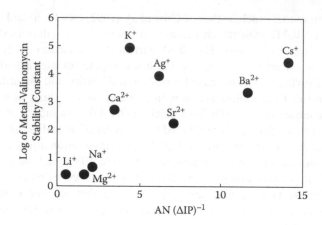

FIGURE 1.2 Ovchinnikov's (1979) estimated stability constants for valinomycin complexes (in methanol solvent) with a series of class (a) metals and one class (b) metal (Ag$^+$) were plotted against the AN(ΔIP)$^{-1}$ metric. The AN(ΔIP)$^{-1}$ metric combines the influences of ion size or inertia and atomic ionization potential (Newman et al. 1998). As is clear in this figure, valinomycin is very selective for K$^+$ and bonding with the one class (b) metal ion is more stable than might have been expected from the class (a) metal ion pattern.

Trends in metal–ligand stability constants for the ionophore, valinomycin, (Ovchinnikov 1979) illustrate several points about class (a) cation movement across membranes (Figure 1.2). The AN(ΔIP)$^{-1}$ metric (Kaiser 1980) in Figure 1.2 combines the atomic number (AN) and the difference in ionization potential (ΔIP) between the ion oxidation number OX and OX-1. Clearly, coordination chemistry trends influence ionophore binding. Another point illustrated in this figure is that extreme specificity might be designed into transport processes that fill an essential biological role: the K$^+$ stability constant is much higher than expected from the general trend. Favored transport can be designed into such structures through a combination of features such as ion size and charge, involved ligand donor atoms, and the configuration of ligand groups (Fraústo de Silva and Williams 1993). As an illustration of the exacting design associated with selectivity, a mutation changing only two amino acid clusters in the K$^+$ channel materially decreases selectivity (Heinemann et al. 1992). Such features are layered onto the general trends predictable from general metal coordination chemistry.

Once a metal enters the cell, a wide range of processes occur that, again, are best understood based on general binding tendencies. By design, the metallothionein proteins and phytochelatin oligopeptides bind intermediate and class (b), but not class (a), metals. Because metallothioneins are synthesized and distributed unevenly among tissues, there will be an associated high accumulation of some metals such as cadmium in organs like the mammalian kidney. This binding to metal-cysteinyl thiolate clusters is crucial to essential element (copper and zinc) and nonessential element (cadmium, mercury, and silver) regulation, sequestration, and elimination in all phyla. More generally, methylmercury forms stable covalent S–Hg bonds with diverse proteins in various tissues (Harris et al. 2003) and will become associated with tissues to a degree related to its delivery rates to potential binding sites and the number of binding sites in the tissue. As another general example, differences in metal binding tendencies produced differential metal accumulation in the various tissues of crayfish (Lyon et al. 1984).

Biomineralization processes play a role in bioaccumulation and are strongly influenced by physicochemical features of metals. Jeffree (1988) found a strong correlation between alkaline earth metal accumulation with age in freshwater mussels (*Velesunio angasi*) and the corresponding metal hydrogen phosphate solubility. Cellular compartmentalization in *Helix aspersa* resulted in intermediate metals (cobalt, iron, manganese, and zinc) being incorporated into sparingly soluble pyrophosphate granules, and class (b) metals (cadmium, lead, and mercury) become associated with sulfur donor atoms of cytosolic proteins (Hopkin and Nott 1979; Simkiss 1981a). Some zinc was also associated with the cytoplasmic proteins. Simkiss (1981b) indicated that another intermediate metal (cobalt) could be found in the calcium/magnesium pyrophosphate granules that also contained protein-rich layers. Some class (a) (barium, strontium) and (b) (silver) metals were also present in the distinct layers of these granules.

Coordination chemistry influences bioavailability, internal transformations, and distribution of metals among internal pools. Fisher (1986) related elemental bioaccumulation in unicellular marine phytoplankton to binding tendencies. He used the logarithm of the corresponding metal hydroxide's solubility product to reflect metal affinity for intermediate ligands such as those with oxygen donor atoms. The Log K_{sp}-MOH was correlated with metal bioaccumulation in phytoplankton, although as illustrated in Figure 1.3, any of a series of other such metrics might have been applied as effectively. In a second excellent paper (Reinfelder and Fisher 1991), the distribution of metals in diatom cells and consequent bioavailability to zooplankton grazers was found to reflect fundamental binding chemistries. The fraction of cellular metal associated with soluble cytoplasmic proteins was higher for class (b) and intermediate metals than for class (a) metals. Moreover, bioavailability to copepods of class (b) and intermediate metals was higher than that for class (a) metals. Similar phenomena occur in metazoans as can be illustrated with the work of Howard and Simkiss (1981), who quantified class (a), intermediate, and class (b) metal binding to hemolymph proteins of the snail, *Helix aspersa*. Based on our previous discussions, a clear trend can be produced to explain intermetal differences in hemolymph protein binding (Figure 1.4). Similar to trends in Figure 1.3, the softer metals (those with outer valence orbitals that readily deform during bonding with ligands) are bound more stably to the plasma proteins than are the harder metals.

As suggested by the above exploration of Howard and Simkiss's data, a metal might be eliminated from the organism at a rate predictable from its binding chemistry. Lyon et al. (1984) explored this theme by introducing a series of class (a), intermediate, and class (b) metals into the open circulatory system of crayfish (*Austropotamobius pallipes*) and measuring elimination from the hemolymph (Figure 1.5). They fit the observed decreases in hemolymph metal concentrations to a first-order elimination model with two mathematical compartments. The percentage of the introduced metal remaining in the hemolymph (C_m) fit the straightforward model

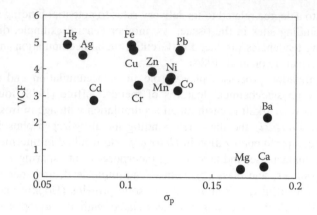

FIGURE 1.3 Phytoplankton volume concentration factor (VCF) (Fisher 1986) is correlated with metal softness index, σ_p. The units for VCF are (Moles of metal $\times \mu m^{-3}$ of cell)/(Moles of dissolved metal $\times \mu m^{-3}$ of ambient water). The softness index reflects the ion's tendency to share electrons during interaction with ligand donor atoms. The three class (a) metals shown are clustered on the bottom right, indicating low bioaccumulation factors compared to the intermediate and class (b) metals in the upper left of the figure. This illustration displays only those metals for which softness parameter values were available in Table 2 of McCloskey et al. (1996). (Data from McCloskey, J.T., M.C. Newman and S.B. Clark. 1996. Predicting the relative toxicity of metal ions using ion characteristics: Microtox® bioluminescence assay. *Environ. Toxicol. Chem.* 15:1730–1737.)

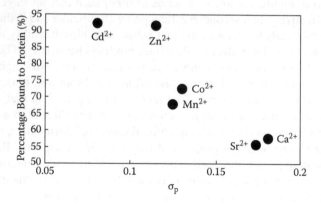

FIGURE 1.4 The percentage of injected metal bound to gastropod hemolymph protein as related to the metal ion softness.

$$C_m = Ae^{-k_{Fast}t} + Be^{-k_{Slow}t} \qquad (1.2)$$

where t is the time elapsed since introduction into the hemolymph, A and B are estimated constants, and k_{Fast} and k_{Slow} are estimated rate constants for elimination from the fast and slow compartments. A general trend emerges if the half-life of a metal in the slow compartment ($T_{1/2} = \ln 2\ (k_{Slow})^{-1}$ is plotted against $\chi^2 r$. Half-life increased

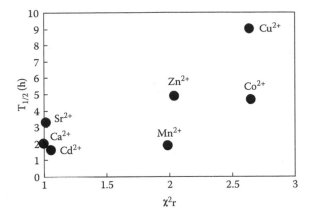

FIGURE 1.5 Elimination of metal ions from crayfish hemolymph is related to metal ion softness. Strong covalent bonding slows elimination. In this illustration, the iron datum was omitted because it was derived from the citrate salt, whereas the other metals were prepared from chloride salts. Based on trends shown here and in Figure 1.3, one could incorrectly assume that metal softness might always be the best descriptor for predicting trends. However, as Ahrland explains, "soft and polarizable are not synonymous; a soft acceptor is certainly always polarizable, but a highly polarizable acceptor need not necessarily be soft, i.e., have (b) properties. For metal ion acceptors, the outer d-electrons are as essential as the polarizability" (Ahrland, S. 1968. Thermodynamics of complex formation between hard and soft acceptors and donors. *Struct. Bond.* 5:118–149).

as the covalent nature of the metal bond with biochemical ligands increased: strong covalent bonding slowed elimination.

1.3.2 BIOMOLECULE-TO-ORGANISM MANIFESTATIONS OF METAL TOXICITY

It should be no surprise to the reader at this point that metal binding differences have also been used to explain intermetal differences in toxicity. The simplest of such effects, in vitro inhibition of enzyme catalysis, can be related to metal affinity to intermediate ligands such as those with oxygen donor atoms (Newman et al. 1998) (Figure 1.6). Examining several enzyme inhibition data sets, Newman et al. (1998) suggested that they could best be modeled with the absolute value of the logarithm of the first hydrolysis constant (i.e., K_{OH} for $M^{n+} + H_2O \rightarrow MOH^{n-1} + H^+$), although clear trends also emerge if plotted against the softness index (σ_p). The $|Log\ K_{OH}|$ reflected metal ion binding affinity for intermediate ligands.

Expanding outward on the biological hierarchy scale from biomolecules to cells, additional mechanisms emerge that could produce differences in metal ion toxicity. Mechanisms include intermetal differences in transport, disruption of ion regulation, binding to and altering protein or nucleic acid functioning, and oxyradical generation. Newman et al. (1998) found coordination chemistry–based trends in metal lethality to cultured cells from fish (Babich et al. 1986; Babich and Borenfreund 1991; Magwood and George 1996) and hamster (Hsie et al. 1984), bacterial bioluminescence suppression (McCloskey et al. 1996; Newman and

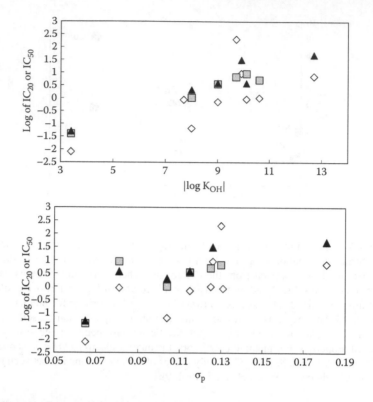

FIGURE 1.6 In vitro inhibition of enzyme activity by metal ions is correlated with the absolute value of the logarithm of the first hydrolysis constant (i.e., K_{OH} for $M^{n+} + H_2O \rightarrow MOH^{n-1} + H^+$) and the softness index (σ_p). Inhibition data for these three enzymes were produced by Christensen (1971/1972) and Christensen and Tucker (1976). White diamonds = catfish carbonic anhydrase IC_{50}, grey squares = white sucker lactic dehydrase IC_{20} and black triangles = white sucker glutamic oxaloacetic transaminase IC_{20}.

McCloskey 1996), and fungal germination (Somers 1961) (Figure 1.7 top panel). Weltje (2002) found similar trends for bacterial bioluminescence inhibition by a series of lanthanides.

Some central themes can be underscored by comparing the McCloskey et al. (1996) and Weltje (2002) data. Modeling trends for bacterial bioluminescence inactivation by divalent metal ions, Newman and McCloskey (1996) found that the $|\log K_{OH}|$ metric produced the best fitting model of a series of candidate models. This suggested that intermetal differences in inactivation were related to differences in affinity for intermediate ligands. This research was expanded (McCloskey et al. 1996) by exploring trends for twenty mono-, di-, and trivalent cations that included class (a), intermediate, and class (b) metals. Although an adequate model was produced by using σ_p alone, the best model for describing the intermetal inactivation trends incorporated both $\chi^2 r$ and $|\log K_{OH}|$. Together, the degree of covalency in metal–ligand bonds and metal affinity for intermediate ligands influenced inactivation.

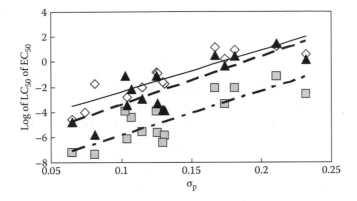

FIGURE 1.7 Lethal and sublethal effects of metal ions on *Daphnia magna* are predictable with the softness index (σ_p). White diamonds = reproductive impairment with chronic exposure (Log EC_{50} after 3 weeks), grey squares = mortality with chronic exposure (Log LC_{50} after 3 weeks), and black triangles = mortality with acute exposure (Log LC_{50} after 48 hours). Effects data were taken from Khangarot and Ray (1989) and Biesinger and Christensen (1972). The regression r^2 values for the linear models fit through the acute and chronic lethality, and chronic reproductive impairment data were 0.83, 0.84, and 0.78, respectively.

Complementing this work, Weltje (2002) took the same approach to study the trivalent, class (a) lanthanide ions. This study was especially interesting for two reasons. The rapidly growing, worldwide use of the rare earth elements (Haxel et al. 2002) makes understanding and predicting lanthanide ion toxicities important (Zhang et al. 1999). But more germane here, how electrons are added to the 4*f* shell of this series of elements permits clear demonstration of how electrostatic interactions influence class (a) metal ion toxicity. The electron occupancy of the seven 4*f* orbitals increases from 0 for lanthanum to 14 for lutetium (see Table 1 of Moeller [1975] for exact configurations). Because the 4*f* electrons are not effective at shielding the nuclear charge, a progressive decrease of ionic radii manifests from lanthanum to lutetium, i.e., the lanthanide contraction (Barrett 2002). Consequently, the charge density (reflected by $Z^2 r^{-1}$) for these trivalent ions increases steadily from lanthanum to lutetium. Weltje (2002) demonstrated a corresponding increase in toxicity: bioluminescence inhibition increased as the energy of the metal ion increased relative to electrostatic interactions with biochemical ligands.

Expanding the biological scale of organization still further outward to the whole metazoan level, it is easy to find published relationships between metal ion qualities and adverse effects. Newman et al. (1998) tabulated a series of such studies. Included effects were acute and chronic lethality, and developmental and reproductive effects. Test species ranged widely from *Planaria* to mice. Walker et al. (2003) also examined the literature addressing this theme, clarifying several important themes. And more such relationships are published each year. As examples, Lewis et al. (1999) produced QSAR models for mice and rats exposed to more than twenty metal ions and Van Kolck et al. (2008) produced models for bioconcentration and lethality for two mollusks.

Barium toxicity by disruption of normal K^+ and Na^+/K^+ channel functioning (Das et al. 1988; Delfino et al. 1988; Taglialatela et al. 1993; Bradberry and Vale 1995) is a particularly informative example of the role played by binding tendencies on effects in metazoans. In this case, it is the class (a) metal ion binding characteristics that emerge as important. Tatara et al. (1998) determined LC_{50} values for nematodes (*Caenorhabditis elegans*) exposed to a series of mono-, di-, and trivalent metal ions including Ba^{2+}. They produced a QSAR with $\chi^2 r$ and |Log of K_{OH}|, but found that Ba^{2+} was much more toxic than predicted from the general QSAR model.* Tatara et al. explained this difference using the relative charge densities of K^+ and Ba^{2+}. The atomic radii of K^+ and Ba^{2+} are 1.38 and 1.36 Å, respectively, but these very similar radii are associated with ions of different charge. The result is very different ion charge densities as reflected in the metric, $Z^2 r^{-1}$. The bonding to the K^+ channel is much more stable for Ba^{2+} than K^+, resulting in a blocking of the essential passage of K^+ through membrane ionophores of excitable tissues. Nervous and muscle tissue could not function properly. This explained the atypical toxicity of Ba^{2+} to the nematode.

In summary, differences in metal coordination chemistries produce differences in the bioaccumulation and effects of single metals. Associated trends can be predicted with basic metrics of metal–ligand interactions.

1.3.3 Metal Interactions in Mixtures

Quantitative means of coping with joint action of metals in mixture have lagged behind those used to quantify effects of single metals. This has produced a body of mixture publications that are more descriptive or graphical than those for single metals. In some extreme cases, they are insufficient for quantifying joint effects despite common use. Typical studies include the toxic unit approach in research such as that implemented by Sprague and Ramsay (1965) and Brown (1968), and also the isobole graphical approach taken by Nash (1981), Broderius (1991), or Christensen and Chen (1991). Some approaches, such as the rudimentary concentration additivity context of toxic units, can be misinformative, tending to confuse as much as advance understanding. The classic quantitative models based on independent and similar joint action provide the best chance of exploring metal ion mixture effects as correlated with coordination chemistry.

Joint action of mixtures is quantified differently depending on whether the mixed toxicants are thought to be acting independently or by a similar mode of action (Finney 1947). In practice, independently acting chemicals are notionally those acting by different modes. In other instances, similar joint action is assumed: the mixed toxicants share a dominant mode of action and display similar toxicokinetics. Mixed toxicants can result in potentiation in which the presence of one chemical at nonlethal levels makes another toxic or more toxic. Mixed toxicants can also be synergistic. In that case, the two or more toxicants together at the specified levels are more toxic than would be predicted by simply summing the effect expected for each alone at those concentrations. The opposite (antagonism) can also occur if the

---

* Lewis et al. (1999) later found similar outlier behavior for barium mouse and rat toxicities.

mixed toxicants produce an effect less than predicted from simply summing the expected effect of each one alone. Functional, dispositional, or receptor-based modes for antagonism exist. Functional antagonism occurs if the toxicants change the process leading to adverse effect in opposite directions, neutralizing each other's effect. Dispositional antagonism occurs if the toxicant influences the uptake, movement, deposition, or elimination of the other(s). Receptor antagonism occurs if the toxicants block or compete in a material way with the other from a receptor. A relevant example might be the competition of metal ions for movement through ion channels, such as that described by Vijverberg et al. (1994). A quick review of the discussions above should reveal the capacity of metals to interact in these manners and for these interactions to be related to intermetal binding trends.

Again, modeling joint action of mixed toxicants is based on whether the mixed chemicals are thought to be predominately independent or similar in action for an adverse effect. The joint effect (P_{M1+M2}) of two independently acting metals ($M1$ and $M2$) combined at concentrations C_{M1} and C_{M2} that alone would produce P_{M1} and P_{M2} proportions (or probabilities) of effect would be predicted with the model,

$$P_{M1+M2}=P_{M1}+P_{M2}(1-P_{M1}) \quad \text{or} \quad P_{M1}+P_{M2}-P_{M1}P_{M2} \tag{1.3}$$

As discussed in Chapter 8, any deviation from ideal independence might be detected by inserting a parameter (ρ) to be estimated into Equation (1.3) in place of the implied 1 and then testing for significant deviation from 1 for the parameter estimate (Newman and Clements 2008).

$$P_{M1+M2}=P_{M1}+P_{M2}-\rho P_{M1}P_{M2} \tag{1.4}$$

Depending on study goals, a general linear modeling approach might be applied to these kinds of data. If there are more metals in the mixture, the independent joint action model can be expanded to Equation (1.5).

$$P_{M1+M2+M3...}=1-(1-P_{M1})(1-P_{M2})(1-P_{M3})... \tag{1.5}$$

The approach is different for mixed metals that are assumed to have a material degree of interaction due to a similar mode of action. Such situations produce toxicant effect–concentration relationships for each of the toxicants alone that have the same slopes ($Slope_{Common}$) (Finney 1947). Extending the notation above to the case of similar joint action,

$$\mathrm{Pr}\,obit(P_{M1})=Intercept_{M1}+Slope_{Common}(Log\,C_{M1}) \tag{1.6}$$

$$\mathrm{Pr}\,obit(P_{M2})=Intercept_{M2}+Slope_{Common}(Log\,C_{M2}) \tag{1.7}$$

The logarithm of the relative potency of these mixed toxicants (e.g., ρ_{M2}) can then be estimated (Equation [1.8]) and used to predict the combined effects of these mixed toxicants (Equation [1.9]).

$$Log \; \rho_{M2} = \frac{Intercept_{M2} - Intercept_{M2}}{Slope_{Common}} \tag{1.8}$$

$$Pr\,obit(P_{M1+M2}) = Intercept_{M1} + Slope_{Common} \, Log \; (C_{M1} + \rho_{M2} \, C_{M2}) \tag{1.9}$$

The relative potency, ρ_{M2}, functions here much like a currency exchange rate functions. It converts concentration of one metal into the equivalent concentration of the other. If one were interested in detecting trends away from perfect similar joint action and toward independent action, the absolute difference between estimated slopes from Equations (1.6) and (1.7), i.e., $|Slope_{M1} - Slope_{M2}|$ could be used as a metric of deviation from similar mode of action.

Newman and Clements (2008) reanalyzed the binary metal mixture effect on bacterial bioluminescence data of Ownby and Newman (2003) using these models (Figure 1.8). A common slope was anticipated if two mixed metals were

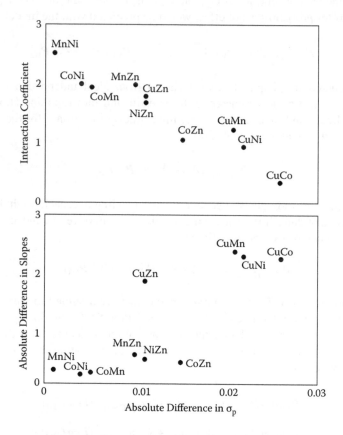

FIGURE 1.8 Deviations from perfect independent (top panel) and perfect similar (bottom panel) action for binary mixtures of metals are predictable from differences in paired metal softnesses. Specific paired metals in mixtures are indicated next to each data point, e.g., MnNi = a binary mixture of Mn^{2+} and Ni^{2+}. This figure is Figure 9.8 from Newman, M.C., and W.H. Clements. 2008. *Ecotoxicology: A Comprehensive Treatment*. Boca Raton, FL: CRC Press.

similar in action. The absolute value of the difference in the probit model slopes for each metal (Equations [1.6] and [1.7]) reflected the degree of deviation from the similar action assumption. As anticipated, a clear trend emerged if this absolute difference were plotted for ten binary mixtures against the differences in the softness indices of the mixed metals (Figure 1.8 bottom panel). The more similar the paired metals' softness indices, the more they tended toward similar action. It is interesting to note that this approach does not require that a complex series of binary metal mixture experiments be conducted. It requires only probit model slopes from single-metal tests. However, mixture experiments are required to explore this approach from the vantage of independent action. Based on metal independent action modeling, the magnitude of deviation of the interaction coefficient (ρ in Equation [1.4]) from 1 suggests the degree of deviation from perfect independent action (Figure 1.8 top panel). The more dissimilar the softness indices of the paired metals, the more the estimated ρ deviated from 1. Independence increased the more dissimilar the coordination chemistry of the paired metals. Clearly, joint action of metals was influenced by coordination chemistry of the combined metals.

1.4 CONCLUSION

[S]everal resolvable issues require attention before the QICAR approach has the same general usefulness as the QSAR approach. These issues include exploration of more explanatory variables, careful evaluation of ionic qualities used to calculate explanatory variables, examination of models capable of predicting effects for widely differing metals (e.g., metals of different valence states), effective inclusion of chemical speciation, examination of more effects, and assessment of the applicability of QICARs to phases such as sediment, soils, and foods.

Newman et al. (1998, p. 1424)

It is now generally accepted that QSAR-like models can be generated for metal ions based on fundamental coordination chemistry trends. Examples ranging from adsorption to biological surfaces, to metal interactions in mixtures, to accumulation and effects in metazoans were used in this short chapter to demonstrate this point. Class (a) metal toxicity was easily related to electrostatic interactions with biological ligands. Trends in effects of class (b) and intermediate metals were related to qualities more closely linked to covalent bonding. Models involving a wider range of class (a), intermediate, and class (b) metals might require more than one explanatory variable based on different binding tendencies.

What is currently needed is a sustained exploration of the approach and further refinement of metrics and methodologies. Progress toward filling the information gaps highlighted in the above quote is evident in the literature, e.g., Can and Jianlong (2007), Kinraide and Yermiyahu (2007), Newman and Clements (2008), Ownby and Newman (2003), Walker et al. (2003), Wolterbeek and Verburg (2001), and Zamil et al. (2009) and Zhou et al. (2011). These models will very likely emerge in the next two decades to the level enjoyed now by QSAR models for organic compounds.

REFERENCES

Ahrland, S. 1968. Thermodynamics of complex formation between hard and soft acceptors and donors. *Struct. Bond.* 5:118–149.

Babich, H., and E. Borenfreund. 1991. Cytotoxicity and genotoxicity assays with cultured fish cells: A review. *Toxicol. in Vitro* 5:91–100.

Babich, H., C. Shopsis, and E. Borenfreund. 1986. In vitro testing of aquatic pollutants (cadmium, copper, zinc, nickel) using established fish cell lines. *Ecotox. Environ. Safe.* 11:91–99.

Barrett, J. 2002. *Atomic Structure and Periodicity*. Cambridge: The Royal Society of Chemistry.

Bichet, D., M. Grabe, Y.N. Jan, and L.Y. Jan. 2006. Electrostatic interactions in the channel cavity as an important determinant of potassium channel selectivity. *P. Natl. Acad. Sci. USA* 103:14355–1460.

Biesinger, K.E., and G.M. Christensen. 1972. Effects of various metals on survival, growth, reproduction, and metabolism of *Daphnia magna*. *Can. J. Fish. Aquat. Sci.* 29:1691–1700.

Blake, J. 1884. On the connection between physiological action and chemical constitution. *J. Physiol.* 5:35–44.

Bradberry, S.M., and A. Vale 1995. Disturbances of potassium homeostasis in poisoning. *Clin. Toxicol.* 33:295–310.

Brezonik, P.L., S.O. King, and C.E. Mach. 1991. The influence of water chemistry on trace metal bioavailability and toxicity to aquatic organisms. In *Metal Ecotoxicology. Concepts & Applications*, edited by M.C. Newman and A.W. McIntosh, 1–31. Chelsea, MI: Lewis Publishers.

Broderius, S.J. 1991. Modeling the joint toxicity of xenobiotics to aquatic organisms: Basic concepts and approaches. In *Aquatic Toxicology and Risk Assessment: Fourteenth Volume, ASTM STP 1124*, edited by M.A. Mayes and M.G. Barron, 107–127. Philadelphia: American Society for Testing and Materials.

Brown, V.M. 1968. The calculation of the acute toxicity of mixtures of poisons to rainbow trout. *Water Res.* 2:723–733.

Can, C., and W. Jianlong. 2007. Correlating metal ion characteristics with biosorption capacity using QSAR model. *Chemosphere* 69:1610–1616.

Carlson, R.M. 1990. Assessment of the propensity for covalent binding of electrophiles to biological substrates. *Environ. Health Persp.* 87:227–232.

Chen, C., and J. Wang. 2007. Influence of metal ionic characteristics on their biosorption capacity by *Saccharomyces cerevisiae*. *Appl. Microbiol. Biot.* 74:911–917.

Christensen, E.R., and C.-Y. Chen. 1991. Modeling of combined toxic effects of chemicals. *Toxic Subst. J.* 11:1–63.

Christensen, E.R., and J.H. Tucker. 1976. Effects of selected water toxicants on the in vitro activity of fish carbonic anhydrase. *Chem.-Biol. Interact.* 13:181–192.

Christensen, G.M. 1971/1972. Effects of metal ions and other chemicals upon the in vitro activity of two enzymes in the blood plasma of the white sucker, *Catostomus commersoni* (Lacủpède). *Chem.-Biol. Interact.* 4:351–361.

Crist, R.H., K. Oberholser, J. McGarrity, D.R. Crist, J.K. Johnson, and J.M. Brittsan. 1992. Interaction of metals and protons with algae. 3. Marine algae, with emphasis on lead and aluminum. *Environ. Sci. Technol.* 26:496–502.

Crist, R.H., K. Oberholser, D. Schwartz, J. Marzoff, D. Ryder, and D.R. Crist. 1988. Interactions of metals and protons with algae. *Environ. Sci. Technol.* 22:755–760.

Das, T., A. Sharma, and G. Talukder. 1988. Effects of barium on cellular systems: A review. *Nucleus* 31:41–68.

Delfino, G., S. Amerini, and A. Mugelli. 1988. Barium cardiotoxicity: Relationship between ultrastructural damage and mechanical effects. *Toxicol. in Vitro* 2:49–55.

Di Toro, D., H.E. Allen, H.L. Bergman, J.S. Meyer, P.R. Paquin, and R.C. Santore. 2001. Biotic ligand model of the acute toxicity of metals. 1. Technical basis. *Environ. Toxicol. Chem.* 20:2383–2396.

Duffus, J.H. 2002. Heavy metals: A meaningless term? *Pure Appl. Chem.* 74:793–807.

Erichsen Jones, J.R. 1940. A further study of the relation between toxicity and solution pressure, with *Polycelis nigra* as test animal. *J. Exp. Biol.* 17:408–415.

Finney, D.J. 1947. *Probit Analysis.* Cambridge: Cambridge University Press.

Fisher, N.S. 1986. On the reactivity of metals for marine phytoplankton. *Limnol. Oceanogr.* 31:443–449.

Fraústo de Silva, J.J. R., and R.J.P. Williams. 1993. *The Biological Chemistry of the Elements: The Inorganic Chemistry of Life.* Oxford: Oxford University Press.

Harris, H.H., I.J. Pickering, and G.N. George. 2003. The chemical form of mercury in fish. *Science* 301:1203.

Haxel, G.B., J.B. Hedrick, and G.J. Orris. 2002. Rare earth elements: Critical resources for high technology. USGS Fact Sheet 087–02, http://geopubs.wr.usgs.gov/fact-sheet/fs087-02/.

Heinemann, S.H., H. Terlau, W. Stühmer, K. Imoto, and S. Numa. 1992. Calcium channel characteristics conferred on the sodium channel by single mutations. *Nature* 356:441–443.

Hopkin, S.P., and J.A. Nott. 1979. Some observations on concentrically structured, intracellular granules in the hepatopancreas of the shore crab *Carcinus maenas* (L.). *J. Mar. Biol. Assoc. UK* 59:867–877.

Howard, B., and K. Simkiss. 1981. Metal binding by *Helix aspersa* blood. *Comp. Biochem. Phys.* A 70:559–561.

Hsie, A.W., R.L. Schenley, E.-L. Tan, S.W. Perdue, M.W. Williams, T.L. Hayden, and J.E. Turner. 1984. The toxicity of sixteen metallic compounds in Chinese hamster ovary cells: A comparison with mice and *Drosophilia*. In *Alternative Methods in Toxicology. Volume 2: Acute Toxicity Testing: Alternative Approaches*, edited by A.M. Goldberg, 117–125. New Rochelle, NY: Mary Ann Liebert, Inc.

Hudson, R.J.M., and C.W. Shade. 2004. In their Brevia "The chemical form of mercury in fish." *Science* 303:763.

Jeffree, R.A. 1988. Patterns of accumulation of alkaline-earth metals in the tissue of the freshwater mussel *Velesunio angasi* (Sowerby). *Arch. Hydrobiol.* 112:67–90.

Jones, M.M., and W.K. Vaughn. 1978. HSAB theory and acute metal ion toxicity and detoxification processes. *J. Inorg. Nucl. Chem.* 40:2081–2088.

Kaiser, K.L. 1980. Correlation and prediction of metal toxicity to aquatic biota. *Can. J. Fish. Aquat. Sci.* 37:211–218.

Khangarot, B.S., and P.K. Ray. 1989. Investigation of correlation between physicochemical properties of metals and their toxicity to the water flea *Daphnia magna* Straus. *Ecotox. Environ. Safe.* 18:109–120.

Kinraide, T.B. 2009. Improved scales for metal ion softness and toxicity. *Environ. Toxicol. Chem.* 28:525–533.

Kinraide, T.B., and U. Yermiyahu. 2007. A scale of metal ion binding strengths correlating with ionic charge, Pauling electronegativity, toxicity, and other physiological effects. *J. Inorg. Biochem.* 101:1201–1213.

Lewis, D.F.V., M. Dobrota, M.G. Taylor, and D.V. Parke. 1999. Metal toxicity in two rodent species and redox potential: Evaluation of quantitative structure-activity relationships. *Environ. Toxicol. Chem.* 18:2199–2204.

Lyon, R., M. Taylor, and K. Simkiss. 1984. Ligand activity in the clearance of metals from the blood of the crayfish (*Austropotamobius pallipes*). *J. Exp. Biol.* 113:19–27.

Magwood, S., and S. George. 1996. in vitro alternatives to whole animal testing: Comparative cytotoxicity studies of divalent metals in established cell lines derived from tropical and temperate water fish species in a neutral red assay. *Mar. Environ. Res.* 42:37–40.

Mathews, A.P. 1904. The relation between solution tension, atomic volume, and the physiological action of the elements. *Am. J. Physiol.* 10:290–323.

McCloskey, J.T., M.C. Newman, and S.B. Clark. 1996. Predicting the relative toxicity of metal ions using ion characteristics: Microtox® bioluminescence assay. *Environ. Toxicol. Chem.* 15:1730–1737.

McKinney, J.D., A. Richard, C. Waller, M.C. Newman, and F. Gerberick. 2000. The practice of structure-activity relationships (SAR) in toxicology. *Toxicol. Sci.* 56:8–17.

Moeller, T. 1975. *The Chemistry of the Lanthanides.* Oxford: Pergamon Press.

Nash, R.G. 1981. Phytotoxic interaction studies: Techniques for evaluation and presentation of results. *Weed Sci.* 29:147–155.

Newman, M.C. 1995. *Quantitative Methods in Aquatic Ecotoxicology.* Boca Raton, FL: Lewis Publishers/CRC Press.

Newman, M.C., and W.H. Clements. 2008. *Ecotoxicology: A Comprehensive Treatment.* Boca Raton, FL: CRC Press.

Newman, M.C., and M.C. McCloskey. 1996. Predicting relative toxicity and interactions of divalent metal ions: Microtox® bioluminescence assay. *Environ. Toxicol. Chem.* 15:275–281.

Newman, M.C., J.T. McCloskey, and C.P. Tatara. 1998. Using metal-ligand binding characteristics to predict metal toxicity: Quantitative ion character-activity relationships (QICARs). *Environ. Health Persp.* 106:1419–1425.

Nieboer, E., and D.H.S. Richardson. 1980. The replacement of the nondescript term "heavy metals" by a biologically and chemically significant classification of metal ions. *Environ. Pollut. B* 1:3–26.

Ovchinnikov, Y.A. 1979. Physico-chemical basis of ion transport through biological membranes: Ionophores and ion channels. *Eur. J. Biochem.* 94:321–336.

Ownby, D.R., and M.C. Newman. 2003. Advances in quantitative ion character-activity relationships (QICARS): Using metal-ligand binding characteristics to predict metal toxicity. *QSAR Comb. Sci.* 22:241–246.

Pearson, R.G. 1963. Hard and soft acids and bases. *J. Am. Chem. Soc.* 85:3533–3539.

Pearson, R.G. 1966. Acids and bases. *Science* 151:172–177.

Pigman, E.A., J. Blanchard, and H.E. Laird II. 1997. A study of cadmium transport pathways using the Caco-2 cell model. *Toxicol. Appl. Pharmacol.* 142:243–247.

Rainbow, P.S., and W.H. Black. 2005. Cadmium, zinc and the uptake of calcium by two crabs, *Carcinus maenus* and *Eriocheir sinesis. Aquat. Toxicol.* 72:45–65.

Raize, O., Y. Argaman, and S. Yannai. 2004. Mechanisms of biosorption of different heavy metals by brown marine macroalgae. *Biotechnol. Bioeng.* 87:451–458.

Reinfelder, J.R., and N.S. Fisher. 1991. The assimilation of elements ingested by marine copepods. *Science* 251:794–796.

Shaw, W.H.R. 1961. Cation toxicity and the stability of transition-metal complexes. *Nature* 192:754–755.

Simkiss, K. 1981a. Cellular discrimination processes in metal accumulating cells. *J. Exp. Biol.* 94:317–327.

Simkiss, K. 1981b. Calcium, pyrophosphate and cellular pollution. *Trends Biochem. Sci.* April 1981: III–V.

Somers, E. 1961. The fungitoxicity of metal ions. *Ann. Appl. Biol.* 49:246–253.

Sprague, J.B., and B.A. Ramsay. 1965. Lethal levels of mixed copper-zinc solutions for juvenile salmon. *Can. J. Fish. Aquat. Sci.* 22:425–432.

Taglialatela, M., J. Drewe, and A. Brown. 1993. Barium blockage of a clonal potassium channel and its regulation by a critical pore residue. *Mol. Pharmacol.* 44:180–190.

Tatara, C.P., M.C. Newman, J.T. McCloskey, and P.L. Williams. 1998. Use of ion characteristics to predict relative toxicity of mono-, di-, and trivalent metal ions: *Caenorhabditis elegans* LC50. *Aquat. Toxicol.* 42:255–269.

Van Kolck, M., M.A.J. Huijbregts, K. Veltman, and A.J. Hendriks. 2008. Estimating bioconcentration factors, lethal concentrations and critical body residues of metals in the mollusks *Perna viridis* and *Mytilus edulis* using ion characteristics. *Environ. Toxicol. Chem.* 27:272–276.

Vijverberg, H.P.M., M. Oortgiesen, T. Leinders, and R.G.D.M. van Kleef. 1994. Metal interactions with voltage- and receptor-activated ion channels. *Environ. Health Persp.* 102:153–158.

Walker, J.D. 2003. *QSARs for Pollution Prevention, Toxicity Screening, Risk Assessment and Web Applications*. Pensacola, FL: SETAC Press.

Walker, J.D., M. Enache, and J.C. Dearden. 2003. Quantitative cationic-activity relationships for predicting toxicity of metals. *Environ. Toxicol. Chem.* 22:1916–1935.

Walker, J.D., M. Enache, and J.C. Dearden. 2007. Quantitative cationic-activity relationships for predicting toxicity of metal ions from physicochemical properties and natural occurrence levels. *QSAR Combin. Sci.* 26:522–527.

Walker, J.D., and J.P. Hickey. 2000. QSARs for metals: Fact or fiction? In *Metal Ions in Biology and Medicine*, edited by J.A. Centeno, P. Colleery, G. Vernet, R.B. Finkelman, H. Gibb, and J.C. Etienne, 401–405. Montrouge, France: John Libbey Eurotext Limited.

Weltje, L. 2002. Bioavailability of lanthanides to freshwater organisms: Speciation, accumulation and toxicity. PhD diss., Delft University.

Williams, M.W., and J.E. Turner. 1981. Comments on softness parameters and metal ion toxicity. *J. Inorg. Nucl. Chem.* 43:1689–1691.

Wolterbeek, H.T., and T.G. Verburg 2001. Predicting metal toxicity revisited: General properties vs. specific effects. *Sci. Total Environ.* 279:87–115.

Xue, H.-B., W. Stumm, and L. Sigg. 1988. The binding of heavy metals to algal surfaces. *Water Res.* 22:917–926.

Zamil, S.S., S. Ahmad, M.H. Choi, J.Y. Park, and S.C. Yoon. 2009. Correlating metal ionic characteristics with biosorption capacity of *Staphylococcus saprophyticus* BMSZ711 using QICAR model. *Bioresource Technol.* 100:1895–1902.

Zhang, H., J. Feng, W. Zhu, C. Liu, S. Xu, P. Shao, D. Wu, W. Yang, and J. Gu. 1999. Chronic toxicity of rare-earth elements on human beings. *Biol. Trace Elem. Res.* 73:1–17.

Zhou, D.M., L.Z. Li, W.J.G.M. Peijnenburg, D.R. Ownby, A.J. Hendriks, P. Wang, and D.D. Li. 2011. A QICAR approach for quantifying binding constants for metal-ligands complexes. *Ecotox. Environ. Safe.* 74:1036–1042.

2 Electronic Structure of Metals and Atomic Parameters

2.1 ABOUT QSAR AND THE DESCRIPTORS OF CHEMICAL STRUCTURE

In the absence of precise information regarding the biological mechanisms that contribute to the production of the biological activity, and in the absence of satisfactory information regarding the state of the metal ions in biological environments (this could be either the free ion state or, most probably, the bonded form in a chemical compound), a suitable quantitative structure-activity relationship (QSAR) approach would be to choose descriptors that represent the long path that the metal ions have to travel up to the biological target (Schwietert and McCue 1999). The choice of descriptors must consider all the physical–chemical processes that happen to the metal ion or metal compound up to the last event, which is the interaction with the biological receptor. Correspondingly, a good correlation of metal biological activity with descriptors depends on a better understanding of the physicochemical or biological mechanisms in which metal ions participate in biological environments.

Many descriptors that can be used for metals are generally common to chemical structures—for example, molar refractivity, electronegativity, and so on. Other descriptors are specific for metals or metal ions, depending on their electronic configuration and their position in the periodic table.

A review of the parameters found in the literature to correlate with metal toxicity shows the use of atomic parameters such as atomic radii and the ionization energies; the use of basic chemical properties such as the valence; the electrochemical character and thermodynamic measures that describe the electrochemical reactions of the metals; the theory of hard and soft acids and bases; and properties of the metal compounds, for example, the solubility of metal compounds or other particularities of the metal compounds (a review is found in Walker et al., 2003). To better understand such descriptors, some important characteristics of metals are discussed in this and the following chapter. In addition to the descriptors relevant for the free metal ions, the array of structural descriptors is greatly enlarged by the use of the descriptors for organometallic complexes (Chapter 4).

2.2 GENERAL PROPERTIES OF METALS

More than 80 chemical elements in the periodic table are classified as metals (~65%): they all have a set of physical and chemical properties that characterize the metallic character. Another 18 elements are classified as nonmetals, while 7 elements that

Group →	1	2	3	4	5	6	7	8	9	10	11	12	13	14	15	16	17	18
↓Period	IA	IIA	IIIB	IVB	VB	VIB	VIIB	VIIIB	VIIIB	VIIIB	IB	IIB	IIIA	IVA	VA	VIA	VIIA	VIIIA
1	1 H																	2 He
2	3 Li	4 Be											5 B	6 C	7 N	8 O	9 F	10 Ne
3	11 Na	12 Mg											13 Al	14 Si	15 P	16 S	17 Cl	18 Ar
4	19 K	20 Ca	21 Sc	22 Ti	23 V	24 Cr	25 Mn	26 Fe	27 Co	28 Ni	29 Cu	30 Zn	31 Ga	32 Ge	33 As	34 Se	35 Br	36 Kr
5	37 Rb	38 Sr	39 Y	40 Zr	41 Nb	42 Mo	43 Tc	44 Ru	45 Rh	46 Pd	47 Ag	48 Cd	49 In	50 Sn	51 Sb	52 Te	53 I	54 Xe
6	55 Cs	56 Ba	*	72 Hf	73 Ta	74 W	75 Re	76 Os	77 Ir	78 Pt	79 Au	80 Hg	81 Tl	82 Pb	83 Bi	84 Po	85 At	86 Rn
7	87 Fr	88 Ra	**	104 Rf	105 Db	106 Sg	107 Bh	108 Hs	109 Mt	110 Ds	111 Rg	112 Cn						

*Lanthanides	57 La	58 Ce	59 Pr	60 Nd	61 Pm	62 Sm	63 Eu	64 Gd	65 Tb	66 Dy	67 Ho	68 Er	69 Tm	70 Yb	71 Lu
**Actinides	89 Ac	90 Th	91 Pa	92 U	93 Np	94 Pu	95 Am	96 Cm	97 Bk	98 Cf	99 Es	100 Fm	101 Md	102 No	103 Lr

Notes:

Metals: alkali metals (group 1 elements), alkaline earth metals (group 2 elements), transition metals (groups 3 to 11), post-transition metals (group 12 elements and Al, Ga, In, Tl, Sn, Pb, Bi), lanthanides and actinides

Metalloids: B, Si, Ge, As, Sb, Se, Te, sometimes Po is also added to the list

Non-metals: H, C, N, P, O, S, halogens (group 17 elements) and the noble gases (group 18 elements)

FIGURE 2.1 The periodic table of the elements and the classification of chemical elements as metals, nonmetals, and metalloids.

have intermediate properties between metals and nonmetals are called *metalloids* (Figure 2.1). The main characteristics that separate metals, nonmetals, and metalloids are presented in Table 2.1.

The metallic character is a result of a particular type of chemical bond (the metallic bonding) that can be found in the solid state and in the liquid (melted) state of a chemical substance, but disappears in the gaseous state.

The metallic bonding holds together the atoms of a metal and determines the formation of the metal crystal structures. All valence electrons of the atoms in the metal form a common electronic cloud in the crystal lattice. In the crystal structures, the metal atoms usually have the coordination number 12 or 8 with lower coordination numbers being found in less-compact crystal structures (In, white Sn, Sb, Bi).

Metals have few electrons in their valence shells, and these electrons are loosely bound; therefore, metals have low ionization energy and a low electronegativity. Within the periodic table groups, the metallic character is increasingly downward, due to the increase of the atomic radius; therefore, the longer the distance between the nucleus and the valence electrons, the less electrical attraction can be manifested between them. Also, within the periodic table periods, the metallic character decreases from left to right, following the decrease of the atomic radius. An electron situated near a positively charged nucleus is more difficult to remove; therefore, such atoms tend to accept electrons to fill the valence shell rather than lose them.

The density of the metals varies between 0.53 g/cm³ for lithium and 22.6 g/cm³ for osmium; most metals have values of 5–9 g/cm³. In general, the higher density of

TABLE 2.1

Some Physical and Chemical Properties of Metals, Nonmetals, and Metalloids

Metals	Metalloids	Nonmetals
Opaque with a metallic luster	Metallic luster	No luster
Solid at room temperature (Hg is liquid)	Solid at room temperature	Gases, liquids, or solids
High thermal conductivity	Good thermal conductivity	Good thermal isolators (excepting C in diamond)
High electrical conductivity	Good electrical conductivity	Poor electrical conductivity (excepting C in graphite)
Malleable and ductile	Brittle	Brittle (when solid)
Metallic bonding; closed packed crystal structures, high coordination numbers	Relatively open crystal structures, medium coordination numbers	Weak molecular forces, low coordination numbers
Oxidation numbers of metals in compounds are positive	Oxidation numbers in compounds are positive or negative	Oxidation numbers in compounds are positive or negative
In solution they tend to lose electrons to form positive ions	They form cations and anions	Tend to gain electrons and form negative ions
Low ionization energy	Intermediate ionization energies	High ionization energy
Low electronegativity	Electronegativity values: from 1.9 to 2.4	High electronegativity
They form ionic compounds with nonmetals	They tend to form covalent bonds in compounds	They form ionic compounds with metals and covalent compounds with other nonmetals

metals as compared to nonmetals is a consequence of the tight metallic packing in the metal crystal structures. The strongest attractive force between atoms in metals can be found in the middle of the transition metals series, in elements where the delocalization of electrons is strongest. Metals that have density values higher than 5 g/cm^3 are called *heavy metals*. However, this term has been used for a variety of transition metals in various biological studies, leading to contradictory definitions and the proposal to abandon its use (for example, Nieboer and Richardson 1980; Phipps 1980; Borovik 1990; Duffus 2002).

2.3   CHARACTERIZATION OF METALS ACCORDING TO THEIR ELECTRONIC CONFIGURATION

The physical and chemical properties of the chemical elements are a consequence of their electronic structure. The chemistry of metals and of their compounds show the existence of analogies for metals that have to fill s, p, d, or f subshells. Therefore, those elements that have the same value of the second quantum number (l) (see Section 2.3.1) are classified in the s, p, d, f blocks of elements (Figure 2.2).

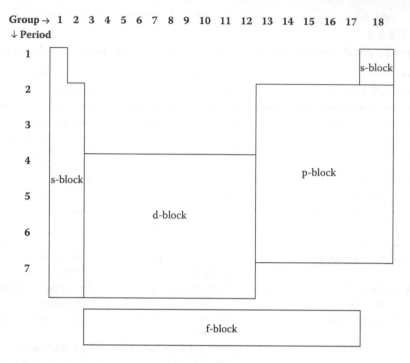

FIGURE 2.2 Periodic table blocks of elements.

2.3.1 Atomic Orbitals: Penetration Effects

The atom is the smallest unit of a substance that has its chemical properties. From the 118 known elements, only 92 are found in nature. The other elements have been obtained artificially and have generally a very short life, with a rapid decay.

The atomic model shows that the atom is made of three fundamental particles: protons, neutrons, and electrons. The protons and neutrons are found in the center of the atom forming a dense, positively charged nucleus. The protons have the electric charge +1 and the weight equal to one atomic mass unit. The neutrons have a weight equal to one atomic mass unit too, but do not have electric charge.

The protons, neutrons, and electrons are always written in lowercase: p, n, and e. The uppercase versions of these letters are not being used since they have other meanings in both chemistry and physics (for example, P is phosphorus and E is energy).

The nucleus of any atom or element is characterized by an atomic number, Z, equal to the number of protons and a mass number, A, equal to the total number of protons and neutrons ($A = Z + n$).

The electrons surround the nucleus like clouds, spinning at high speeds on their paths called *orbitals*. They have the same electrical charge as protons, but have the opposite sign, namely negative.

The atom, as such, cannot be observed directly; therefore, different models have been imagined to represent it as realistically as possible and ease the explanation of

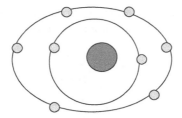

FIGURE 2.3 Electrons rotate around the nucleus like planets around the sun.

most of the chemical transformations. The most common model of the atom is the classic model imagined by Rutherford-Bohr, also known as the *solar system model* (Figure 2.3). This model is a graphic representation that is similar to the real representation of an atom, but it is not identical to the atom; however, it is very useful for viewing and explaining different chemical phenomena.

From this representation it can be seen that, unlike the solar system where on an orbit only one planet is rotating, in an atom an *orbital* (orbit) is in fact an electronic level or electronic layer capable of hosting a certain number of electrons, as represented in Figure 2.3. Each level has a certain energy and is placed at different mean (median) distances from the nucleus. Because of the energy that the electronic layers carry, they are also called *energy levels*.

The electronic structure of the atom is the key to understanding the properties of the elements, of the compounds that they form their chemical reactions, and the molecules that they shape.

The electrons, similar to protons, are electrically charged particles. Their charge is the same size as the protons, but is negative. The number of electrons in an atom is the same as the number of protons, so that an atom is neutral from an electric point of view. The mass of an electron is 1836 times smaller than that of a proton, and this is why in chemical calculations it is usually not considered (it is neglected).

In his wave mechanics studies, Erwin Schrödinger used mathematics to assess the probability that an electron will be in some point in the atomic space. The mathematical relationship that expresses this is similar to the equation used to represent wave propagation in acoustics. Thus, the motion of the electron, the probability of finding it in a certain position within the atom, is characterized by a wave function ψ or orbital that actually indicates the charge density or electronic cloud. The Heisenberg uncertainty principle shows that it is impossible to know the exact position of an electron at a certain point in space at a certain moment in time. In line with this principle, the probability that an electron would find itself at a certain point in space at a certain moment in time is given by the square of the wave function ψ^2. The internal motion of the electron around the nucleus is calculated according to the Schrödinger equation:

$$-\frac{\hbar}{2\mu}\nabla^2\psi + V\psi = E\psi \qquad (2.1)$$

where V is the potential coulombian energy of the electron, E is the total energy of the system, $\hbar$ is a convenient modification of Planck's constant, and μ is its reduced mass:

$$V = -\frac{Ze^2}{4\pi\varepsilon_0 r} \tag{2.2}$$

$$\hbar = \frac{h}{2\pi} \tag{2.3}$$

$$\frac{1}{\mu} = \frac{1}{m_e} + \frac{1}{m_N} \tag{2.4}$$

In the potential energy equation, r is the distance between the electron and the nucleus, and ε_0 is the permittivity in vacuum (Atkins 1994; Atkins and Beran 1990; Massey 1990).

A simpler representation of Schrödinger's equation is:

$$H\psi = E\psi \tag{2.5}$$

In this equation, H is the operator of the kinetic and potential energy of the electron–nucleus system and it describes the position of the electron in the three-dimensional Cartesian space (Equation [2.6]).

$$H = -\frac{h^2}{8\pi^2 m}\left(\frac{\partial^2}{\partial x^2} + \frac{\partial^2}{\partial y^2} + \frac{\partial^2}{\partial z^2}\right) + V(x,y,z) \tag{2.6}$$

where $\frac{\partial^2}{\partial x^2}, \frac{\partial^2}{\partial y^2}$ and $\frac{\partial^2}{\partial z^2}$ are partial differentials, and V(x, y, z) is the potential energy in the Cartesian system.

Schrödinger's equation cannot be solved exactly, especially for multi-electron atoms, mainly because of the effects determined by the attraction and rejection of electrons phenomena. However, by using different methods of approximation, one can obtain satisfactory results.

The wave function that indicates the charge density or electronic cloud provides a space model of the movement of the electrons. Specific electron clouds are positioned like concentric layers, at different energy levels around the nucleus.

Each atomic orbital is described by a unique set of three quantum numbers: n, l, and m_l. Also, each electron is described by a fourth unique quantum number m_s, which gives information regarding the spin quantum number.

The principal quantum number n indicates on which layer the electron is positioned and is represented by a positive integer number: 1, 2, 3, 4, 5, ... ∞. It can also be represented with the capital letters K, L, M, N, O, P, and Q according to Bohr's system. The energy value of these levels increases with distance from the nucleus; therefore, the lowest energy level is the level with the principal quantum number 1.

The next quantum number (l) is called the secondary quantum number, orbital, or azimuthal quantum number. For each principal quantum number there are values of integer numbers of l from 0 to (n–1) (Lide 2008). Consequently, each energy layer corresponding to a principal quantum number is made of a specific number of subshells corresponding to the values of those orbitals. These subshells are represented with the letters s, p, d, f, g corresponding to the secondary quantum numbers as follows:

l =	0	1	2	3	4
	s	p	d	f	g

To specify the principal level, the letters s, p, d, f, and g are preceded by the number that indicates the level to which that substrate belongs. The first energy level contains a single subshell marked as 1s; the second principal energy level includes two subshells marked as 2s and 2p; and so on.

The subshells also include a certain number of stable orbitals and a maximum number of electrons. The difference in energy between different subshells determines the shape of the orbitals, and the orbital quantum number describes the orbital shape.

For n = 1, we have l = 0 and the orbit is spherical. For n ≠ 1 and the values l = 1, 2, 3 ... the orbits are ellipses that look like nodal planes with each part forming a lobe. These planes are similar, but perpendicular to each other for degenerate orbitals.

Since electrons are charged particles, they produce a magnetic field as they move. Thus, the orbitals are, in turn, characterized by a magnetic quantum number (m_l). This number takes $2l + 1$ values from $-l$ to $+l$ and reflects how many electrons can exist on each orbital on each subshell, as well as their orientation in an exterior magnetic field. For example, there are four orbitals for n = 2, one in the sublevel s and three in the sublevel l = 1 with m_l equal to +1, 0 and –1. Therefore, the s subshell has only one orbital, the subshell p has three orbitals, the subshell d has five orbitals, and the subshell f has seven orbitals.

The orbitals in a subshell have the same average (median) energy and are called *degenerate orbitals*. Consequently, the three p orbitals (p_x, p_y, and p_z) have the same energy; this also holds true for the five d orbitals or the seven f orbitals.

The fourth quantum number characterizes the movement or autorotation around the electron's own axis. It is called the *spin motion* and is characterized by an angular momentum and an inherent magnetic momentum, both of which can be quantified. Accordingly, an electron has two spin states only, one in the clockwise state (↑ state) and one in the counterclockwise state (↓ state). This quantitation is represented by the magnetic quantum number of the spin and has only two values for an electron.

The number of electrons that rotate on a certain level around the nucleus is determined by precise quantum laws. The filling of orbitals with electrons occurs according to the Aufbau principle (in German the "construction principle") so that the electronic configuration leads to the lowest energy state. This principle uses the rule of Hund and the rule of Pauli.

There are a maximum of 2 electrons for an orbital. Pauli's exclusion principle shows that 2 electrons cannot exist in the same quantum state, that is, with 4 identical numbers. They must differ at least by one quantum number. That means that there will be a maximum of 2 electrons in each orbital that have different values of the magnetic quantum number.

Hund's rule shows the order of occupation of degenerate orbitals—that is, the pairing of electrons on the orbitals will not take place until after each orbital contains one electron. These (single) electrons have the same spin; 1 that is, they are parallel.

The filling of orbitals with electrons starts with the lowest energy. The orbitals of the p, d, and f subshells have the same energy for the corresponding subshell. Therefore, these are filled with one electron first, and then receive the pair. On one orbital the two electrons are antiparallel; their magnets are oriented in opposite directions ($\uparrow\downarrow$). Using these rules, the distribution of electrons on the orbitals of an element can be defined easily. For example, chlorine has 17 electrons. They will be placed on the following orbitals: $1s^2$, $2s^2$, $2p^6$, $3s^2$, $3p^5$. The following schematic representation is obtained:

$$\uparrow\downarrow \quad \uparrow\downarrow \quad \uparrow\downarrow \quad \uparrow\downarrow \quad \uparrow\downarrow \quad \uparrow\downarrow \quad \uparrow\downarrow \quad \uparrow\downarrow \quad \uparrow$$

$$1s \quad 2s \quad 2p_x \quad 2p_y \quad 2p_z \quad 3s \quad 3p_x \quad 3p_y \quad 3p_z$$

In this context, we note that an energy level can only have 2, 8, 18, or 32 electrons (Table 2.2).

The distances between the energy levels, or subshells, are not uniform. The filling of orbitals with electrons is done from the lowest energy level, and the next one starts to fill only after the previous level is full.

From the structure of polyelectronic atoms, we would expect that electrons would fill the orbitals according to the following order: $1s < 2s < 2p < 3s < 3p < 3d < 4s < 4p < 4d < 4f$ and so on. In fact, data has shown that the energies of the orbitals do not increase progressively according in this order. The order of filling of the orbitals is dependent on the shielding effects (Z^*) and the electronic penetration of the orbitals.

Due to their charges, the electrons are attracted by the nucleus and repel each other. Thus, polyelectronic atoms have three types of energy:

- the kinetic energy of the electron motion around the nucleus,
- the energy of attraction of the electron to the nucleus, and
- the energy of rejection from the other electrons.

The electrons in lower (inner) layers more strongly shield the electrons in the outer layers than do the electrons on the same subshell or orbital shield, or they reject each

---

TABLE 2.2
The Number of Electrons in Layers and Sublayers

	n	l	m_l	Orbitals	Electrons
K	1	0	0	s	2
L	2	0,1	0,−1,0,+1	s, p	2+6=8
M	3	0,1,2	0,−1,0,+1;−2,−1,0,+1,+2	s, p, d	2+6+10=18
N	4	0,1,2,3	0,−1,0,+1;−2,−1,0,+1,+2; 0,−1,0,+1;−3,−2,−1,0,+1,+2,+3	s, p, d, f	2+6+10+14=32

other. The levels of energy increase as the atomic number increases. The presence of some internal electron levels reduces the attraction exerted by the nucleus on an electron. Thus, the effective nuclear energy that acts upon an electron in the multi-electronic nucleus is lower than the charge of the nucleus and is given by the formula:

$$Z_{eff} = Z - \sigma \tag{2.7}$$

where Z_{eff} is the effective nuclear charge and is the nuclear constant that can be calculated using Slater's empirical rules:

- the electrons can occupy the following groups of orbitals: 1s, 2s2p, 3s3p, 3d, 4s4p, 4d, 4f, 5s5p, 5d, 5f, 5g, etc.,
- the electrons placed on the orbitals on the right-hand side of the electron of interest are not considered,
- each electron in the same group with the electron being considered contributes 0.35 to the shielding, except for the electrons in group 1s, in which the contribution of an electron is 0.30,
- if the electron of interest is in an nsp group, the electrons in the penultimate level $(n-1)$ will each contribute 0.85 to the shielding, and those in the lower layers from $n=1$ to $n-2$ will contribute 1, and
- if the electron of interest is part of a d or f group, the contribution of each electron in the lower layers is 1, while for those in the same layer the contribution is 0.35.

Consider some examples:

1. The oxygen atom has $Z=8$; therefore, it has the electron configuration $1s^2 2s^2 2p^4$. One electron in the nsp group is shielded by the three others by 0.35 and the electrons in the layer 1s by 0.85. Therefore, the effective nuclear charge will be:

$$Z_{eff} = 8 - (5*0.35 + 2*0.85) = 4.55 \quad \text{with } \sigma = 3.45 \tag{2.8}$$

2. The vanadium atom has $Z=23$ and the electron configuration $1s^2 2s^2 2p^6 3s^2 3p^6 3d^3 4s^2$. For one electron in the 3d orbital, we have 18 electrons in the lower layers, 2 in the same layer, and those in the top layer are not considered. Therefore, the effective nuclear charge will be:

$$Z_{eff} = 23 - (18*1 + 2*0.35) = 4.30 \quad \text{with } \sigma = 18.70 \tag{2.9}$$

3. The europium atom has $Z=63$ and the electronic configuration $1s^2 2s^2 2p^6 3s^2 3p^6 3d^{10} 4s^2 4p^6 4f^7 5s^2 5p^6 6s^2$. The effective nuclear charge of one electron in layer 6s will be:

$$Z_{eff} = 63 - (53*1 + 8*0.85 + 1*0.35) = 2.85 \quad \text{with } \sigma = 60.15 \tag{2.10}$$

It can be seen from these examples that, as the atom size increases, the effective nuclear charge becomes bigger. Therefore, the attraction of an electron by the nucleus decreases as the electron moves farther away from the nucleus.

The possibility that an electron (an electron s in particular) is found in the inner layers of an atom—that is, as close as possible to the nucleus—is defined as *penetration of the electron*.

The wave function ψ^2 is proportional to the density of an electron at a certain point in space at a certain moment in time. This function consists of an angular and a radial distribution function. With the help of the last one, the probable distance (r) from an electron to the nucleus can be determined. (The notation R (r) refers to the radial wave function.) The radial distribution function is

$$4\pi r^2 R(r)^2 \tag{2.11}$$

Thus, the graphical representation of the radial distribution function leads to a maximum probability that an electron of the $n = 1$ layer, that is the 1s layer, to be very close to the nucleus. For the $n = 2$ layer, the graphical representation shows two maxima for the 2s orbital and a curve with a maximum for the 2p orbital. For $n = 3$, the graphical representation shows 3 maxima for 3s, 2 maxima for 3p, and one maximum for 3d (Figure 2.4) (Pop 2003; Brezeanu et al. 1990; Whitten et al. 1988).

It can be seen that the radial functions $R(r)^2$ are equal to zero between maximum values. These points are radial nodes. Their number increases with the principal

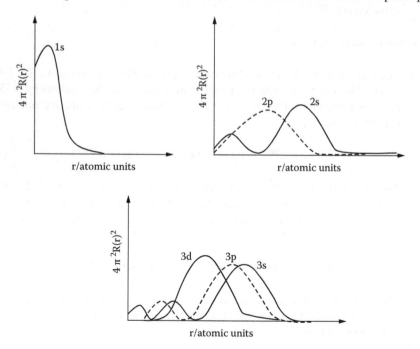

FIGURE 2.4   Graphs of radial electron density function for 1s, 2s, 2p, 3s, 3p, and 3d orbitals.

quantum number. In fact, the presence of these nodes shows that the ns orbitals are composed of concentric layers like an onion, and that at least one electron is much closer to the nucleus than the electron on the 2p orbital. Therefore, the 2s orbitals achieve more penetration than the 2p orbitals. In fact, the s orbitals are much more penetrating than the p or d of the same layer, that is, an electron on an s orbital has a higher probability of being near the nucleus than an electron on a p or d orbital. This is also the reason that s electrons have a higher shielding power than the electrons of other orbitals. It must be emphasized that the relative energies of the ns atomic orbitals are smaller than the energies of the corresponding $(n-1)$d orbitals.

Together, shielding and penetration phenomena, and the laws of Hund and Pauli, provide a partial explanation for the way the orbitals are filled by electrons. In a more simplistic way, Kleshkowski formulated the rule of the minimum $n+l$ sum that says, "the subshells are being arranged in ascending order of the $n+l$ sum, and for equal values of the sum, in an order corresponding to increasing principal quantum number n" (Wong 1979). Thus, if we write the sequence of symbols of the subshells arranged in ascending order of n and l, and the value of the $n+l$ sum, we get:

n	1s	2s	2p	3s	3p	3d	4s	4p	4d	4f	5s	5p	5d	5f	5g	6s	6p	6d	..	7s	7p	..
n+l	1	2	3	3	4	5	4	5	6	7	5	6	7	8	9	6	7	8	..	7	8	

With Kleshkowski's law we obtain the real order in which electrons fill the atomic subshells: 1s, 2s, 2p, 3s, 3p, 4s, 3d, 4p, 5s, 4d, 5p, 6s, 4f, 5d, 6p, 7s, 5f, 6d, 7p.... This law is illustrated in Figure 2.5 where the order for filling the orbitals of an atom is indicated by the orientation of the arrow.

If the maximum number of electrons in orbitals is included, the order for filling the orbitals of an atom can be written as follows:

$$1s^2\, 2s^2\, 2p^6\, 3s^2\, 3p^6\, 4s^2\, 3d^{10}\, 4p^6\, 5s^2\, 4d^{10}\, 5p^6\, 6s^2\, 4f^{14}\, 5d^{10}\, 6p^6\, 7s^2\, 5f^{14}\, 6d^{10}...$$

where the figure in front of the orbital is the principal quantum number and the superscript represents the number of electrons in the orbital. There are deviations from this rule. For example, known exceptions include lanthanum and actinium. Thus, before the filling with electrons of the 4f subshell, 5f subshell, respectively, the distinctive electron of La (respectively Ac) is placed on an orbital of the 5d subshell (respectively 6d subshell). Such exceptions also occur in the other d-block elements. Spectroscopic data show that d-block elements have the form $3d^n4s^2$ with the 4s orbitals filled, but chromium has the electronic structure $1s^2 2s^2 2p^6 3s^2 3p^6 4s^1 3d^5$ instead of $1s^2 2s^2 2p^6 3s^2 3p^6 4s^2 3d^4$. The energy decreases by putting electrons in odd orbitals. Such behavior is also found in copper, silver, and gold atoms.

It is obvious from the above that the orbital energies, or more precisely the order by which orbitals are filled, depends on the shielding effect, the effect of electron penetration, and the filling level of the neighboring orbitals.

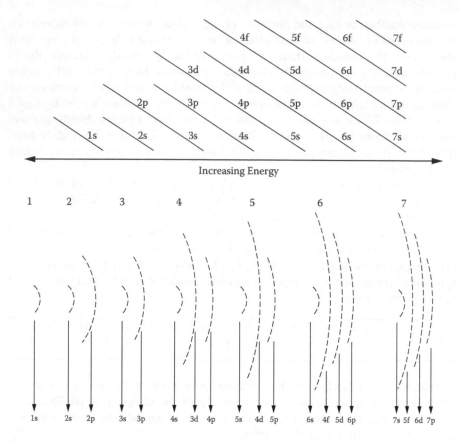

FIGURE 2.5   Order of filling of the shells by electrons.

2.3.2   BLOCKS OF METALLIC ELEMENTS

Elements are arranged in the periodic table set up by Mendeleev in 1875 according to their atomic number, forming the most important method for grouping related elements. The horizontal rows (periods) indicate the total number of electron shells in the atoms of the elements and the vertical columns (groups) include elements that have similar configurations of the outermost electron shells of their atoms.

The periodic table is divided into four blocks of elements (Figure 2.2):

- The s-block includes group 1 (the alkali metals) and group 2 (the alkaline earth metals) plus hydrogen and helium. The element hydrogen is placed in group 1 for convenience (International Union of Pure and Applied Chemistry [IUPAC] 2011) due to its electron configuration (ns^1) but it is not an alkali metal.
- The p-block is formed of the groups 13–18 (the nonmetals, metalloids, and a few metals). The elements in the s-block and p-block are called *representative elements*.

- The d-block includes groups 3–12 (the transition metals and post-transition metals).
- The f-block comprises the lanthanides and the actinides (the inner transition metals).

The representative elements have the valence electrons on their outermost electron shell (valence shell) (Tables 2.3 and 2.4), the transition elements have the valence electrons on a subshell in the penultimate electron shell (Table 2.5) and the inner transition elements have the valence electrons on a subshell that belongs to the ante-penultimate energy shell (Table 2.6).

In the s-block and p-block, there are differences between the first element and the rest of the elements in each group in the context of the number of electrons present in the penultimate electron shell (from 2 to 8 electrons). A similar situation can be seen between Al and Ga in group 13 (8 and 18 electrons, respectively, on the penultimate electron shell). Such differences in the electronic configuration of the ionic state are reflected in a certain variability in the properties of the elements and of their combinations, and is responsible for the diagonal analogies between some elements, such as Li–Mg, Be–Al, and B–Si.

Group 1 (I A) includes the most electropositive and reactive metals of the periodic table; they have large atomic volumes. Group 2 (II A) elements have much higher ionization energies compared to group 1 elements, but calcium, strontium, and barium are very electropositive and reactive. However, beryllium tends to form predominantly covalent bonds due to its high ionization potential and small ionic radius. The last element in both groups (francium and radium) is radioactive. They are homogeneous in their behavior and properties with differences being correlated to the variation of atomic and ionic size.

Due to their tendency to readily lose the outermost electrons, the alkali metals and the alkaline earth metals are considered some of the most powerful reducing agents. In nature, the alkali metals always exist in chemical combinations.

The reducing power of these elements in aqueous solution decreases in the order $Cs > Li > Rb > K > Ba > Sr > Ca > Na > Mg > Be$. Based on their ionization potentials,

TABLE 2.3
Electron Configuration of the Atoms of the Elements in s-Block

Period	Group 1		Group 2	
		1		2
2	Li	$[He]2s^1$	Be	$[He]2s^2$
3	Na	$[Ne]3s^1$	Mg	$[Ne]3s^2$
4	K	$[Ar]4s^1$	Ca	$[Ar]4s^2$
5	Rb	$[Kr]5s^1$	Sr	$[Kr]5s^2$
6	Cs	$[Xe]6s^1$	Ba	$[Xe]6s^2$
7	Fr	$[Rn]7s^1$	Ra	$[Rn]7s^2$

TABLE 2.4

Electron Configuration of the Atoms of the Elements in p-Block

Period	Group				
	13	14	15	16	17
2	B $[He]2s^22p^1$	C $[He]2s^22p^2$	N $[He]2s^22p^3$	O $[He]2s^22p^4$	F $[He]2s^22p^5$
3	Al $[Ne]3s^23p^1$	Si $[Ne]3s^23p^2$	P $[Ne]3s^23p^3$	S $[Ne]3s^23p^4$	Cl $[Ne]3s^23p^5$
4	Ga $[Ar]3d^{10}4s^24p^1$	Ge $[Ar]3d^{10}4s^24p^2$	As $[Ar]3d^{10}4s^24p^3$	Se $[Ar]3d^{10}4s^24p^4$	Br $[Ar]3d^{10}4s^24p^5$
5	In $[Kr]4d^{10}5s^25p^1$	Sn $[Kr]4d^{10}5s^25p^2$	Sb $[Kr]4d^{10}5s^25p^3$	Te $[Kr]4d^{10}5s^25p^4$	I $[Kr]4d^{10}5s^25p^5$
6	Tl $[Xe]4f^{14}5d^{10}6s^26p^1$	Pb $[Xe]4f^{14}5d^{10}6s^26p^2$	Bi $[Xe]4f^{14}5d^{10}6s^26p^3$	Po $[Xe]4f^{14}5d^{10}6s^26p^4$	At $[Xe]4f^{14}5d^{10}6s^26p^5$

TABLE 2.5
Electron Configuration of the Atoms of Elements in the d-Block

Group		Period 4		5		6
3	Sc	$[Ar]3d^14s^2$	Y	$[Kr]4d^15s^2$	La	$[Xe]5d^16s^2$
4	Ti	$[Ar]3d^24s^2$	Zr	$[Kr]4d^25s^2$	Hf	$[Xe]4f^{14}5d^26s^2$
5	V	$[Ar]3d^34s^2$	Nb	$[Kr]4d^45s^1$	Ta	$[Xe]4f^{14}5d^36s^2$
6	Cr	$[Ar]3d^54s^1$	Mo	$[Kr]4d^55s^1$	W	$[Xe]4f^{14}5d^46s^2$
7	Mn	$[Ar]3d^54s^2$	Tc	$[Kr]4d^55s^2$	Re	$[Xe]4f^{14}5d^56s^2$
8	Fe	$[Ar]3d^64s^2$	Ru	$[Kr]4d^75s^1$	Os	$[Xe]4f^{14}5d^66s^2$
9	Co	$[Ar]3d^74s^2$	Rh	$[Kr]3d^84s^1$	Ir	$[Xe]4f^{14}5d^76s^2$
10	Ni	$[Ar]3d^84s^2$	Pd	$[Kr]4d^{10}5s^0$	Pt	$[Xe]4f^{14}5d^96s^1$
11	Cu	$[Ar]3d^{10}4s^1$	Ag	$[Kr]4d^{10}5s^1$	Au	$[Xe]4f^{14}5d^{10}6s^1$
12	Zn	$[Ar]3d^{10}4s^2$	Cd	$[Kr]4d^{10}5s^2$	Hg	$[Xe]4f^{14}5d^{10}6s^2$

TABLE 2.6
Electron Configuration of the Atoms of the Elements in the f-Block

Period 6		7	
Ce	$[Xe]4f^15d^16s^2$	Th	$[Rn]5f^06d^27s^2$
Pr	$[Xe]4f^35d^06s^2$	Pa	$[Rn]5f^26d^17s^2$
Nd	$[Xe]4f^45d^06s^2$	U	$[Rn]5f^36d^17s^2$
Pm	$[Xe]4f^55d^06s^2$	Np	$[Rn]5f^46d^17s^2$
Sm	$[Xe]4f^65d^06s^2$	Pu	$[Rn]5f^66d^07s^2$
Eu	$[Xe]4f^75d^06s^2$	Am	$[Rn]5f^76d^07s^2$
Gd	$[Xe]4f^75d^16s^2$	Cm	$[Rn]5f^76d^17s^2$
Tb	$[Xe]4f^95d^06s^2$	Bk	$[Rn]5f^96d^07s^2$
Dy	$[Xe]4f^{10}5d^06s^2$	Cf	$[Rn]5f^{10}6d^07s^2$
Ho	$[Xe]4f^{11}5d^06s^2$	Es	$[Rn]5f^{11}6d^07s^2$
Er	$[Xe]4f^{12}5d^06s^2$	Fm	$[Rn]5f^{12}6d^07s^2$
Tm	$[Xe]4f^{13}5d^06s^2$	Md	$[Rn]5f^{13}6d^07s^2$
Yb	$[Xe]4f^{14}5d^06s^2$	No	$[Rn]5f^{14}6d^07s^2$
Lu	$[Xe]4f^{14}5d^16s^2$	Lr	$[Rn]5f^{14}6s^17s^2$

the calculated reducing power would decrease according to the order: $Cs > Rb > K > Na > Ba > Li > Sr > Ca > Mg > Be$. These two series coincide if lithium and sodium are excluded.

Group 13 (III A) includes boron (a metalloid) that has a nonmetal behavior, and aluminum, gallium, indium and thallium (poormetals). These elements have lower electropositive character than the s-block metals. Gallium, indium, and thallium are post-transitional metals, so their nuclear effective charges are higher than those of boron and aluminum. Their first ionization energy is higher than that of aluminum. Due to the very high ionization energies for Al^{3+}, Ga^{3+}, and In^{3+}, the compounds of these elements in the oxidation state +3 are mainly covalent, yet these ions are formed easily in solution due to their high enthalpy of hydration.

Group 14 (IV A) (the carbon group) includes a nonmetal (carbon), silicon, and germanium, which are metalloids (Si behaves as a nonmetal mostly, but Ge behaves as a metal) and the typical metals, tin and lead. These elements have 4 electrons in the outer energy level, 2 of which are in the last p orbitals (p^2) (Table 2.4). These 4 electrons are usually shared; silicon and germanium can form +4 ions, tin and lead can form +2 ions, and only carbon can form −4 ions.

Group 15 (V A) (the nitrogen group) has an electronic configuration ns^2np^3, with five valence electrons, three of which are unpaired electrons in the p subshell. These are sometimes shared, in double or triple bonds of very stable compounds. Nitrogen and phosphorus are nonmetals. The group includes two metalloids (As behaves as a nonmetal mostly, while Sb as a metal), and one metal (Bi). The gradual variation of the chemical character in the group is linked to the decrease of the energy of ionization and to the increase of the atomic radii.

Similar to the other groups that have been presented, group 16 (VI A) (the chalcogens or the oxygen family) starts with an element whose physical properties and chemical reactivity do not resemble those of the rest of the group. Here also, the metallic character increases down the group: oxygen and sulphur are nonmetals, selenium and tellurium are considered metalloids (they are referred to as a metal when in elemental form), while radioactive polonium is classified either as a post-transitional metal or metalloid (Hawkes 2010; Bentor 2011).

The properties and behavior of p-block metals are less homogenous than those of the s-metals because of their atomic and ionic sizes and other characteristics such as their unusual crystalline structures and the presence of the pair of electrons on the s orbital (s^2) of the outermost electron shell. These s^2 orbital electrons do not participate in chemical interactions (covalent or ionic) (*inert electron pair*). This influence increases with atomic number within a group.

The d-block is situated in the middle of the periodic table and includes groups 3 to 12 (III B, IV B, V B, VI B, VII B, VIII B, IX B, X B, I B, II B) (Figure 2.2). Within the d-block there are metals known as *transition metals* or *transition elements*, and also some post-transition metals (group 12). The term *post-transition metal* is sometimes used to describe the elements that have metallic character and are situated at the right of the transition elements.

The d-block transition metals/elements are considered those elements that have, as their name suggests, the d-orbital subshells partly occupied in either the neutral atom or ionic state (IUPAC 2004). Accordingly, the following metals can be included in this group:

- Copper, silver, and gold, which have a completely filled $(n-1)d^{10}ns^1$ subshell in atomic state, but have a partly filled d orbital in ionic state: Cu(II) $3d^9$, Ag(II) $4d^9$ and Au(III) $5d^8$.

- The elements in the group III B: scandium, yttrium, lanthanum and actinium that have an incompletely filled d subshell in their atomic state: $(n-1)d^1$ ns^2. Although both lanthanum and actinium could be included in the d transition metal series, they are very similar physically and chemically to the elements in the f-block and therefore are considered to be f-type transition elements (4f-, 5f-type transition elements, respectively). The last element of the lanthanides series, lutetium, also has a partly filled d orbital (Table 2.6) and could also be included in the d transition metal group. However, it has similar properties to the 4f-type transition metals, where it is usually grouped with lanthanum and the rest of the lanthanides series.

Zinc, cadmium, and mercury are considered d-block elements, but are not included in the transition metals category because their d-shell is full in the atomic state (the electronic configuration is $d^{10}s^2$) and they only have the oxidation state two with the electronic configuration of their ions: $d^{10}s^0$. Because they have a similar chemical behavior to the transition metals, they are studied along with the elements in the d-block.

The d-block elements can have variable chemical reactivities compared to the other metallic elements, either representative elements or f-type transition metals. They can have several oxidation states (Tables 2.7 and 2.8), and have a higher tendency to form metal complexes. Metal cations can form coordination complexes with different types of electron-donating groups, and these complexes are the active species in biological environments (Krantzberg 1989; Lepădatu et al. 2009).

The f-block elements are the lanthanides and the actinides. Each of these series consists of 14 elements with similar properties in both atomic state and in their combination. They are situated in the 6th and 7th period of the periodic table, respectively, but are customarily separated from the main body of the table. These elements fill their 4f and their 5f orbital, respectively (Table 2.6). The lanthanides are more reactive than the d-block metals, behaving more like the alkali earth metals. Actinides are all very radioactive and therefore have high lethality. The actinides, with atomic numbers higher than 92, have been synthesized during nuclear reactions.

TABLE 2.7
Characteristic Oxidation States of s, p, d and f Metals

s-Type Metals	p-Type Metals	d-Type Metals	f-Type Metals
Have just one oxidation state	Have two oxidation states, 2 units difference between them	The +2 oxidation state is present; there are also several higher oxidation states, one unit difference between them	The lanthanides: +3 almost exclusively
Group 1: +1			The actinides: +3, however the first 5
Group 2: +2	Group 13: +1; +3		elements can have
	Group 14: +2; +4		several higher
	Group 15: +3; +5		oxidation states

TABLE 2.8
Some Essential Trace Elements and Their Oxidation States

Transition Metal	Oxidation Number
Molybdenum	III
	IV
	V
	VI
Manganese	II
	III
	IV
	VI
	VII
Iron	II
	III
	IV
	VI
Cobalt	II
	III
	IV
Nickel	II
	III
Copper	I
	II
	III
Zinc	II

2.3.3 ATOMIC RADII

The atomic radii describe the characteristic size of the neutral and isolated atoms that are not involved in any kind of bonds. The radius of the free atom is considered the distance from the nucleus to the point of maximum electron radial density of the occupied atomic orbital.

The atomic radius cannot be determined directly. The electron cloud is diffuse due to the undulatory waveform nature of the electron, making it impossible to assign a well-defined outer structure to the atom. One indirect method for the determination of the relative atomic radius is the use of the formula derived from Bohr's atomic model. The formula for the hydrogen atom is the following:

$$r_H = \frac{\varepsilon_0 h^2}{\pi m e^2} = 0.53 \ \text{Å} \tag{2.12}$$

where all measures are universal constants, namely:

- ε_0 is the permittivity of vacuum ($8.85*10^{-12}$ F/m),
- h is Planck's constant ($6.63*10^{-34}$ Js),

- m is the mass of the electron ($9.1*10^{-31}$ kg), and
- e is the charge of the electron ($1.6*10^{-19}$ C).

The radius of the hydrogen atom, known as the Bohr radius, is used as a unit of atomic radius. For the other atoms, the calculation of the atomic radius of the free atom involves the relationship:

$$r = \frac{(n_{eff})^2}{Z_{eff}} r_H \tag{2.13}$$

where n_{eff} is the effective principal quantum number, and Z_{eff} is the effective nuclear charge.

These measures are relative sizes of atoms because they cannot be measured accurately as might be done for some tangible objects such as a ball or wheel. The relative atomic radii according to the corresponding atomic numbers are shown in Figure 2.6.

It is very interesting that the atomic radii decrease within a period according to the atomic number and increase within groups. The resulting tendency is an increase in the size of atoms from the right to the left of the periodic table, and from the top to the bottom (of groups) (Figure 2.7).

Although with increasing atomic number in period the number of electrons and number of occupied electrons increase, the atomic radius decreases. This results from the increase in the effective nuclear charge that produces a "contraction" in periods of the orbitals toward the nucleus. In those cases where the outer electron layer is the same valence layer, the increasing attraction of electrons by the nucleus leads to a decrease in atomic radius. In groups, the atomic radius increases due to the increasing number of electronic layers.

A special case is that of lanthanides where the 4f subshell is occupied progressively before the filling of the 5d orbital from cerium (Z=58) to lutetium (Z=71) (Table 2.9). The f orbitals have weak shielding and penetration properties so that the electrons are

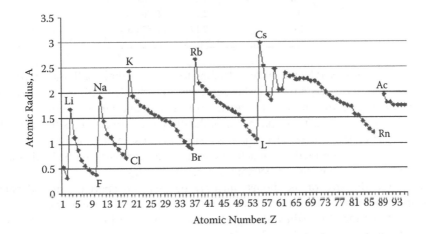

FIGURE 2.6 The variation of the atomic radius according to the atomic number.

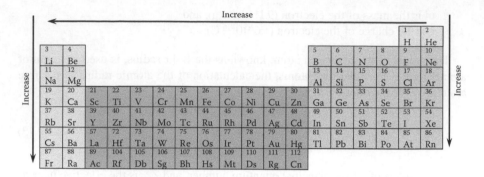

FIGURE 2.7   General trend of the increase of atomic radii in the periodic table.

TABLE 2.9

The Lanthanide Contraction

Element and Atomic Number	Å	Xe
[58]Ce	1.85	$4f^1 5d^1 6s^2$
[59]Pr	2.47	$4f^3 6s^2$
[60]Nd	2.06	$4f^4 6s^2$
[61]Pm	2.05	$4f^5 6s^2$
[62]Sm	2.38	$4f^6 6s^2$
[63]Eu	2.31	$4f^7 6s^2$
[64]Gd	2.33	$4f^7 5d^1 6s^2$
[65]Tb	2.25	$4f^9 6s^2$
[66]Dy	2.28	$4f^{10} 6s^2$
[67]Ho	2.26	$4f^{11} 6s^2$
[68]Er	2.26	$4f^{12} 6s^2$
[69]Tm	2.22	$4f^{13} 6s^2$
[70]Yb	2.22	$4f^{14} 6s^2$
[71]Lu	2.17	$4f^{14} 5d^1 6s^2$

more strongly attracted by the nucleus, leading to smaller atomic radii than expected. Moreover they have very similar, and in some cases nearly identical, sizes. This effect in the two series is known as the *lanthanide contraction* or *actinide contraction*.

Similarly, the contractions that occur in the 4th period, in elements that have the d orbital completely filled (d^{10}), occur as a result of poor shielding of the electrons in the d orbital. This contraction is much less pronounced compared to the lanthanides. The elements that show this effect are Ga, Ge, As, Se, and Br.

However, the size of the atoms can only be assessed approximately depending upon the bonds that form between two adjacent atoms. Several measures are described: the metallic radius, the covalent radius, and the ionic radius. The most straightforward, the metallic radius, is equal to half of the distance between two adjacent atoms in the metallic network.

The covalent radius of nonmetallic element is calculated according to the homonuclear or heteronuclear nature of the molecules. Thus, in the case of homonuclear molecules (Cl_2, Br_2, graphite) in solid state, this radius is equal to half the experimentally measured distance between the nuclei of two neighboring atoms. For heteronuclear molecules the distance between the two covalent bound atoms is experimentally determined. From that value, the value of the known radius is subtracted in order to obtain the unknown radius. For example, the length of the covalent Cl–Cl bond is 1.998 Å, therefore the covalent radius of the chlorine atom is 0.994 Å. If the covalent C–Cl bond is 1.766 Å, the value of the known covalent radius of the chlorine atom is subtracted and so the covalent radius of C is obtained, that is 0.722 Å.

Besides the covalent radius, there is also the Van der Waals radius, which is characteristic for atoms in covalent compounds. This radius represents the shortest distance between the atoms that are not chemically bound, that is, the distance at which they can approach without the electronic clouds repelling each other.

The ionic radius is difficult to measure because it varies depending on features such as the environment surrounding the ion and the number of ions bound to a particular ion. However, there are different methods that can be used to measure this radius. The most used are those of Landé and of Pauling. The ionic radius of an element is the contribution of each ion to the distance between the nuclei of two neighboring ions or the distance between the nucleus of a cation and its neighboring anion is the sum of the radii of the two ions. The radius of the positive ions is always smaller than the radius of the atoms of origin because the change to the cation state is made by the removal of one or more electrons from the outer layer, which increases the effective nuclear charge. It follows that the anion radius is always larger than that of the corresponding atoms because electron addition decreases the electric nuclear charge.

Also, the ionic radii change with the oxidation state of the ion, with the increase of the oxidation state leading to a decrease in ionic radius. Figures 2.8, 2.9, and 2.10 compare the atomic radius, ionic radius, and the covalent radius for periods 1 and 2, and period 3 and period 4 elements, respectively (Brezeanu et al. 1990; Whitten et al. 1988; Housecroft and Constable 1997). It can be seen from these figures that covalent radii follow the same general trend as the ionic radii: they have even smaller values than the ionic radii for metals and higher values for the nonmetals.

The d-block elements have similar properties from this point of view, and because they are found together in natural minerals, they are difficult to separate.

In the case of lanthanides and actinides, the ionic radii decrease greatly and the differences between the atomic radii and ionic radii are significant. They are much bigger than those of other elements. If other elements display 10–20% decreases, in lanthanides and actinides the ionic radii would decrease by 50–60%. Also, among lanthanide elements and actinide elements, respectively, the differences between the atomic and the ionic radii are quite small. They are practically equal in most instances (Figures 2.11 and 2.12).

2.3.4 IONIZATION ENERGIES

The ease with which an atom forms positive ions by electron transfer is dependent on its ionization energy (IE). The *ionization energy*, or the *ionization potential*, is the

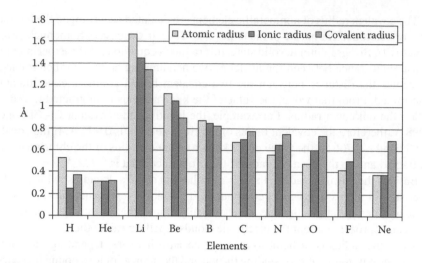

FIGURE 2.8 The variation of atomic radii, ionic radii, and covalent radii for the elements in periods 1 and 2.

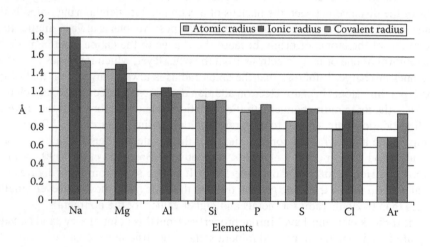

FIGURE 2.9 The variation of atomic radii, ionic radii, and covalent radii for the elements in period 3.

minimum quantity of energy required for the removal of one or more electrons from an atom or from an isolated ion that exists in a gaseous state. For this endothermic process, the ionization energy is expressed in Joules or electron volts ($1eV = 1.602 *10^{-19}$ J). For example, for the hydrogen atom to lose an electron requires 1310kJ/mol:

$$H(g) + 1310 \text{ kJ} \rightarrow H^+(g) + e^- \tag{2.14}$$

Each electron in an atom corresponds to a specific ionization energy. Thus, an initial energy IE_1 is required to remove the first electron, which is the weakest bound in

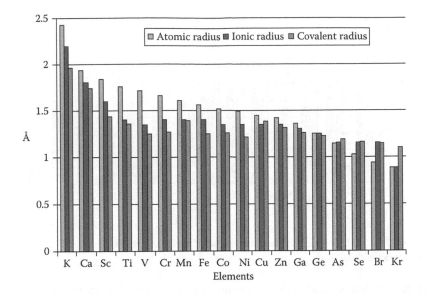

FIGURE 2.10 The variation of atomic radii, ionic radii, and covalent radii for the elements in period 4.

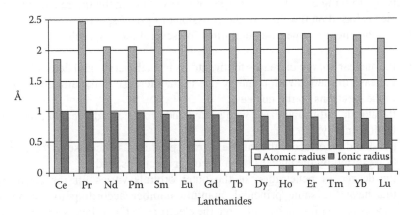

FIGURE 2.11 The variation of atomic and ionic radii for the lanthanides.

the neutral atom, and a secondary energy IE_2 is used to remove the next electron, and so on. For example, IE_1 is 737 kJ/mol, and for IE_2 it is 1449 kJ/mol for magnesium:

$$Mg(g) + 737 \text{ kJ} \rightarrow Mg^+(g) + e^- \qquad (2.15)$$

$$Mg^+(g) + 1449 \text{ kJ} \rightarrow Mg^{2+}(g) + e^- \qquad (2.16)$$

Following the removal of one electron, the electrostatic repulsion among the rest of the electrons decreases. The remaining electron cloud becomes a lot more compact so

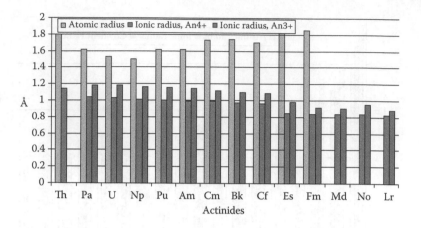

FIGURE 2.12 The variation of atomic and ionic radii for the two oxidation states 3+ and 4+ of the actinides.

that the removal of another electron requires more energy. The secondary ionization energy is bigger than the initial energy, the third is bigger than the second, and so on.

The ionization energy is one of the few properties of the atom that can be measured directly by studying electrical discharges of gases. Among the indirect methods, the most used is based on the study of the emission of optical spectra.

Ionization energy values are determined by several factors, such as:

- the nuclear charge increases for the atoms in each period and determines an increase in the ionization energy due to the increase in the attraction force of the electrons by the nucleus,
- the atomic radius increases with period and produces a decrease of the ionization energy,
- the shielding effect of the inner electrons produces a decrease of the attraction exerted by the nucleus upon the outer electrons, and
- the degree of penetration of the outer orbitals varies for the electrons that have the same principal quantum number according to the order $ns > np > nd > nf$, which means that the electrons in the ns layers are the most strongly retained, followed by the np electrons, nd electrons, and so on.

The ionization energy is dependent on all these factors and they must be considered in order to obtain the correct evaluation.

Figure 2.13 shows the modification of the first ionization energy according to the atomic number. It can be seen that for each period, the first ionization energy increases to maximum values in rare gases, which have very stable ns^2np^6 structures.

Minimum values are found in the elements in group 1 because the distinctive electron that is lost is situated on a new energetic layer. The atomic radius increases compared to the radius of the atoms in the previous period, which determines a stronger shielding compared to the electron of interest. There are remarkably high values for ionization energies of the elements situated at the end of the periods in

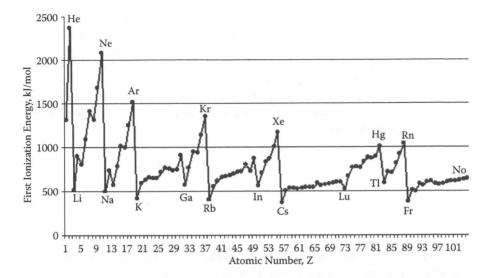

FIGURE 2.13 The variation of the first ionization energy according to the atomic number Z.

the d-block, i.e., zinc, cadmium and mercury. This situation is a consequence of the more stable electron configurations of the type: $(n-1)d^{10}ns^2$. The ionization energy is decreasing per group in general from period 2 to 7. This is a result of increasing atomic radius and a concurrent increase in the shielding effect.

With some exceptions, the first ionization energy increases generally from left to right in periods in the periodic table and decreases downward in groups. The nuclear effective charge (Z_{eff}) increases in periods, and as a consequence, it is natural that the valence electrons are more strongly linked to the nucleus (Housecroft and Constable 1997). The number of layers in groups increases with each layer that is further from the nucleus. The electrons become more amenable to removal.

The ionization energy increase is not uniform in periods. In each period there are some values that are bigger or smaller than expected. In group 13 (B, Al, Ga, In, Tl) for example, as we expect, the first ionization energy decreases sharply from boron to aluminum, but it is slightly higher for gallium than for aluminum. The explanation is that there is the first row of d-block elements before gallium, such that gallium has 10 more positive charge units and 10 electrons on the 3d orbitals. Although the d electrons shield less than the s or p electrons, the outer electrons are very strongly attracted by the nucleus so that the ionization energy is slightly higher than that of Al despite the larger size of the atom. The same phenomenon is found for indium, which has slightly lower ionization energy than gallium. For thallium there is a significant increase of the ionization energy and Tl has the highest value in its group after boron. This can be explained by the presence of the row of elements that fill the 4f subshell before thallium, resulting in an increase of the effective nuclear charge and simultaneous decrease of the energy of the 6p orbitals.

The elements whose atoms have low ionization energy values have metallic character and are situated on the bottom left side of the periodic table.

The *electron affinity* (EA) is defined as the energy that is absorbed or released by an atom or ion, in the gaseous state, to capture an electron of zero kinetic energy. It is expressed in Joules (J) similar to the ionization energy or electron volts (eV). Unlike the ionization energy, electron affinities are measured by indirect methods, and relative to the ionization energy, have less accurate values. If energy is absorbed during electron attachment to an atom, the value of the energy is designated as positive. If energy is released, the energy is negative. For example, the electron affinity of magnesium and fluoride are:

$$Mg(g) + e^- + 231\, kJ \rightarrow Mg^- \qquad\qquad EA = +231\, kJ \qquad\qquad (2.17)$$

$$F(g) + e^- \rightarrow F^- + 322\, kJ \qquad\qquad EA = -322\, kJ \qquad\qquad (2.18)$$

Therefore, the first reaction is endothermic and the second is exothermic.

Atoms, such as those of the halogens, have high electron affinity in absolute value and in negative value are considered highly electronegative.

The value of the electron affinity depends on numerous factors of which we can mention the atomic radius and the nuclear effective charge, in particular. Generally, the value of the electron affinity decreases with an increase in atomic radius, and increases with a decrease in shielding effect. Consequently, it will depend largely on the type of orbital accepting the electron. Thus, the energy released when the first electron is accepted is greater when it is accepted to the ns orbital. The energy released decreases in order of np, nd, and nf orbitals.

Group 16 and 17 elements have the highest electron affinities. The alkaline earth elements (group 2) and the noble gases (group 18) do not form stable negative ions. But, as shown in Figure 2.14, some d-block elements also have electron affinity equal to zero (Mn, Zn, Cd, Hf, and Hg).

The ionization energies and the electron affinities of the atoms are used to determine the electrochemical character of the elements (their electropositivity and electronegativity). The electronegativity (tendency of an atom to accept electrons) and the electropositivity (the tendency of an atom to give electrons) are complementary properties, so that the electrochemical character of the elements can be defined by only one of these two properties. The higher the electron affinity is, the higher the ionization energy. The more electronegative elements are nonmetals. The elements with low ionization energy and electron affinity are metals that are electropositive. Elements that show both metallic and nonmetallic character are metalloids.

The electronegativity increases in periods as Z_{eff} increases and atomic radius decreases. In the main groups, the electronegativity decreases as the atomic radius increases and Z_{eff} decreases. In the secondary groups, this property is irregular. In elements in the d-block and f-block in lower oxidation states (where ns or $(n-1)$ d^1ns^2 electrons participate only) the electropositive (metallic) character prevails. This character is less pronounced than that of the typical metallic elements because of the contraction of the atomic orbitals. In higher oxidation states these elements have a weak electropositive character.

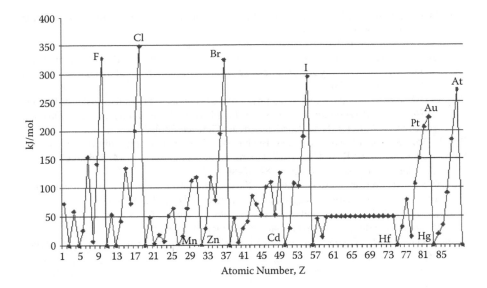

FIGURE 2.14 The electron affinity of the elements.

REFERENCES

Atkins, P.W. 1994. *Physical Chemistry*, 5th edition. Oxford: Oxford University Press.

Atkins, P.W., and J.A. Beran. 1990. *General Chemistry*, 2nd edition. New York: Scientific American Books.

Bentor, Y. 2011. Polonium. Chemical Elements.com, http://www.chemicalelements.com/elements/po.html (accessed December 4, 2011).

Borovik, A.J. 1990. Characterizations of metal ions in biological systems. In *Heavy Metal Tolerance in Plants: Evolutionary Aspects*, edited by A.J. Shaw, 1–5. Boca Raton, FL: CRC Press.

Brezcanu, M., E. Cristurean, D. Marinescu, A. Antoniu, and M. Andruh. 1990. *Chimia metalelor.* Bucharest: Editura Academiei Romane.

Duffus, J.H. 2002. "Heavy metals" a meaningless term? (IUPAC Technical Report). *Pure Appl. Chem.* 74 (5):793–807.

Hawkes, S.J. 2010. Polonium and astatine are not semimetals. *J. Chem. Educ.* 87 (8):783–783.

Housecroft, C.E., and Constable, E.C. 1997. *Chemistry: An Integrated Approach*. New York: Addison Wesley Longman.

International Union of Pure and Applied Chemistry (IUPAC). 2004. *Provisional Recommendations for the Nomenclature of Inorganic Chemistry*, IUPAC (online draft of an updated version of the *Red Book* IR 3-6), http://old.iupac.org/reports/provisional/abstract04/connelly_310804.html.

International Union of Pure and Applied Chemistry (IUPAC). 2011. Periodic Table of the Elements, IUPAC, http: //old.iupac.org/reports/periodic_table/ (accessed May 1, 2011).

Krantzberg, G. 1989. Accumulation of essential and nonessential metals by chironomid larvae in relation to physical and chemical properties of the elements. *Can. J. Fish. Aquat. Sci.* 46:1755–1761.

Lepădatu, C., M. Enache, and J.D. Walker. 2009. Toward a more realistic QSAR approach to predicting metal toxicity. *QSAR Comb. Sci.* 28 (5):520–525.

Lide, D.R. (ed.). 2008–2009. *Handbook of Chemistry and Physics*, 89th edition. Boca Raton, FL: CRC Press, Taylor & Francis Group.

Massey, A.G. 1990. *Main Group Chemistry*, New York: Ellis Horwood Limited.

Nieboer, E., and D.H.S. Richardson. 1980. The replacement of the nondescript term "heavy metals" by a biologically and chemically significant classification of metal ions. *Environ. Pollut.*, Ser. B, 1:3–26.

Phipps, D.A. 1980. Chemistry and biochemistry of trace metals in biological systems. In *Effect of Heavy Metal Pollution on Plants. Vol 1. Effects of Trace Metals on Plant Function*, edited by N.W. Lepp, 1–54. London: Applied Science Publishers.

Pop, V. 2003. *Métaux et leurs composés*. Bucharest: Editura Universitatii din Bucuresti.

Schwietert, C.W., and J.P. McCue. 1999. Coordination Compounds in Medicinal Chemistry. *Coordination Chemistry Reviews* 184:67–89.

Walker, J.D., M. Enache, and J.C. Dearden. 2003. Quantitative cationic-activity relationships for predicting toxicity of metals. *Environ. Toxicol. Chem.*, 22:1916–1935.

Whitten, K.W., K.D. Gailey, and R.E. Davis. 1988. *General Chemistry with Qualitative Analysis*, 3rd edition. New York: Saunders College Publishing.

Wong, D.P. 1979. Theoretical justification of Madelung's rule. *J. Chem. Educ.*, 56(11):714–718.

3 Properties of Metals and Metal Ions Related to QSAR Studies

3.1 PROPERTIES OF METALS AND METAL IONS AS TOOLS IN QUANTITATIVE STRUCTURE-ACTIVITY RELATIONSHIP (QSAR) STUDIES

A complex set of interrelationships involving physical, chemical, biological, and pharmacological factors are involved when a metal ion interacts with a biological system. Therefore, it is more important to characterize the metal–biological system than to characterize the species in a simple chemical system. In a comprehensive work, Walker et al. (2003) reviewed approximately 100 diverse contributions dating from 1835 to 2003 to evaluate the relationships between about 20 physicochemical properties of cations and their potential to produce toxic effects in different organisms.

Data are presented in the form of ionic properties, surrogate metal ion characteristics described to reflect metal–ligand binding tendencies, thermodynamic considerations, and equilibrium constants of metal ions with a variety of inorganic and organic ligands of biological significance. Some of these physicochemical characteristics described in this review are atomic number (AN), atomic weight (AW), melting point (m.p.), boiling point (b.p.), density (ρ), molar refractivity (MR), electric dipole polarizability for ground-state atoms (α), atomic radius, atomic electron affinity (EA), electronegativity (X, X_{AR}, or X_m), ionization energy (I), total ionization potential (ΣI_n), ionization potential differential (ΔIP), electrochemical potential, standard reduction-oxidation potential or absolute value of the electrochemical potential between the ion and its first stable reduced state (ΔE°), or standard reduction potential or standard electrode potential (E°), first hydrolysis constant (|log KOH|), softness parameter (σ_P; σ_A; σ_W), stability constant of a metal ion with ethylenediamine-tetraacetic acid (EDTA) and sulfate, heat of atomization of the elements (ΔH_{atom}), enthalpy of formation of the elements (ΔH°_f), enthalpy of fusion (ΔH°_{298}), enthalpy and Gibbs energy of hydration (ΔH°_h; ΔG°_h), coordinate bond energy of halides (CBE F; CBE Cl; CBE Br; CBE I), crystal lattice energy of the elements and lattice energy of metal inorganic compounds (LE), enthalpy of formation of oxides and sulfides (ΔH_o, ΔH_s), ion charge (Z), ionic potential (charge to radius ratio; Z/r), the ionic index (Z^2/r), the covalent index ($X^2_m r$), AN/ΔIP, log AN/ΔIP, and finally, ZX_{III}.

Qualitative and, especially, quantitative correlation between toxicity and physicochemical properties of metal ions may be useful in predicting toxicity to biologically important organisms, and to identifying and interpreting modes of toxic action. The qualitative correlations are expressed by structure-activity relationships SARs), which relates a (sub)structure to the presence or absence of a property or activity. The quantitative correlations were expressed as quantitative structure-activity relationships (QSARs), quantitative ion character-activity relationships (QICARs), and quantitative cationic activity relationships (QCARs).

QSARs include statistical methods to relate biological activities (most often expressed by logarithms of equipotent molar activities) with structural elements (Free Wilson analysis), physicochemical properties (Hansch analysis), or fields (3D QSAR). The parameters used in a QSAR model are also called (molecular) *descriptors*. Classical QSAR analyses (Hansch and Free Wilson analyses) consider only 2D structures. Their main field of application is in substituent variation of a common scaffold. 3D-QSAR analysis (CoMFA) has a much broader scope. It starts from 3D structures and correlates biological activities with 3D-property fields (McKinney et al. 2000).

QICARs use the metal–ligand bonding characteristics to predict metal ion toxicity (Newman et al., 1998). In general, the models developed for metals with the same valence were better than those combining mono-, di-, and trivalent metals. The metal ion characteristics included a softness parameter and the absolute value of the log of the first hydrolysis constant. The first stable reduced state also contributed to several two-variable models. Since most metals can interact in biological systems as cations and because toxicity of metals depends on cationic activity, the term (quantitative) *cationic-activity relationships* or (Q)CARs also describes the qualitative and quantitative relationships for predicting the bioconcentration, biosorption, or toxicity of metals, from their physicochemical properties and natural occurrence levels.

3.2 ELECTROCHEMICAL CHARACTERISTICS OF METALS

Electrochemical characteristics of metals include properties such as ionization potential, electron affinity, electropositivity, electronegativity, oxidation state, and standard electrode potential.

The ionization potential IP (or ionization energy IE) is a measure of capability of a metal to form the corresponding positive ion. The electron affinity (EA) is a measure of the capability of an ion to form the corresponding negative ion. Both properties are discussed in Chapter 2. Unlike the ionization potential and the electron affinity, which are atomic parameters characteristic of the isolated atoms, electronegativity is a chemical characteristic of bonded atoms. Ionization potential, electron affinity, and electronegativity are considered energy-base descriptors (Todeschini and Consonni 2008). The standard electrode potential is a characteristic of bulk metal reflecting the capacity of a metal to generate positive ions in aqueous solution.

3.2.1 ELECTRONEGATIVITY

Electronegativity is a chemical property that describes the power of an atom in a molecule to attract electrons to itself (Pauling 1932). Unlike ionization potential and electron affinity, characterized by associated energies that are determined experimentally, electronegativity is a qualitative concept that is not amenable to direct measurement. Electronegativity was evaluated by various indirect methods that resulted in several electronegativity scales named after their originator. Experience has shown that they lead to scales that have similar values, although different methods were used.

The Pauling scale, named after the chemist Linus Pauling, is based on thermodynamic data associated with bond energy, the energy that must be provided to break a bond between two atoms. Considering formation of a polar molecule AB from nonpolar molecules A_2 and B_2, the bond A–B contains a supplementary energy relative to the average of energies A–A and B–B. Pauling considers this energy as an ionic contribution to the covalent bond, correlated with the electronegativity difference between atoms A and B:

$$E_{A-B} = \frac{1}{2}\left(E_{A-A} + E_{B-B}\right) + \Delta \qquad (3.1)$$

where Δ is a bonding extra-ionic energy.

If χ_A and χ_B are the electronegativities of the atoms A and B, Pauling proposed the following formula:

$$|\chi_A - \chi_B| = 0.102\sqrt{\Delta} \qquad \text{with } \Delta \text{ calculated in kJ·mol}^{-1}. \qquad (3.2)$$

The scale is chosen so that the relative electronegativity of hydrogen (χ_H) would be 2.1. With the exception of the noble gases, dimensionless values of relative electronegativity are calculated for all chemical elements relative to the hydrogen value. Variation of the electronegativity (Pauling scale) is dependent on the atomic number Z for the elements from the first three periods (Figure 3.1).

Electronegativity shows the same variation as ionization potential; 1 that is, it increases in a period from left to right and decreases from top to bottom in a group.

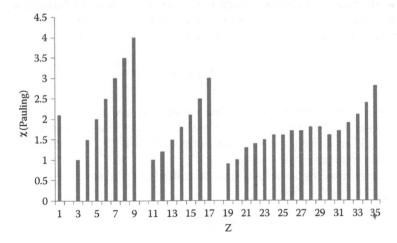

FIGURE 3.1 Variation of electronegativity for the elements from the first three periods (Pauling scale).

It follows that fluorine is the most electronegative of the elements and cesium is the least electronegative. Metallic elements are characterized by low electronegativities, generally lower than 2.1. The exceptions are gallium and germanium, which have higher electronegativities than aluminum and silicon, respectively, because of the d-block contraction.

Allred and Rochow suggested a scale of electronegativity (Allred and Rochow 1958) based upon the electrostatic force of attraction between the nucleus and the valence electrons:

$$F = \frac{e^2 Z_{eff}}{4\pi \varepsilon_0 r^2} \tag{3.3}$$

where

- r is the distance between the electron and the nucleus (covalent radius),
- e is the charge on an electron,
- Z_{eff} is the effective nuclear charge at the electron due to the nucleus and its surrounding electrons (calculated from Slater's rules), and
- ε_0 is vacuum permittivity ($\varepsilon_0 \approx 8.854 \times 10^{-12}\, F \cdot m^{-1}$).

The larger the effective nuclear charge (Z_{eff}) and the smaller the atomic radius, the more electronegative the element will be.

The quantity Z_{eff}/r^2 correlates well with Pauling electronegativities and the two scales can be made to coincide by expressing the Allred-Rochow electronegativity as

$$\chi = 0.744 + 0.359 Z_{eff}/r^2 \tag{3.4}$$

where r is the covalent radius expressed in picometers (1 pm = 10^{-12} m).

The Mulliken scale considers electronegativity as a measure of the tendency of an atom to attract electrons (Mulliken 1934). Electronegativity is calculated as the arithmetic mean between the first ionization energy (IE) and the electron affinity (EA),

$$\chi_M = \frac{1}{2}(EI + EA). \tag{3.5}$$

Because this definition is not dependent on an arbitrary relative scale, it has also been termed *absolute electronegativity*, with the units of kilojoules per mole or electron volts. If an element has low values of both EI and EA, as is the case of most metals, electronegativity is also low.

It is more usual to use a linear transformation of these absolute values into the Pauling scale (Bratch 1988),

$$\chi_P = 1.35\sqrt{\chi_M} - 1.37 \pm 0.14. \tag{3.6}$$

The use of Mulliken electronegativity is limited by the determination of the electron affinity, which is difficult to carry out for some elements.

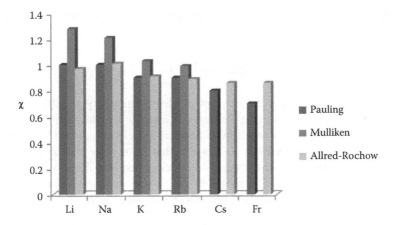

FIGURE 3.2 The electronegativities of the elements in group 1A calculated by Pauling, Mulliken, and Allred-Rochow methods (Mulliken values are not determined for Cs and Fr).

The chemical potential (defined below) is equal, but has an opposite sign, to the Mulliken electronegativity,

$$\mu = -\chi_M = -\frac{EI + EA}{2}. \qquad (3.7)$$

When electronegativity values obtained by the Allred-Rochow and Mulliken formula are converted to the Pauling scale, they produce values comparable to those from the Pauling scale (Figure 3.2).

The Sanderson method introduced a new parameter, the stability ratio, which is calculated with the following formula (Sanderson 1955),

$$S = \frac{\rho_e}{\rho_i} \qquad (3.8)$$

where ρ_e is the medium electronic density, and ρ_i represents the most stable electronic density for that respective number of electrons. The values of ρ_i are obtained by linear extrapolation of ρ_i values of the preceding and succeeding noble gases for the considered atom. For example, aluminum is found between the rare gases, neon ($\rho_i = 1.06$) and argon ($\rho_i = 0.82$). If the eight supplementary electrons that appear at the argon atom are compared with the neon atom, the density difference ($\Delta\rho_i$) is equal to 0.24 ($\rho_{iNe} - \rho_{iAr}$). The aluminum atom, which possesses three supplementary electrons compared to the neon atom, ρ_i, will be calculated as

$$\rho_{iAr} = \rho_{iNe} - \frac{8}{3} \cdot 0.24 = 1.6 - 0.09 = 0.97. \qquad (3.9)$$

The empirical relationship that relates the stability ratio with the Pauling electro-negativity is the following:

$$\sqrt{\chi} = 0.21S + 0.77.$$ (3.10)

This method also underlies the concept of electronegativity equalization, which suggests that electrons distribute themselves around a molecule to minimize or equalize the Mulliken electronegativity. Sanderson's electronegativity equalization principle (Geerlings and De Proft 2002) states that, upon molecule formation, atoms (or other more general portions of space of the reactants) with initially different elec-tronegativities χ_i^0 ($i = 1, ..., M$) combine in such a way that their *atoms-in-molecule* electronegativities are equal. The corresponding value is termed the *molecular electronegativity* (χ_M).

This can be expressed symbolically:

$$\chi_1^0, \chi_2^0, ..., \chi_M^0 \rightarrow \chi_1 = \chi_2 = ... = \chi_M$$ (3.11)

isolated atoms   molecule

Electron transfer takes place from atoms with lower electronegativity to those with higher electronegativity. In this way, the electronegativity of the donor atom increases, whereas the electronegativity of the acceptor atom decreases (Figure 3.3).

The Allen method introduced spectroscopic electronegativity as related to the average energy of the valence electrons in a free atom (Allen 1989),

$$\chi = \frac{n_S \varepsilon_S + n_P \varepsilon_P}{n_S + n_P}.$$ (3.12)

where ε_s and ε_p are the one-electron energies of s- and p-electrons in the free atom and n_s and n_p are the number of s- and p-electrons in the valence shell.

If a scaling factor of 1.75×10^{-3} is applied for energies expressed in kilojoules per mole or 0.169 for energies measured in electronvolts, we can obtain values that are

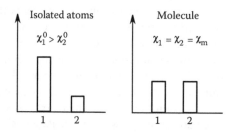

FIGURE 3.3 Sanderson's electronegativity equalization principle. (Adapted from P. Geerlings and F. De Proft. "Chemical Reactivity as Described by Quantum Chemical Methods." *Int. J. Mol. Sci.* 3, no. 4 (2002):276–309.)

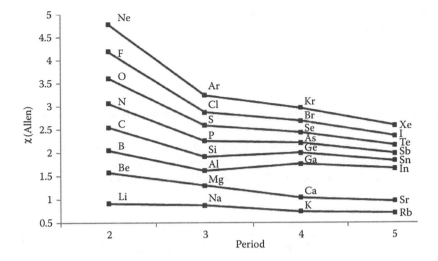

FIGURE 3.4 Spectroscopic electronegativities for groups I–VIII of representative elements. (Adapted from L.C. Allen, "Electronegativity Is the Average One-Electron Energy of the Valence-Shell Electrons in Ground-State Free Atoms. *J. Am. Chem. Soc.* 111, no. 25 (1989):9003–9014.)

numerically similar to Pauling electronegativities. The one-electron energies can be determined directly from spectroscopic data that are available for almost all elements. This method allows the estimation of electronegativities for elements that cannot be generated otherwise. However, it is unclear what should be considered the valence electrons for the d- and f-block elements. This leads to an ambiguity for their electronegativities calculated by the Allen method.

For the Allen electronegativity, several general trends can be noted:

- Noble gases are the most electronegative elements in their period (Figure 3.4).
- Neon has the highest electronegativity of all elements, followed by fluorine, helium, and oxygen.
- Elements with $n=2$ valence orbitals are significantly more electronegative than the other elements in their respective groups.
- Electronegativity decreases down a group, except: $B > Al < Ga$ and $C > Si < Ge$.

This last trend is called the *alternation effect* and is caused by the increased nuclear charge that accompanies the filling of the 3d orbitals. The 3d electrons shield the 4p electrons poorly, making Ga and Ge more electronegative.

Electronegativity was used as descriptor in some QSAR studies related to the metal ion toxicity (Somers 1959; Biesinger and Christensen 1972; Khangarot and Ray 1989; Enache et al. 2003).

Modern definitions of electronegativity were proposed on the basis of the density functional theory (DFT).

For a molecule or atom of N electrons, the electronic chemical potential μ is defined as follows (Parr et al. 1978):

$$\mu = \left(\frac{\partial E}{\partial N}\right)_{v(r)} \tag{3.13}$$

where E is the ground-state energy and v(r) is the composite nuclear potential at the point r.

Chemical potential is correlated with ionization energy and electron affinity, and as a result, also with Mulliken electronegativity (Equation [3.7]).

Molecular electronegativity (Iczowski and Margrave 1961; Parr et al. 1978) is defined as follows:

$$\chi = -\left(\frac{\partial E}{\partial N}\right)_{V(r)} = -\mu \quad or \quad \mu = \left(\frac{\partial E}{\partial N}\right)_{V(r)} = -\chi \tag{3.14}$$

where μ is the electronic chemical potential.

In DFT, the most important descriptors are hardness and softness indices (Parr and Yang 1989, Geerlings et al. 1996a, Geerlings et al. 1996b).

3.2.1.1 Hardness Indices

The second derivative of energy with respect to the number of electrons is called *absolute hardness* (η) or *chemical hardness* (Parr and Pearson 1983). For a molecule with N electrons, the absolute hardness is defined by

$$\eta = \frac{1}{2}\left(\frac{\partial^2 E}{\partial N^2}\right)_{V(r)} = \frac{1}{2}\left(\frac{\partial \mu}{\partial N}\right)_{V(r)} = \int h(r)\,dr = \frac{1}{2S} \tag{3.15}$$

where μ is the electronic chemical potential, h(r) is called *local hardness* (or *hardness density*), and S is the total softness (defined below). Absolute hardness is the resistance of the chemical potential to change the number of electrons. The harder a chemical species, the more difficult it will be to change its oxidation state.

3.2.1.2 Softness Indices

The total softness is defined by

$$S = \frac{1}{2}\left(\frac{\partial N}{\partial \mu}\right)_{V(r)} = \int s(r)\,dr = \frac{1}{2\eta} \tag{3.16}$$

where η represents the absolute hardness, μ is the electronic chemical potential, and s(r) is the local softness (or softness density).

Based on molecular orbital energies, especially the frontier orbital energies (Clare 1994; Huang et al. 1996), some main descriptors can be defined and information can be obtained about reactivity/stability of specific regions of the molecule.

- Highest occupied molecular energy (ε_{HOMO}) is related to the EI and is a measure of nucleophilicity.
- Lowest unoccupied orbital energy (ε_{LUMO}) is related to the EA and it is a measure of electrophilicity.
- The energy gap between the highest occupied molecular orbital (HOMO) and the lowest unoccupied molecular orbital (LUMO) represents the difference between HOMO and LUMO energies:

$$GAP = \varepsilon_{LUMO} - \varepsilon_{HOMO}. \tag{3.17}$$

GAP is an important stability index. A large GAP means a high stability and low reactivity in chemical reaction. The HOMO–LUMO energy fraction ($f_{H/L}$) is a stability index defined as the ratio of HOMO and LUMO energies,

$$f_{H/L} = \frac{\varepsilon_{HOMO}}{\varepsilon_{LUMO}}. \tag{3.18}$$

From the frontier orbitals energies, an approximated absolute hardness is also obtained,

$$\eta = \frac{EI - EA}{2} = \frac{GAP}{2} = \frac{\varepsilon_{LUMO} - \varepsilon_{HOMO}}{2} \tag{3.19}$$

where EI is the ionization energy, EA is the electron affinity, GAP denotes the HOMO–LUMO energy difference, and ε_{LUMO} and ε_{HOMO} are the energies of the lowest unoccupied orbital and the highest occupied orbital, respectively.

The total softness can be also calculated by

$$S = \frac{1}{EI - EA}. \tag{3.20}$$

3.2.2 ELECTROPOSITIVITY

Electropositivity is a measure of an element's ability to donate electrons, and consequently, to form positive ions with a stable configuration. Electropositivity reflects a tendency that is opposite that of electronegativity. Metals are electropositive elements: the greater the metallic character of an element, the greater the electropositivity. The alkali metals are the most electropositive of all as is related to their low ionization energies. Electropositivity decreases along periods (from left to right) and increases down groups.

3.2.3 OXIDATION NUMBER (OXIDATION STATE)

Each atom tends to have its outer energy level complete, and also to gain, lose, or share its valence electrons to achieve a stable configuration. The absolute number of electrons gained, lost, or shared is referred to as the valence of the atom. When valence electrons are lost or partially lost by an atom, the valence number is assigned a plus (+) sign, whereas its valence is assigned a minus (–) sign when the valence electrons are gained or partially gained by an atom.

Oxidation numbers, sometimes called *oxidation states*, are signed numbers given to atoms in molecules and ions to define their positive or negative character. The oxidation number refers to the number of formal charges that an atom would have in a molecule or polyatomic ion in the case of electrons being completely transferred in the direction indicated by the electronegativity difference between atoms.

The oxidation state of a metal ion can be defined as the charge that the metal ion would have in the case of a purely ionic model for the complex (Yatsimirskii 1994). The oxidation numbers are written as +1, +2, and so on, whereas charges on ions are written 1+, 2+, and so on. In themselves, oxidation numbers have no physical meaning; they are used to simplify tasks that are more difficult to accomplish without them.The oxidation numbers are calculated taking into account some basic rules:

1. All pure elements are assigned the oxidation number of zero.
2. All monatomic (single-element) ions are assigned oxidation numbers equal to their charges.
3. Certain elements usually possess a fixed oxidation number in compounds.
   - The oxidation number of O in most compounds is –2, except in peroxides where it is –1; in combination with fluorine, it is +2.
   - The oxidation number of H in most compounds is +1, except in metal hydrides, where it is –1.
   - The oxidation number of F in all compounds is –1.
   - The oxidation numbers of alkali metals (Group 1) and alkaline earth metals (Group 2) are +1 and +2, respectively.
4. The sum of all oxidation numbers in a compound equals zero, and the sum of oxidation numbers in a polyatomic ion equals the charge of the ion.

Table 3.1 presents the oxidation numbers of some common chemical elements from the periodic table (Bailar 1984, Greenwood and Earnshaw 2006).

Metals have positive oxidation numbers (Table 3.1). Generally, the representative metals have oxidation numbers equal to the group A number. The metals from the p-block can have a second oxidation number, two units smaller than the group A number. In the case of these elements (e.g., Tl, Sn, and Pb) the stability of the highest oxidation state decreases down a group due to the *ns* inert electron pair. The penetration effect is greater for *ns* orbitals than *np* orbitals. As a consequence, the *ns* electrons are more attracted to the nucleus and more inert.

The transition metals are characterized by wide variability in oxidation numbers that differ by one for the same metal. The highest oxidation state is less than or equal to the group B number. The only exceptions are metals in group 1 B (11), of which the highest oxidation number is +3. In the iron group (8, VIII B), the first element, iron, does not achieve the maximal group oxidation state (+8), and in the cobalt group (9, VIIIB), no metal achieves an oxidation state equal to the group number.

TABLE 3.1
Common Oxidation Numbers of the Chemical Elements in the Periodic Table

1	2	3	4	5	6	7	8	9	10	11	12	13	14	15	16	17	18
H −1 −1																	2 He
3 Li +1	4 Be +2											5 B +3	6 C +4 +2 −4	7 N +5 +4 +3 +2 −3	8 O −1 −2	9 F −1	10 Ne
11 Na +1	12 Mg +2											13 Al +3	14 Si +4 −4	15 P +5 +3 −3	16 S +6 +4 −2	17 Cl +7 +5 +3 +1 −1	18 Ar
19 K +1	20 Ca +2	21 Sc +3	22 Ti +4 +3	23 V +5 +4 +3	24 Cr +6 +3 +2	25 Mn +7 +2	26 Fe +3 +2	27 Co +3* +2	28 Ni +2	29 Cu +2 +1	30 Zn +2	31 Ga +3	32 Ge +4 −4	33 As +5 +3 −3	34 Se +6 +4 −2	35 Br +7 +5 +1 −1	36 Kr +4 +2

(Continued)

TABLE 3.1 (Continued)
Common Oxidation Numbers of the Chemical Elements in the Periodic Table

37	38	39	40	41	42	43	44	45	46	47	48	49	50	51	52	53	54
Rb	Sr	Y	Zr	Nb	Mo	Tc	Ru	Rh	Pd	Ag	Cd	In	Sn	Sb	Te	I	Xe
+1	+2	+3	+4	+5	+6	+7	+4	+3	+4	+1	+2	+3	+4	+5	+6	+7	+8
				+3	+4	+4	+3	+2	+2				+2	+3	+4	+5	+6
							+2							-3	-2	+1	+4
																-1	+2

55	56	57	72	73	74	75	76	77	78	79	80	81	82	83	84	85	86
Cs	Ba	La	Hf	Ta	W	Re	Os	Ir	Pt	Au	Hg	Tl	Pb	Bi	Po	At	Rn
+1	+2	+3	+4	+5	+6	+7	+4	+4	+4	+3	+2	+3	+4	+5	+4	-	
					+4	+4	+3	+3	+2	+1	+1	+1	+2	+3			
							+2										

* in complexes or insoluble compounds

Ce	Pr	Nd	Pm	Sm	Eu	Gd	Tb	Dy	Ho	Er	Tm	Y	Lu
+4	+4	+3	+3	+3	+3	+3	+4	+3	+3	+3	+3	+3	+3
+3	+3			+2	+2		+3				+2	+2	

Th	Pa	U	Np	Pu	Am	Cm	Bk	Cf	Es	Fm	Md	No	Lr
+4	+5	+6	+7	+7	+6	+4	+4	+3	+3	+3	+3	+3	+3
+3	+4	+5	+6	+6	+5	+3	+3			+2	+2	+2	+2
		+4	+5	+5	+4								
		+3	+4	+4	+3								
			+3	+3									

In the case of transition metals, unlike the representative metals, stability of the maximum oxidation state increases down a group. This behavior is explained by reduction of the effective nuclear charge on the d orbitals that makes them more available for bonding. This effect is enhanced by atomic volume.

Oxidation state is correlated with the redox properties of metal ions. Thus, the metals in the highest oxidation state are oxidizing agents, whereas the metals in the lowest oxidation state are reducing agents. Between these extreme situations, the metals in intermediate oxidation states are redox ampholytes.

The toxicity of some metals depends on both their oxidation state and the rapidity with which the metal ion can undergo oxidation and reduction. Some compounds of metals such as arsenic and antimony are more toxic in their lower oxidation state than the higher oxidation state. This feature can be explained by the tendency of these compounds to become more stable in higher valence states, disrupting cellular processes. The possibility of transition elements to exist in several oxidation states is an attribute that makes them particularly suitable for biological functions, but simultaneously redox changes in vivo can have a strong influence on overall metal toxicity (Hoeschele et al. 1991).

3.2.4 STANDARD ELECTRODE POTENTIAL (E^o)

The electrode potential is determined using a galvanic cell, a device that produces an electric current based on the transfer of electrons between two half-cells where an oxidation and a reduction process take place. A galvanic cell is designed so that the reductant and the oxidant are physically separated, but connected by an external circuit made of a conductor (to carry electrons) and a salt bridge (to carry charged ions in solution). A galvanic cell is thus composed of two half-cells, a reductant half-cell, and an oxidant half-cell. This arrangement ensures that electrons cannot go directly from the reductant to the oxidant, but they will move through the external circuit.

Current flows from one half-cell to the other based on the difference in the electrical potential energy between the two electrodes. The difference in electrical potential energy per unit charge, as measured with a potentiometer, is called *electromotive force* (EMF), or cell potential. Because it is measured in volts, it is also referred to as the *cell voltage*. Note that potential difference is always measured, not the individual potentials of any half-cell. By convention, the term *electrode potential* always means the reduction potential, corresponding to a reduction reaction, such as,

$$M^{n+}_{(aq)} + ne^- \rightarrow M_{(s)}. \tag{3.21}$$

An oxidation potential is the negative of the reduction potential,

$$E_{ox} = -E_{red}. \tag{3.22}$$

The cell potential is noted with ΔE, with the Δ symbol included because the measured voltage is a difference between two electrode potentials:

$$\Delta E = \begin{pmatrix} \text{potential at the electrode} & - & \text{potential at the electrode} \\ \text{at which reduction occurs} & & \text{at which oxidation occurs} \end{pmatrix} \tag{3.23}$$

or:

$$\Delta E = \text{cathode potential} - \text{anode potential} \qquad (3.24)$$

or:

$$\Delta E = E_{cathode} - E_{anode} \qquad (3.25)$$

If all substances involved in the reaction are in their standard states (all concentrations are specified to be 1 M and the H_2 gas is 1-atm pressure, temperature is 25°C), the cell is referred to as a standard cell, its voltage being denoted as superscript zero, or ΔE^0.

In order to determine the standard electrode potential for a metal, the galvanic cell is designed so that a half-cell is formed by a piece of metal immersed in a solution that contains 1.00 M of ions of that metal, and one half-cell with potential convention defined to be exactly zero volts. This electrode is called the *standard hydrogen electrode* (SHE) and it consists of a platinum electrode over which H_2 gas at 1-atm of pressure is bubbled, immersed in a solution that contains 1.00 M of hydronium ion at 25°C.

The notation of the cell formed by the half-cells described above can be:

$$M(s) \,|\, M^{n+}(1M) \,||\, H_3O^+(1M) \,|\, H_2(1atm) \,|\, Pt(s) \qquad (3.26)$$

or

$$Pt(s) \,|\, H_2(1atm) \,|\, H_3O^+(1M) \,||\, M^{n+}(1M) \,|\, M(s) \qquad (3.27)$$

where a single vertical bar represents a boundary between two phases; a double vertical bar represents a salt bridge or some other device used to maintain electrical contact between two different solutions that cannot be allowed to mix. The situations described by Relations (3.26) and (3.27) correspond to two main groups of metals:

1. Metals with an oxidation tendency higher than that of H_2. These metals oxidize before H_2 and represent the anode in a coupling with SHE. For these metals:

$$\Delta E^0 = E^0{}_{H^+|H2|Pt} - E^0{}_{M^{n+}|M} \Rightarrow E^0{}_{M^{n+}|M} = -\Delta E^0 \qquad (3.28)$$

2. Metals with an oxidation tendency lower than that of H_2. These metals oxidize after H_2 and represent the cathode in a coupling with SHE. For these metals:

$$\Delta E^0 = E^0{}_{M^{n+}|M} - E^0{}_{H^+|H2|Pt} \Rightarrow E^0{}_{M^{n+}|M} = \Delta E^0 \qquad (3.29)$$

According to their standard electrode potentials, all metals are arranged in an order called an *electrochemical series* (Figure 3.5).

Li K Ba Sr Ca Na Mg Al Mn Zn Cr Fe Ni Sn Pb $\quad\quad$ H$_2$ Sb Cu Ag Hg Pd Pt Au

$\quad\quad\quad\quad$ E° < 0 $\quad\quad\quad\quad\quad\quad\quad\quad\quad\quad\quad$ E° > 0

$\quad\quad\quad\quad\quad\quad\quad\quad\quad\quad\quad\quad\quad\quad$ Not displace H$_2$ from H$_2$O$_{(l)}$,

$\quad$ Displace H$_2$ from H$_2$O$_{(l)}$, steam or acid $\quad\quad\quad\quad$ steam or acid

$\quad\quad\quad\quad\quad\quad\quad\quad$ Ease of oxidation increases

FIGURE 3.5 Electrochemical series of metals.

TABLE 3.2
Table of Standard Reduction Potential

	Oxidizing Agents			Reducing Agents		E° (V)
Strong	F$_2$ (g)	+ 2e$^-$	→	2F$^-$	Weak	2.87
	MnO$_4^-$ + 8H$^+$	+ 5e$^-$	→	Mn^{2+} + 4H$_2$O		1.51
	Ag$^+$	+ e$^-$	→	Ag$_{(s)}$		0.799
	Cu^{2+}	+ 2e$^-$	→	Cu$_{(s)}$		0.337
	2H$^+_{(aq)}$	+ e$^-$	→	H$_{2(g)}$		0.000
	Pb^{2+}	+ 2e$^-$	→	Pb$_{(s)}$		−0.126
	Zn^{2+}	+ 2e$^-$	→	Zn$_{(s)}$		−0.763
Weak	Li$^+$	+ e$^-$	→	Li$_{(s)}$	Strong	−3.05

Source: Adapted from B.G. Segal, *Chemistry: Experiments and Theory* (New York: Wiley, 1989).

By devising various galvanic cells and measuring their electromotive forces, tables of values of standard electrode potentials can be constructed. A table that lists the value of electrode potential for any half-cell in which all concentrations are 1M and all gases are at 1-atm pressure is a Table of Standard Reduction Potential (Table 3.2). By convention, the tabulated values are standard reduction potentials relative to the potential of the standard hydrogen electrode, which is defined as exactly zero volts. The analysis of the data from Table 3.2 highlights some important aspects.

1. Oxidizing agents are tabulated at the left of the table in the decreasing order of the oxidizing power.
2. The strongest oxidizing agent is fluorine gas, F$_2$.
3. The reducing agents are written at the right of the table in the increasing order of the reducing power.
4. The strongest reducing agent is lithium metal.
5. The more positive the reduction potential, the weaker the reducing agent and conversely, the more negative the reduction potential, the stronger the reducing agent.

It can be surprising that lithium is the stronger reducing agent, although cesium is the most electropositive metal. This behavior can be explained if all the processes are included in the reduction process (Figure 3.6). The standard reduction potential is related to the free energy change: $\Delta G° = -nFE°$, and $\Delta G° = \Delta H° - T\Delta S°$.

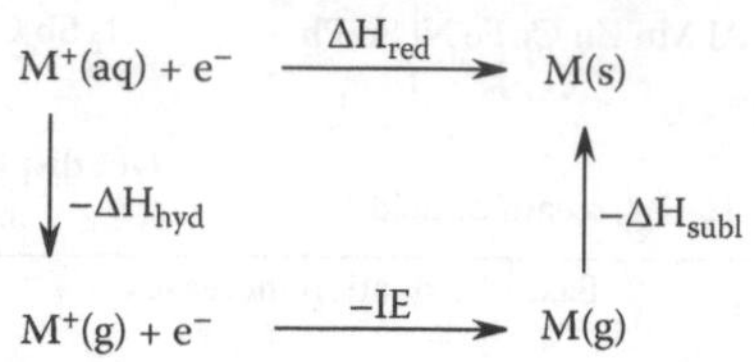

$$M^+(aq) + e^- \xrightarrow{\Delta H_{red}} M(s)$$

$$\downarrow -\Delta H_{hyd} \qquad \qquad \uparrow -\Delta H_{subl}$$

$$M^+(g) + e^- \xrightarrow{-IE} M(g)$$

FIGURE 3.6　Alternative path for the reduction half-reaction.

TABLE 3.3
Enthalpy Changes ($kJ\ mol^{-1}$) for Electrode
Half-Reaction $M^+(aq) \rightarrow M(s)$ for Alkali Metals

Element	$-\Delta H_{sub}$	$-IE$	$-\Delta H_{hyd}$	ΔH_{red}
Li	−159	−519	506	−172
Na	−107	−498	397	−208
K	−89	−418	318	−189
Rb	−81	−402	289	−194
Cs	−76	−377	259	−194

Knowing the energies for all alkaline metals, the enthalpy change for the reduction process ($\Delta H_{red} = -\Delta H_{subl} - IE - \Delta H_{hydr}$) can be calculated (Table 3.3). The enthalpy change for lithium is the smallest, meaning that the tendency of its ion to be reduced is also the smallest. The crucial factor involved is the high hydration energy of the lithium ion. The second factor that influences the ΔG° values is the entropy change. It is a much larger entropy change for the lithium electrode reaction due to the more severe disruption of the H_2O structure by the lithium ion.

The standard electrode potential has proved a useful parameter to correlate with metal toxicity (Turner et al. 1983; Lewis et al. 1999). Regarding the interaction of metals in a biological milieu, it can be stated generally that the moderate reduction potential of a biological system tends to convert metals to lower oxidation states.

Another electrochemical parameter, the standard reduction-oxidation potential (ΔE^0) represents the absolute difference in electrochemical potential between an ion and its first stable reduced state, or in other terms, the ability of an ion to change its electronic state. This parameter is seldom used alone in the studies concerning the toxicity of metal ions, but is usually combined with AN/ΔIP (where AN = atomic number, ΔIP = the difference in the ionization potential (in eV) between the actual oxidation number (O.N.) and the next lower one (O.N. −1) or with log AN/ΔIP (Kaiser 1980; Kaiser 1985; McCloskey et al. 1996; Enache et al. 2003).

3.3　METAL IONS IN A COORDINATION ENVIRONMENT

The description of some basic aspects regarding the coordination chemistry of metal ions is especially important for understanding their behavior in biological media. In the predominantly aqueous environment found in biological systems, the metal ions do not exist as free ions, but are present in a bound form through the formation of coordination complexes (Lepădatu et al. 2009). In aqueous environments, metal ions are probably surrounded by water molecules bounded coordinatively, as $[M(H_2O)_n]^Z$ species. Thus, the interaction between the metal ion and a molecular target is of the type: $[M(H_2O)_n]^Z$ – Receptor or Ligand. Many factors affect metal–ligand interaction in biological systems. Some factors depend on the metal ion characteristics, such as ion size and charge, or preferred metal coordination geometry. Ligand chelation also affects the binding. Hard–soft acid–base matching is an important consideration, as well as various competing equilibria such as solubility product, complexation, and acid–base equilibrium. In the following paragraphs, some aspects regarding formation, structure, thermodynamic stability, kinetic reactivity/lability, and redox properties of coordination complexes will be discussed, especially focused on hydrated metal ions. Moreover, the need to use molecular descriptors that characterize the molecular character of the coordinated metal ions in QSAR studies will be highlighted.

3.3.1 Coordinating Capacity of Metal Ions: HSAB Theory

A metal complex is formed by two main components: a metal ion, called the *generator of complex*, and one or more anions or molecules called *ligands*. The number of coordinated ligands represents the coordination number (C.N.). In the simplest case, when the charges of both the metal ion and the ligands are not considered, the formation of a metal complex can be represented as

$$M + nL \rightleftharpoons ML_n. \tag{3.30}$$

Most theories of metal–ligand bonding consider the Lewis behavior of the species involved. In a Lewis acid–base reaction, a pair of electrons from a species is used to form a covalent bond with another species. A Lewis base is the species that donates an electron pair, and a Lewis acid is the species that accepts the electron pair. The reaction can be written,

$$A \quad + \quad :B \quad \rightarrow \quad A-B \tag{3.31}$$

acid   base   adduct

(electron acceptor,   (electron donor,

electrophile)   nucleophile)

A generalized theory elaborated by R.G. Pearson that explains the differential complexation behavior of cations and ligands is the *hard and soft acid and bases* (HSAB) principle or theory. Prior to the Pearson generalized theory, some ideas brought attention to the different behaviors of metal ions toward the same ligand, and on the binding preference of metal ions to certain ligands. Irving and Williams pointed out that for a given ligand, the stability of dipositive metal ion complexes increases in the following order (Irving and Williams 1953):

$$Mn^{2+} < Fe^{2+} < Co^{2+} < Ni^{2+} < Cu^{2+} > Zn^{2+}.$$

This series is known as the Irving-Williams stability series.

It was also known that certain ligands formed their most stable complexes with metal ions like Al^{3+}, Ti^{4+}, and Co^{3+} while others formed stable complexes with Ag^+, Hg^{2+}, and Pt^{2+}.

Ahrland, Chatt, and Davies proposed that metal ions could be described as class (a) if they formed stronger complexes with ligands whose donor atoms are N, O, or F and class (b) if they prefer binding to ligands whose donor atoms are P, S, or Cl (Table 3.4). Ligands were classified as type (a) or type (b) depending upon whether they formed more stable complexes with class (a) or class (b) cations (Ahrland et al. 1958).

Based on the classification of cations and ligands proposed by Ahrland, Pearson (Pearson 1963, Pearson 1966) developed the type (a) and type (b) by explaining the differential complexation behavior of cations and ligands in terms of electron pair–donating Lewis bases and electron pair–accepting Lewis acids:

$$\text{Lewis acid} + \text{Lewis base} \rightarrow \text{Lewis acid/base complex} \tag{3.32}$$

Pearson classified Lewis acids as hard, borderline, or soft (Table 3.5). The hard acids are generally characterized by a small ionic radius, high positive charge, and no electron pairs in their valence shells. The soft acids have a large radius, low or partial δ^+ positive charge, and electron pairs in their valence shells. They are also easy to polarize and oxidize.

TABLE 3.4
Classification of Cations and Ligands According to Ahrland

Class (a) of Cations	Class (a) of Ligands	Class (b) of Cations	Class (b) of Ligands
Alkali metal cations: Li$^+$ to Cs$^+$		Heavier transition metal cations in lower oxidation states: Cu$^+$, Ag$^+$, Cd^{2+}, Hg$^+$, Ni^{2+}, Pd^{2+}, Pt^{2+}.	
Alkaline earth metal cations: Be^{2+} to Ba^{2+}	N >> P > As > Sb > Bi		N << P > As > Sb > Bi
Lighter transition metal cations in higher oxidation states: Ti^{4+}, Cr^{3+}, Fe^{3+}, Co^{3+}	O >> S > Se > Te		O << S ~ Se ~ Te
The proton, H$^+$	F > Cl > Br > I		F < Cl < Br < I

TABLE 3.5
Pearson Classification of Lewis Acids
Represented by Metal Ions

Hard (Class [a])	Soft (Class [b])
H^+, Li^+, Na^+, K^+	Cu^+, Ag^+, Au^+, Tl^+, Hg^+, Cs^+
Be^{2+}, Mg^{2+}, Ca^{2+}, Sr^{2+}, Mn^{2+}	Pd^{2+}, Cd^{2+}, Pt^{2+}, Hg^{2+}
Al^{3+}, Sc^{3+}, Ga^{3+}, In^{3+}, La^{3+}	Tl^{3+}, Au^{3+}, Te^{4+}, Pt^{4+}
Cr^{3+}, Co^{3+}, Fe^{3+}, As^{3+}, Ce^{3+}	
Si^{4+}, Ti^{4+}, Zr^{4+}, Th^{4+}, Pu^{4+}	
Ce^{4+}, Ge^{4+}, VO^{2+}	
UO_2^{2+}	

Borderline

Fe^{2+}, Co^{2+}, Ni^{2+}, Cu^{2+}, Zn^{2+}, Pb^{2+}, Sn^{2+}

Sb^{3+}, Bi^{3+}, Rh^{3+}, Ir^{3+}, Ru^{2+}, Os^{2+}

TABLE 3.6
Pearson Classification of Lewis Bases

Hard	Soft
H_2O, OH^-, F^-	R_2S, RSH, RS^-
$CH_3CO_2^-$, PO_4^{3-}, SO_4^{2-}	I^-, SCN^-, $S_2O_3^{2-}$
Cl^-, CO_3^{2-}, ClO_4^-, NO_3^-	R_3P, R_3As, $(RO)_3P$
ROH, RO^-, R_2O	CN^-, RNC, CO
NH_3, RNH_2, N_2H_4	C_2H_4, C_6H_6
	H^-, R^-

Borderline

$C_6H_5NH_2$, C_5H_5N, N_3^-, Br^-, NO_2^-, SO_3^{2-}, N_2

In a similar manner, the bases are classified as hard, borderline, or soft (Table 3.6). A hard base has a small, electronegative atomic center ($\chi = 3.0 - 4.0$) and is a weakly polarizable species that is difficult to oxidize. A soft base has a large atomic center of intermediate electronegativity ($\chi = 2.5 - 3.0$) and it is easy to polarize and oxidize. Between the two main categories of acids and bases are species with intermediate character.

Besides the classification of acids and bases, the Pearson principle introduced the idea of binding preference, namely: hard acids prefer to bind to hard bases, and soft acids prefer to bind to soft bases.

The knowledge of coordinative binding preference of metal ions is important relative to the capacity of metal ions to form stable and irreversible coordination complexes with various donor atoms of the macromolecules, and therefore notionally, metal toxicity. In biological systems, hard metal cations prefer oxygen donor atoms that are present in groups such as $-OH$, $-COOH$, $-PO_4^{3-}$ or nitrogen donor atoms that are present in groups such as amines ($R-NH_2$). Soft cations prefer donor sulfur atoms such as those present in $-SH$, RSH, or RS^- (Roat-Malone 2003).

3.3.2 Crystal Field Theory of Bonding in Transition Metal Complexes

Bethe (1929) initiated the crystal field theory with which van Vleck, Ilse, and Hartman explained the color and magnetic properties of metal complexes. The crystal field theory (CFT) constitutes a foundation for predicting the structure, stability, kinetic lability, and redox properties, and of metal complexes. It also accounts for certain trends in the physicochemical properties of metal complexes (Orgel 1952).

In CFT, the interaction between central metal ions and ligands is approached exclusively from the electrostatic vantage point, the ligands being considered as point negative charges. According to this theory, degeneration of d atomic orbitals of a metal ion in a spherical field is raised when the ion is in a crystal lattice. The set of d orbitals split in two groups that are energetically nonequivalent depending on the crystal field symmetry. The most common in the ionic compounds is the cubic symmetry field, and the groups of cubic symmetry are the cube, octahedron, and tetrahedron.

3.3.2.1 The d Orbitals in an Octahedral Field

When a metal ion is surrounded by six ligands, these ligands will position on an octahedron whose vertices coincide with the center of the cube's faces. Between metal ions and ligands, two types of electrostatic interactions will appear. The first is an attraction between the positive ion and the electronic pairs of the ligands, and the second is the repulsion between the d electrons of the metal ion and the electronic pairs of ligands. The magnitude of this repulsion depends on the type of d orbital. Thus, if the metal ion has an electron in one of the orbitals, dx^2-y^2 or dz^2, with lobes that lie along the Cartesian coordinates, the repulsion will be greater than that corresponding to the electron found in one of the orbitals, dxy, dxz, dyz, with lobes directed between the axes (Figure 3.7).

As a consequence, the energy of orbitals dx^2-y^2 or dz^2 increases, while the energy of orbitals dxy, dxz, dyz decreases. The result of the metal–ligand interactions is the degeneration raising of the five d orbitals with their splitting into two groups: triple degenerated of t_{2g} symmetry and lower energy formed by dxy, dyz, dxz, and the second one double degenerated of e_g symmetry and higher energy formed by dx^2-y^2 or dz^2 orbitals.

The energy difference between the two sets of d orbitals, labeled 10Dq or Δ_o is called *crystal field splitting in octahedral field*. The quantity Dq or $1/10\Delta_o$ is called the *crystal field parameter in octahedral field*. The magnitude of this parameter depends on a) the nature of central metal ion; b) the charge of metal ion; and c) the nature of the ligand.

The crystal field stabilization energy (CFSE) is the energy with which a certain electrostatic configuration is stabilized upon d orbitals splitting.

The relative energy of the e_g and t_{2g} compared to d orbitals before splitting can be calculated based on an ion with d^{10} configuration. When this ion is introduced in an octahedral field created by 6 ligands, the five d orbitals are split in t_{2g} and e_g levels. Because the energetic level e_g is higher than the energetic level t_{2g}, the four electrons

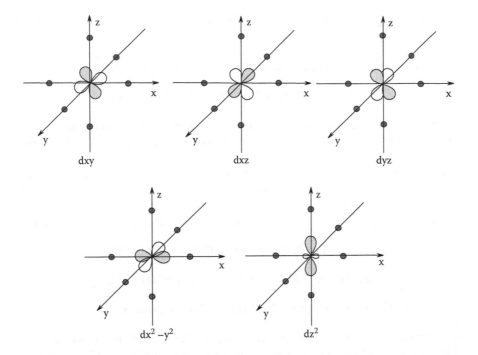

FIGURE 3.7 Orbitals d in an octahedral surrounding.

from e_g must balance the energy decrease of the six electrons from t_{2g} according to the energetic gravity center rule. Through preferential occupying of t_{2g} orbitals with electrons, the system's energy decreases with 2/5 Δ_0 multiples, and by completion of e_g orbitals with electrons, the energy increases with 2/5 Δ_0 multiples. In other terms, each electron from t_{2g} contributes $-4Dq$ to the crystal field stabilization energy, while each electron from e_g contributes with $+6Dq$ to the CFSE.

In the case of d^1, d^2, and d^3 configurations, the first three electrons successively occupy the orbitals from the t_{2g} group, according to the Hund rule. The values of CSFE are $-4Dq$, $-8Dq$, and $-12Dq$, respectively. The d^4, d^5, d^6, d^7 configurations of the associated electrons involve two possible position modes: either in the e_g orbitals, with uncoupled spin, or in the t_{2g} orbitals with coupled electronic spins.

From these different arrangements, two types of complexes can be formed: high-spin complexes with uncoupled d electrons, and low-spin complexes with coupled d electrons. The crucial factor that occurs in these situations is the spin-pairing energy (P) as follows:

a. When $\Delta_0 < P$, a low field is generated. For a d^4 configuration, for example, the fourth electron is placed in an e_g orbital resulting in the overall configuration $t_{2g}^3 e_g^1$ (and $t_{2g}^3 e_g^2$, $t_{2g}^4 e_g^2$, $t_{2g}^5 e_g^2$, respectively).

b. When $\Delta_0 > P$, a strong field is generated. In a d^4 configuration, the fourth electron is placed in a t_{2g} orbital resulting in the overall configuration t_{2g}^4 (and t_{2g}^5, t_{2g}^6, $t_{2g}^6 e_g^1$, respectively).

For the d^8, d^9, d^{10} configurations, there is just one possibility for the arrangement of electrons, namely: $t_{2g}^6 e_g^2$, $t_{2g}^6 e_g^3$, $t_{2g}^6 e_g^4$.

In the general case of an electronic configuration represented as $t_{2g}^m e_g^n$ with x electronic pairs, CFSE can be calculated with the general formula:

$$CFSE = -4mDq + 6nDq + xP = -(4m - 6n)Dq + xP. \qquad (3.33)$$

CFSE can be spectroscopically measured: it is **inversely** proportional with the light absorbed when an excited electron promotes from the t_{2g} level in e_g level: $10Dq = hc/\lambda$.

For octahedral transition metal complexes, Δ_o varies depending on the nature of the ligands, which can be arranged in increasing order of field strength forming a series called the *Fajans-Tsushida series* or *spectrochemical series* (so called, because the series can be determined using UV-visible absorption spectroscopy):

$$I^- < Br^- < Cl^- < SCN^- < N_3^- < (EtO)_2PS_2^- < F^- < (NH)_2CO$$

$$< OH^- < C_2O_4^{2-} < H_2O < NCS^- < H^- < C\underline{N}^- < NH_2CH_2CO_2^-$$

$$< NH_3 < C_5H_5N < en < SO_3^{2-} < NH_2OH < NO_2^- < phen < CH_3^- < \underline{C}N^-$$

In general, the predicted order of decreasing tendency of donor ligands to cause spin pairing is $C > N > O > S > F > Cl > Br$. The actual order found is $S > O$ and $Cl > Br > F$.

The Δ_o depends not only on the nature of the ligands, but also on the metal and its oxidation state. The approximate spectrochemical series for metal ions is shown here:

$$Mn^{2+} < Ni^{2+} < Co^{2+} < Fe^{2+} < V^{2+} < Co^{3+} < Mn^{4+} < Mo^{3+} <$$

$$Rh^{3+} < Ru^{3+} < Pd^{4+} < Ir^{3+} < Pt^{4+}$$

This series is not quite as regular as the spectrochemical series of ligands, but some general regularities are seen:

1. For a given metal and ligand set, Δ_o increases with increasing oxidation state ($Co^{2+} < Co^{3+}$, etc.).
2. For a given oxidation state and ligand set, Δ_o increases down a group ($Co^{3+} << Rh^{3+} < Ir^{3+}$).
3. For a given oxidation state and ligand set, Δ_o varies irregularly across the transition metals (groups 3 to 12).

Relative to an octahedral field some general conclusions can be drawn:

- CFSE has higher values in a strong field than in a weak field.
- CFSE is zero for the d^0, d^5 (high spin) and d^{10} configurations.
- The d^3 and d^8 configurations are the most stable configurations in a weak field, while the d^6 low-spin configuration is the most stable in a strong field.

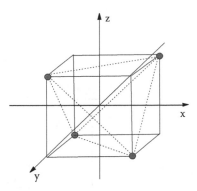

FIGURE 3.8   A tetrahedral arrangement of ligands.

3.3.2.2   The d Orbitals in a Tetrahedral Field

When a metal ion situated in the center of a cube is surrounded by four ligands, these occupy alternatively 4 of the 8 cube corners to form a tetrahedron. The ligands' arrangement leads to interactions with dxy, dyz, and dxz orbitals of metal ions, placed on the coordinate axes bisectors (Figure 3.8). These orbitals will constitute a triple-degenerated group noted with t_2, which is energetically destabilized. The dx^2-y^2 and dz^2 orbitals will form a double-degenerated group noted with e_2, energetically stabilized. (The indices g are missing because the tetrahedron does not have a center of symmetry).

The splitting parameter in a tetrahedral field is denoted by $\Delta_T = 10$ Dq, so the e orbitals are stabilized with $3/5\ \Delta_T$ or 6Dq, and the t_2 orbitals increase in energy with $2/5\ \Delta_T$ or 4Dq.

The splitting parameter Δ_T is always smaller than the parameter Δ_o because the ligand number is smaller, and the ligands are not pointed toward the direction of electronic density of the d orbitals:

$$\Delta_T \cong -4/9\ \Delta_o \qquad (3.34)$$

The minus sign indicates that the splitting is reversed in the two geometries.

Since the splitting parameter in the tetrahedral field is smaller than in the octahedral field, the tetrahedral field is always a weak field, $\Delta_T < P$. The electrons occupy the e and t_2 orbitals by the Hund's rule. In the tetrahedral field, the highest values of CFSE correspond to d^2 and d^7 configurations. Figure 3.9 presents the comparative crystal field splitting of d orbitals of the central ion in complexes of geometry: tetrahedral, octahedral, tetragonal, and square-planar.

3.3.2.3   Applications of CFT

Crystal field stabilization energy is a factor that contributes to the thermodynamic stability of complexes with predominantly ionic metal–ligand interactions, and also to the variation in properties of d metals and their compounds. Some of these properties are the size of di- and trivalent ions, hydration enthalpies, lattice energies, and stability of oxidation states.

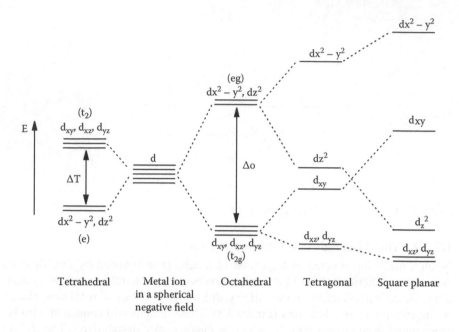

FIGURE 3.9   Crystal field splitting of d orbitals of central ion in complexes with geometries: tetrahedral, octahedral, tetragonal, and square-planar.

3.3.2.4   Size of Transition 3d-Metal Ions

Variation with Z of the size of di- and trivalent $3d$ cations in MO oxides of an NaCl-type lattice presents a maximum for the d^5 configuration and two minimums for the d^3 configurations (V^{2+}, Cr^{3+}), and d^8 configurations (Ni^{2+}).

In a weak field and for a given oxidation state, the ionic radius decreases when going from left to right in a transition series. The magnitude of this decrease is not uniform, being greater for d^4, d^5, d^9, and d^{10}. That is explained by differences in repulsion between the distinctive electron and the anionic ligands. Thus, for ions with configurations $d^1 \equiv t_{2g}^1$, $d^2 \equiv t_{2g}^2$, $d^3 \equiv t_{2g}^3$, $d^6 \equiv t_{2g}^4 e_g^2$, $d^7 \equiv t_{2g}^5 eg^2$, $d^8 \equiv t_{2g}^6 e_g^2$, the distinctive electron that occupies the t_{2g} orbitals is more weakly repulsed by the negative charges of anionic ligands. As a consequence, it is subjected only to the influence of the increasing nuclear effective charge. Ions with configurations $d^4 \equiv t_{2g}^3 e_g^1$, $d^5 \equiv t_{2g}^3 eg^2$, $d^9 \equiv t_{2g}^6 e_g^3$, and $d^{10} \equiv t_{2g}^6 e_g^4$ have the distinctive electron in the e_g orbitals and more interaction with anionic ligands, resulting in the electron being more weakly attracted by the nucleus.

In a strong field, the radii of M^{2+} and M^{3+} ions decreases with Z up to the configuration t_{2g}^6. At that configuration, there is an increase with stepwise occupation of e_g orbitals being more influenced by the repulsing interaction with the anionic ligands.

Figure 3.10 represents the variation of divalent transition metal ions in combination with anions generating a weak and a strong field, respectively.

3.3.2.5   Hydration Enthalpy of Transition Metal Ions

Hydration enthalpy is extrapolated to infinite dilution when a metal ion coordinates only six water molecules forming hexaaquacomplexes of type $[M(H_2O)_6]^{2+,3+}$.

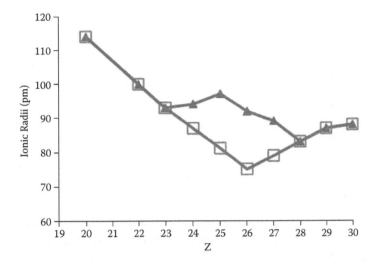

FIGURE 3.10 Size of transition $3d$-metal ions (□ low spin; ▲ high spin). (Data from J.E. Huheey, E.A. Keiter, and R.L. Keiter, *Inorganic Chemistry: Principles of Structure and Reactivity,* 4th edition [New York: Harper Collins, 1993].)

Hydration enthalpies are closely related to formation enthalpies of octahedral aqua complexes. Because the water molecule is a weak ligand, the resulting configurations are of the high-spin type.

The variation of enthalpy of M^{2+} ions corresponds to the process:

$$M^{2+}(g) + 6 H_2O(l) \rightarrow [M(H_2O)_6]^{2+}(aq) \tag{3.35}$$

Ca^{2+}, Mn^{2+}, and Zn^{2+} have d^0, d^5, and d^{10}, so CFSE is 0. Other metal ions deviate from the expected line due to extra CFSE, with two maxima for the configuration d^3 ($[V(H_2O)_6]^{2+}$ species) and d^8 ($[Ni(H_2O)_6]^{2+}$ species). Figure 3.11 presents the hydration enthalpies of divalent transition metal ions.

3.3.2.6 Lattice Energy

The lattice energy variation for some ionic compounds with C.N. 6, for example, halogenide MX_2 from CaX_2 to ZnX_2, is influenced by CFSE (octahedral field). In Figure 3.12, the lattice energies of the difluoride of the $3d$ metal ions are plotted as a function of the atomic number.

3.3.2.7 Stability of Some Oxidation States

In aqueous solutions, the Co^{3+} ion is not stable. It is reduced by water to the Co^{2+} ion. However, if there are strong field ligands present in solution, the Co^{3+} ion is stabilized, due to the CFSE (Oh) that are highest for a high-spin configuration t_{2g}^6 ($-24Dq + 3P$). The oxidation of Co^{2+} to Co^{3+} is accomplished by configuration changes from low spin to high spin and it is considered to take place in two stages:

I. Redistribution of electrons in the low spin state

$$Co^{2+}(t_{2g}^5 e_g^2) \rightarrow Co^{2+}(t_{2g}^6 e_g^1) \tag{3.36}$$

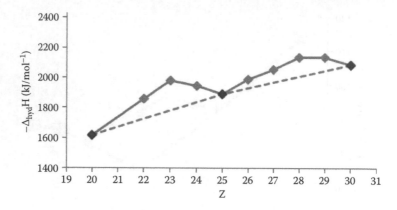

FIGURE 3.11   Hydration enthalpies of transition 3*d*-metal ions. (Data from J. Barrett, *Inorganic Chemistry in Aqueous Solution* [Cambridge, UK: The Royal Society of Chemistry, 2003].)

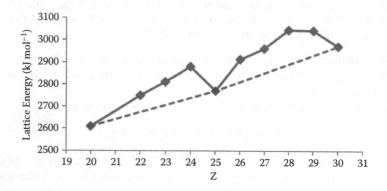

FIGURE 3.12   Lattice energy for the MF_2 of first row transition metals (F = weak field ligand).

II. Release of the 7th electron:

$$Co^{2+}(t_{2g}{}^6 e_g{}^1) \rightarrow Co^{3+}(t_{2g}{}^6) \tag{3.37}$$

3.3.3   Coordination Number and Stereochemistry

An important application of the crystal field theory is prediction of the geometry corresponding to different coordination numbers based on the magnitude of Δ and the nature of the bound ligands (Table 3.7).

The use of the coordination number as a parameter in prediction of metal toxicity has potential significance, although the coordination number is difficult to determine in biological systems. Moreover, two or more coordination numbers are possible for some metal ions.

TABLE 3.7
Coordination Numbers (CN), Geometries, and Preferred Ligands for Selected Metal Ions in Biological Systems

Metal Ion	Coordination Number	Geometry	Ligand Atom Donor
Na^+	6	Octahedral	O
K^+	6–8	Flexible	O
Mg^{2+}	6	Octahedral	O
Ca^{2+}	6–8	Flexible	O
Mn^{2+} (d^5)	6	Octahedral	O, N
Mn^{3+} (d^4)	6	Tetragonal	O
Fe^{2+} (d^6)	4	Tetrahedral	O
	6	Octahedral	O, N
Fe^{3+} (d^5)	4	Tetrahedral	S
	6	Octahedral	O
Co^+ (d^8)	6	Octahedral (usually missing the 6th ligand)	O, N
Co^{2+} (d^7)	4	Tetrahedral	S, N
	6	Octahedral	O, N
Co^{3+} (d^6)	6	Octahedral	O, N
Ni^{2+} (d^8)	4	Square planar	S, N
	6	Octahedral	Uncommon
Cu^+ (d^{10})	4	Tetrahedral	S, N
Cu^{2+} (d^9)	4	Tetrahedral	S, N
	4	Square planar	O, N
	5	Square pyramidal	O
	6	Tetragonal	N
Zn^{2+} (d^{10})	4	Tetrahedral	O, S, N
	5	Square pyramidal	O, N
Mo^{4+} (d^2)	6	Octahedral	O
Mo^{5+} (d^1)	6	Octahedral	O
Mo^{6+} (d^0)	6	Octahedral	S

Source: Adapted from J.A. Cowan, *Inorganic Biochemistry: An Introduction.* (New York: Wiley-VCH Inc., 1996) and R.M. Roat-Malone, *Bioinorganic Chemistry: A Short Course.* Hoboken, NJ: Wiley, 2003).

3.3.4 THERMODYNAMIC STABILITY OF METAL COMPLEXES

The thermodynamic stability of a metal complex may be represented by a stepwise formation constant (or stability constant) K_n or an overall stability constant β_n. The formation of a complex between a metal ion, M, and a ligand, L, is usually a substitution reaction. For example, metal ions in aqueous solution will be present as aqua ions. The reaction for the formation of the first complex could be written as:

$$[M(H_2O)_n] + L \rightleftharpoons [M(H_2O)_{n-1}L] + H_2O \tag{3.38}$$

The equilibrium constant for this reaction is given by:

$$K = [M(H_2O)_{n-1}L][H_2O]/[M(H_2O)_n][L] \qquad (3.39)$$

The expression can be simplified by removing constants. The number of water molecules attached to each metal ion is constant, and in dilute solutions the concentration of water is essentially constant. The expression becomes

$$K = [ML]/[M][L]. \qquad (3.40)$$

The stepwise coordination of a ligand L to a metal M to form a complex ML_n, can be represented as:

$$M + L \rightleftharpoons ML \qquad K_1 = [ML]/[M][L] \qquad (3.41)$$

$$ML + L \rightleftharpoons ML_2 \qquad K_2 = [ML_2]/[ML][L] \qquad (3.42)$$

$$ML_2 + L \rightleftharpoons ML_3 \qquad K_3 = [ML_3]/[ML_2][L] \qquad (3.43)$$

$$\cdots$$

Or, in the general form,

$$ML_{n-1} + L \rightleftharpoons ML_n \qquad K_n = [ML_n][ML_{n-1}][L] \qquad (3.44)$$

where the K_1, K_2 ..., Kn represent individual stability constants.
   An alternative representation may be the following:

$$M + L \rightleftharpoons ML \qquad \beta_1 = [ML]/[M][L] \qquad (3.45)$$

$$M + 2L \rightleftharpoons ML_2 \qquad \beta_2 = [ML_2]/[M][L]^2 \qquad (3.46)$$

$$M + 3L \rightleftharpoons ML_3 \qquad \beta_3 = [ML_3]/[M][L]^3 \qquad (3.47)$$

$$\cdots$$

$$M + nL \rightleftharpoons ML_n \qquad \beta_n = [ML_n]/[M][L]^n \qquad (3.48)$$

where the β_1, β_2, ..., β_n represent overall stability constants.
   The relationship that connects β_n and K_n:

$$\beta_1 = K_1 \qquad (3.49)$$

$$\beta_2 = K_1 \cdot K_2 \qquad (3.50)$$

$$\beta_3 = K_1 \cdot K_2 \cdot K_3 \qquad (3.51)$$

$$\cdots$$

$$\beta_n = K_1 \cdot K_2 \dots K_n \qquad (3.52)$$

TABLE 3.8

Overall Stability Constants for Some Complexes with Mono- and Bidentate Ligands in Aqueous Solution at 298 K

Complex ML_n	$\beta_n = [ML_n]/[M][L]^n$	Complex ML_n	$\beta_n = [ML_n]/[M][L]^n$
	$L = NH_3$		$L = H_2NCH_2CH_2NH_2$
$Mn(NH_3)_6^{2+}$	20	$Mn(en)_3^{2+}$	$5 \cdot 10^5$
$Fe(NH_3)_6^{2+}$	$1.6 \cdot 10^2$	$Fe(en)_3^{2+}$	$4 \cdot 10^9$
$Co(NH_3)_6^{2+}$	$1.3 \cdot 10^5$	$Co(en)_3^{2+}$	$8 \cdot 10^{13}$
$Ni(NH_3)_6^{2+}$	$5.5 \cdot 10^8$	$Ni(en)_3^{2+}$	$4 \cdot 10^{18}$
$Cu(NH_3)_4^{2+}$	$1 \cdot 10^{12}$	$Cu(en)_2^{2+}$	$1.6 \cdot 10^{20}$
$Zn(NH_3)_4^{2+}$	$5 \cdot 10^8$	$Zn(en)_2^{2+}$	$1.2 \cdot 10^{13}$

The value of an overall stability constant β_n is the product of the stepwise stability constants K_1 to K_n:

$$\beta_n = \prod_1^n K_n \qquad (3.53)$$

The magnitude of log K reflects the tendency toward formation of metal complexes in aqueous solution and gives a quantitative measure of the relative stabilities of metal complexes. The larger the magnitude of the equilibrium constant, the more stable the complex ML is in solution.

The stability is greatly increased in the case of complexes formed by the multidentate ligands. The bi- or polydentate ligands form coordination compounds in the shape of a heteroatomic ring. This process of ring formation is called *chelation*, the ligands involved are called chelating agents, and the resulting complex is called a *chelate*. The increase in stability is due to a predominantly entropic effect. Table 3.8 presents the increase of overall stability constants of ethylenediamine chelates of some divalent metal ions compared to the corresponding ammonia complexes.

Considering that each metal possesses its own spectra of affinity constants for different biological ligands, the values of stability constants have been used as descriptors for QSAR studies. Some examples include the log of the equilibrium constant (log K_{eq}) of a metal–ATP complex (Biesinger and Christensen 1972), the stability constant of metal–ion complexes with NH_3, the stability constants of metal ion complexes with EDTA, and the stability constants of divalent metal ions with AMP (Enache et al. 2003).

3.3.5 KINETIC REACTIVITY/LABILITY OF METAL COMPLEXES

The coordination complexes that undergo rapid ligand exchange reactions are referred to as *labile complexes*; those that do not are called *inert complexes*. Metal complexes vary in lability from extremely labile to essentially substitutionally inert. The intrinsic nature of the metal ion largely determines the reactivity of the metal complex. In this respect, the rates of replacement of ligand of aqua complexes by

TABLE 3.9
Rate Constants for Water Exchange in Aqua Complexes

Metal ion	d^n	Hydration Number[a]	Water Exchange Rate (s^{-1})[a]
Group 1A			
Li^+	—	4	$\sim 10^9$
Na^+	—	6	$\sim 10^9$
K^+	—	6	$\sim 10^9$
Group 2A			
Mg^{2+}	—	6	$\sim 10^6$
Ca^{2+}	—	8	$\sim 10^8$
Sr^{2+}	—	6	$\sim 10^9$
Ba^{2+}	—	6	$\sim 10^9$
1st Row Transition Metals			
Cr^{2+}	d^4	6	$\sim 10^9$
Cr^{3+}	d^3	6	2.4×10^{-6}
Mn^{2+}	d^5	6	2.1×10^7
Fe^{2+}	d^6	hs 6	4.4×10^6
Fe^{3+}	d^5	hs 6	$\sim 10^{-3}$
Co^{2+}	d^7	6	3.2×10^6
Co^{3+}	d^6	hs 6	$\sim 10^{-1}$
Ni^{2+}	d^8	6	3×10^4
Cu^{2+}	d^9	6	4.4×10^9
Zn^{2+}	d^{10}	6	$\sim 10^7$
2nd Row Transition Metals			
Ru^{3+}	d^5	6	1.8×10^{-2}
Pd^{2+}	d^8	4	5.6×10^2
Cd^{2+}	d^{10}	6	$\sim 10^8$
3rd Row Transition Metals			
Pt^{2+}	d^8	4	3.9×10^{-4}
Hg^{2+}	d^{10}	6	$\sim 10^9$

Source: Adapted from A.L. Feig and O.C. Uhlenbeckv. "The Role of Metal Ions in RNA Biochemistry, in *The RNA World.* 2nd edition, 287–319 (New York: Cold Spring Harbor Laboratory Press, 1999).

other ligand (Table 3.9) vary over a wide range, from very high rates ($k \approx 10^9 \ M^{-1} s^{-1}$) to very low rates ($k \approx 10^{-9} \ M^{-1} s^{-1}$).

The analysis of the data in Table 3.9 shows that:

1. Nontransition metal aqua complexes are extremely labile.
   The size of the central ion influences ligand replacement. The small central ions are held tightly by the ligand, resulting in more inert complexes. For example, the rate constant for water exchange in the three aqua

complexes, $[Mg(H_2O)_6]^{2+}$, $[Ca(H_2O)_8]^{2+}$, $[Ba(H_2O)_6]^{2+}$ increases from Mg^{2+} to Ca^{2+} with increasing ion size.

The charge of the metal ion is also important in the complexes involving a higher charge being more inert. For example Mg^{2+}(aq) is smaller than Na^+(aq) and is more inert.

The high lability of essential cations (Na^+, K^+, Ca^{2+}, Mg^{2+}) suggests that these ions are predominantly "free" or unbound in the body. They are associated with catalytic sites of enzymes but do not have material inhibitory activity (Hoeschele et al. 1991).

2. The reactivity order for first-row divalent cations is $Mn > Fe > Co > Ni \ll Cu$.

In general, the divalent metal ions of the first-row transition elements are all very labile with rate constants ranging from 10^4 to 10^9 s^{-1}.

The crystal field theory can be used to predict the relative labilities of octahedral transition metal complexes. The very slow rates of exchange of the aqua cations $Cr(H_2O)^{3+}$ and $Co(H_2O)_6^{3+}$ can be explained by conjunction of maxima in CFSE with a high effective positive charge on the metal ion (Jolly 1991).

3. The inertness to substitution increases from first- to third-row metals (as Z increases) for metal oxidization states with the same electronic configuration (e.g., the order of exchange rate $Ni^{2+} > Pd^{2+} > Pt^{2+}$).

The inertness suggests the reason that Mo is the only essential transition metal found in biological systems that is not a first-row transition metal. Most second- and third-row transition metals appear to be too strongly bound and inert to engage in metabolic processes (Hoeschele et al. 1991). However, the inertness of some second- and third-row elements, such as Pt(II), can be useful for the chemotherapeutic activity of these complexes. The inertness probably plays a role also in the mutagenicity and mild carcinogenicity observed for cisplatin and the mutagenicity of other inert complexes, especially Cr(III) complexes.

The kinetic reactivity/lability of metal complexes is especially relevant to the reaction products of aquated cations with biological substrates (ligands) and also for the use of metal complexes as therapeutic agents.

3.3.6 Redox Properties of Metal Complexes

There are two types of electron transfer mechanisms for transition metal species, outer- and inner-sphere electron transfers. The outer-sphere electron transfer occurs when the outer coordination sphere (or solvent) of the metal centers is involved in transferring electrons. This type of transfer does not imply reorganization of the inner coordination sphere of either reactant. An example of this reaction is given in Equation (3.54):

$$[Fe(II)(CN)_6]^{4-} + [Rh(IV)Cl_6]^{2-} \rightarrow [Fe(III)(CN)_6]^{3-} + [Rh(III)Cl_6]^{3-}. \quad (3.54)$$

Inner-sphere electron transfer involves the inner coordination sphere of the metal complexes, and normally takes place through a bridging ligand. A classic example studied and explained by Taube (1953) is given in equation 3.55:

$$[CoCl(NH_3)_5]^{2+} + [Cr(H_2O)_6]^{2+} \rightarrow [Co(NH_3)_5(H_2O)]^{2+} + [CrCl(H_2O)_5]^{2+}. \quad (3.55)$$

In this reaction, the chloride that was initially bound to Co(III), the oxidant, becomes bound to Cr(III) in complexes that are kinetically inert. The bimetallic complex $[Co(NH_3)_5(\mu\text{-}Cl)(Cr(H_2O)_5]^{4+}$ is formed as an intermediary, wherein "μ-Cl" indicates the chloride bridges between the Cr and Co atoms, serving as a ligand for both. The electron transfer occurs across a bridging group from Cr(II) to Co(III) to produce Cr(III) and Co(II).

Redox changes occurring in a biological environment can have a pronounced influence on the overall toxicological (biological) response elicited by a metallic complex.

A typical example is mercury, which can exist in two oxidation states and the free state. These species exhibit marked differences in uptake, distribution, and toxicological effects. The three forms are governed by the following disproportion reaction (Equation [3.56]):

$$Hg_2^{2+} \rightleftharpoons Hg^0 + Hg^{2+} \quad (3.56)$$

The intrinsic lability/reactivity can be modified by redox changes in vivo. Thus, an inert reactant can become a labile product and vice versa. The classical redox reaction of Taube (Equation [3.54]) implies reduction of the inert Co(III) species by the labile Cr(II), producing labile Co(II) and an inert Cr(III) species. Changing of intrinsic lability/reactivity can lead to the stronger bonds between biological ligands (e.g., nucleic acids) and very inert complexes. As a consequence, the metal might not be removed easily by the normal substitution reactions or repair processes.

3.4 PROPERTIES OF METAL IONS RELEVANT TO IONIC AND COVALENT BONDING TENDENCIES

3.4.1 IONIC POTENTIAL (Z/r)

The ionic potential (Z/r) of a metal ion is given by the charge/radius ratio (Z is ion charge and r is ionic radius). It incorporates the distance between an ion and another charge, and the size of the electrostatic force created. It reflects the metal ion tendency to form ionic bonds.

3.4.2 THE IONIC INDEX (Z²/r)

The polarizing power (Z^2/r) is a measure of electrostatic interaction strength between a metal ion and a ligand (Turner et al. 1981). The quotient Z^2/r is a surrogate characteristic indicating ionic bond stability.

3.4.3   THE COVALENT INDEX $(X_m^2 r)$

The $X_m^2 r$ (X_m is electronegativity and r is ionic radius) reflects the degree of covalent interactions in the metal–ligand complex relative to ionic interactions (Nieboer and Richardson 1980). In QSAR studies, the covalent index is commonly used in combination with ionic index Z^2/r or with the constant for the first hydrolysis |log KOH| (see Section 3.5). The subscript "m" refers to the most common (Mulliken) measure of electronegativity. Sometimes other measures are used in the literature and sometimes the "m" is omitted, e.g., the covalence index in Chapter 8 does not include the "m" subscript because the Pauling, Mulliken or Allred-Rochow scales for electronegativity might be pertinent.

3.4.4   COMBINATION OF COVALENT INDEX $X_m^2 r$ AND IONIC INDEX Z^2/r

The combined use of the covalent and ionic indices was proposed by Nieboer and Richardson (1980) to correlate with metal bioactivity. Based on the ionic indices and covalent indices, Nieboer and Richardson (1980) classified metals and metalloids into three classes: A, B, and borderline (see Table 3.1). This classification is related to atomic properties and the solution chemistry of metal ions, and demonstrates the potential for grouping metal ions according to their binding preferences.

The hard acceptors, or class A group ions (oxygen-seeking), are expected to interact with oxygen-containing ligands. The soft acceptors, or class B group ions (nitrogen/sulfur-seeking), form stable bonds with S- and N-containing ligands. The metals from class B (e.g., Ag^+, Tl^+, Hg^{2+}, Cd^{2+}) are often very toxic. Between the two major classes are the borderline elements represented primarily by metal ions from the first series of transition metals.

The combined use of the two indices has been applied in QSAR studies to predict the relative toxicity of metal ions (McCloskey et al. 1996) or to predict biosorption capacity (Can and Jianlong 2007).

3.4.5   POLARIZABILITY AND CHEMICAL SOFTNESS PARAMETER

A widely applied concept in the evaluation of metal ion interaction is that of hardness and softness. In general terms, *hard* and *soft* suggest the resistance to deformation in response to electric forces. Thus, hard ions have greater resistance to deformation of the electron cloud and soft ions have lesser resistance to deformation. Quantitative scales for metal ion hardness or softness were developed in the 1960s, starting with the HSAB theory developed by Pearson (see Section 3.3.1).

HSAB theory is rather qualitative and summarizes the general affinity of a Lewis acid (the metal ion) for a Lewis base (the ligand) in a manner independent of the acidity or basicity of the species. The general trend is that hard metal ions bind preferentially to hard ligands, whereas the soft metal ions bind to soft ligands. This principle reflects the degree of covalency in the metal–ligand bond. The combinations between strong species are predominantly ionic, while the combinations between soft species are predominantly covalent.

Klopman attempted to quantify Pearson's HSAB principle using frontier molecular orbital (FMO) theory (Klopman 1968), with the following equation:

$$\Delta E = \underbrace{-q_r q_s \frac{\Gamma}{\varepsilon}}_{\substack{\text{charge/charge} \\ \text{interaction}}} + \Delta solv. + \underbrace{\sum_m \sum_n \left[\frac{2(c_r^m)^2 (c_s^n)^2 \beta^2}{E_m - E_n} \right]}_{\substack{\text{FMO/FMO} \\ \text{interactions}}} \qquad (3.57)$$

where

ΔE = energy change of interaction of species r with species s,

Q = total initial charges,

Γ = Coulomb repulsion term,

ε = local dielectric constant,

$\Delta solv.$ = solvation or desolvation,

c_r = coefficient of orbital r,

β = resonance integral,

E_m = energy of occupied frontier molecular orbital m, and

E_n = energy of unoccupied frontier molecular orbital n.

A hard acid is likely to be strongly solvated and have a high LUMO energy, whereas a soft acid has a low LUMO energy and large-magnitude LUMO coefficients. A hard base has a high HOMO energy, and a soft base is characterized by low HOMO energy, but large magnitude HOMO coefficient (Figure 3.13).

Klopman also proposed the following bonding principle: Hard acids bind to hard bases to give charge-controlled (ionic) complexes. Such interactions are dominated by the charges on the Lewis acid and Lewis base species. Soft acids

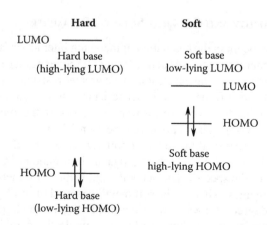

FIGURE 3.13   Relative energies of frontier molecular orbitals for hard and soft acids and bases.

bind to soft bases to give FMO-controlled (covalent) complexes. These interactions are dominated by the energies of the participating FMOs, the HOMO, and the LUMO.

In other words, hard acid–hard base interactions are predominantly electrostatic and soft acid–soft base interactions are predominantly covalent.

The main contribution of Klopman theory is that the contributing aspects of charge-controlled and FMO-controlled Lewis acid–base complexation are separated and quantified.

Based on the research of Klopman, and Parr and Pearson (Klopman 1968; Parr and Pearson 1983), Pearson described the absolute hardness (η) quantitatively as being proportional to the difference between I (ionization potential) and A (electron affinity) of the species (Pearson 1988). Absolute softness is defined as η^{-1}. The absolute electronegativity (χ) and the absolute hardness (η) are applied quantitatively to any given acid–base reaction. Table 3.10 presents χ and η values for some representative metal ions.

A recent contribution to the HSAB principle is the application of concepts of local softness and local hardness (Torrent-Sucarrat et al. 2010). Instead of global hardness (η) and global softness (S) (see Equations [3.15] and [3.16]), which are called *global descriptors* because they describe the properties of a molecule as a whole, the local descriptors are used.

TABLE 3.10
Absolute Hardness and Absolute Electronegativity for Some Metal Ions

Metal	Absolute Hardness η	Absolute Electronegativity χ	Metal	Absolute Hardness η	Absolute Electronegativity χ
	Group 1A			Group 2A	
Li^+	35.12	40.52	Mg^{2+}	47.59	32.55
Na^+	26.21	21.08	Ca^{2+}	19.52	31.39
K^+	17.99	13.64	Sr^{2+}	27.3	16.3
	1st Row Transition Metals			2nd Row Transition Metals	
Cr^{2+}	7.23	23.73	Ru^{3+}	10.7	39.2
Cr^{3+}	9.1	40.0	Pd^{2+}	6.75	26.18
Mn^{2+}	9.02	24.66	Cd^{2+}	10.29	27.20
Fe^{2+}	7.24	23.42			
Fe^{3+}	12.08	42.73		3rd Row Transition Metals	
Co^{2+}	8.22	25.28	Pt^{2+}	8.0	27.2
Co^{3+}	8.9	42.4	Hg^{2+}	7.7	26.5
Ni^{2+}	8.50	26.67		Miscellaneous	
Cu^+	6.28	14.01	Pb^{2+}	8.46	23.49
Cu^{2+}	8.27	28.56	Tl^+	7.16	13.27
Zn^{2+}	10.88	28.84			

Source: Data from R.G. Pearson, "Absolute Electronegativity and Hardness: Application to Inorganic Chemistry. *Inorg. Chem.* 27, no. 4 (1988):734–740.

The local hardness $\eta(r)$ and local softness $s(r)$ are defined as

$$\eta(r) = \left(\frac{\delta\mu}{\delta\rho(r)} \right)_{v(r)} \text{ and} \tag{3.58}$$

$$s(r) = \left(\frac{\delta\rho}{\delta\mu(r)} \right)_{v(r)}. \tag{3.59}$$

The local hardness and local softness are interconnected through this inverse relationship:

$$\int \eta(r)s(r)\,dr = 1. \tag{3.60}$$

The most commonly cited scale producing the most successful correlations (i.e., highest correlation coefficients, r) is the one computed by Pearson and Mawby (Pearson and Mawby 1967). This scale involves the so-called softness parameter (σ_P) defined for a metal ion in terms of the coordinate bond energies of its metal fluoride, CBE(F), and metal iodide, CBE(I).

$$\sigma_P = \frac{CBE(F) - CBE(I)}{CBE(F)} \tag{3.61}$$

The softness parameter (σ_P) is a measure of the ability of a metal ion to give up its valence electrons (Jones and Vaughn 1978; Williams and Turner 1981). A hard metal ion retains its valence electrons very strongly and is not readily polarized, whereas a soft metal ion is relatively large, does not retain its valence electrons firmly, and is easily polarized (Vouk 1979). For ions of a given charge, softness increases as σ_P decreases. The σ_P is probably the most widely used parameter in studies of metal ion toxicities (Williams et al. 1982; Turner et al. 1983; Tan et al. 1984; Babich et al. 1986; Khangarot and Ray 1989; Segner et al. 1994; Magwood and George 1996; McCloskey et al. 1996; Enache et al. 1999; Ownby and Newman 2003). These and additional studies are discussed in Chapter 5.

Other softness parameters used by various studies are the following:

- Ahrland softness parameter (σ_A) (Ahrland 1968) is correlated with total ionization potential for the formation of $M^{n+}(g)$ and the dehydration energy $-\Delta H^\circ$. The larger the difference between the total ionization potential for the formation of $M^{n+}(g)$ and the dehydration energy $-\Delta H^\circ$ are, the softer the ion is.
- The softness parameter of Misono and Saito (Y') (Misono and Saito 1970) is based on the use of a regression model where the log of stability constants of halogeno-complexes of metal ions (Log k) is a function of two parameters. One parameter corresponds to hardness (X) and the other to softness (Y),

$$\log K = \alpha'X + \beta'Y' + \gamma' \tag{3.62}$$

(α' and β' are the dual basicity parameters of a ligand corresponding to X and Y', respectively, and γ' is a constant determined for each ligand considered.) This parameter was used in some QSAR studies (Enache et al. 2003).

- The Williams softness parameter (s_W, R or R (W)) is defined as the ratio of the ionization potential to the ionic function, Z^2/r where Z is the ionic charge and r the ionic radius of the cation (Williams and Hale 1966; Williams et al. 1982):

$$R = \frac{\Sigma I}{\frac{Z^2}{r}} = \frac{\Sigma I \times r}{Z^2}. \tag{3.63}$$

Based on this relation, the behavior of class B metal ions in water can be explained by the high polarizing power expressed by the ionization potential of these cations, relative to their size and charge.

3.4.6 OTHER PARAMETERS

The covalent bond stability parameter ($\Delta\beta$) was described as the difference between the logarithm of the stability constants for the metal fluoride and the metal chloride at infinite dilution (Turner et al. 1981):

$$\Delta\beta = \log \beta^\circ_{MF} - \log \beta^\circ_{MCl} \tag{3.64}$$

where Log β°_{MF} = Log of the stability constant for the metal fluoride and Log β°_{MCl} = Log of the stability constant for the metal chloride. The tendency to form covalent bonds with soft ligands decreases with $\Delta\beta$.

The combination of $\Delta\beta$ and Z^2/r was also used (Turner et al. 1981). Plotting Z^2/r (vertical axis) as a function of $\Delta\beta$ (horizontal axis), it was possible to create a complexation field diagram for metal ions, where the stability constants of metal ion complexes with intermediate ligands increases across the diagram (Walker et al. 2003).

3.5 BRÖNSTED ACIDITY OF METAL IONS

The metal ions are found in aqueous solution surrounded by water molecules as aqua complexes (or aqua acids). The hydrated ions can undergo stepwise hydrolysis with delivery of one or more H^+ to the bulk solvent. This process occurs due the attraction of the metal ion to the electron cloud of its water hydration, and the weakening of the OH bond in the H_2O molecule. The process can be represented by the reaction,

$$MOH_2^{n+} + H_2O = MOH^{(n-1)+} + H_3O^+. \tag{3.65}$$

Any positive ion can participate in this reaction to some extent, as expressed as the equilibrium constant (K_a) for the above equation.

Metal ions with a single positive charge have only one hydrolysis step, while the divalent or trivalent cations undergo stepwise hydrolysis. For comparison only, the first step of hydrolysis, characterized by the pK_{a1} is considered for each element.

Each step in the hydrolysis reduces the charge as hydroxycations are formed. For any step of this reaction, a standard hydrolysis constant can be written:

$$pK_a = pH - \log_{10} \frac{[M^{(n-1)+}]}{[M^{n+}]}. \tag{3.66}$$

Finally, a neutral hydrated metal hydroxide is formed. Usually, such neutral species will be insoluble and precipitate. Metal hydroxide usually starts to precipitate when the pH of the solution is approximately equal to the pK_a of the metal ion (Table 3.11). If the effective electronegativity of the element is high enough, the hydrolysis process can continue to produce hydroxyanions and even oxoanions. This is the case of metal ions in their highest oxidation states, which have extremely high effective electronegativity. For example, $Mn(VII)$ forms the MnO_4^- oxoanion in

---

TABLE 3.11

Acid Hydrolysis Constants (pK_a's) for Some Aqua Ions at 298 K and pH Precipitation Range of Hydroxides or Hydrous Oxides

Metal Ion	pK_a	Precipitation Range	Metal Ion	pK_a	Precipitation Range
Tl^+	13.2		Sn^{2+}	3.9	
Hg_2^{2+}	5.0		Pb^{2+}	7.8	
Mg^{2+}	11.4	>10	Al^{3+}	5.1	
Ca^{2+}	12.6		Sc^{3+}	5.1	
Ba^{2+}	13.2		In^{3+}	4.4	
				($pK_{a2} = 3.9$)	
Mn^{2+}	10.6	8–10	Tl^{3+}	1.1	
				($pK_{a2} = 1.5$)	
Fe^{2+}	9.5	6–8	Bi^{3+}	1.6	
Co^{2+}	8.9	6–8	V^{3+}	2.8	
Ni^{2+}	10.6	6–8	Ti^{3+}	2.2	0–2
Cu^{2+}	6.8	4–6	Cr^{3+}	3.8	4–6
Zn^{2+}	8.8	6–8	Co^{3+}	0.7	
Cd^{2+}	9.0		Fe^{3+}	2.2	2–4
				($pK_{a2} = 3.3$)	
Hg^{2+}	3.7				
($pK_{a2} = 2.6$)					

Source: Adapted from F. Basolo and E.G. Pearson. *Mechanisms of Inorganic Reactions: A Study of Metal Complexes in Solution,* 2nd edition (New York: John Wiley, 1967); and H.F. Walton, *Principles and Methods of Chemical Analysis* (New York: Prentice Hall, 1952).

aqueous solution. The constant for the first hydrolysis ($|\log K_{OH}|$) is a parameter often used in QSAR studies, based on the premise that the metal ions are found as aqua complexes in biological systems (Newman and McCloskey1996; Tatara et al. 1997; Enache et al. 2003).

Quantitative correlations between a cation's acidity and electrostatic parameter can be made (Wulfsberg 1991). A good correlation can be found by plotting the pK_a versus the electrostatic parameter (Z^2/r where r is expressed in pm) for the elements with an electronegativity of $\chi \leq 1.5$. The following empirical equation was obtained;

$$pK_a = 15.14 - 88.16\frac{Z^2}{r}. \tag{3.67}$$

For the more electronegative metals ($\chi > 1.5$ on the Pauling scale), the points deviate substantially from the line, so Equation (3.67) was corrected with an empirically derived factor:

$$pK_a = 15.14 - 88.16\left\{\frac{Z^2}{r} + 0.096\left(\chi_{Pauling} - 1.50\right)\right\}. \tag{3.68}$$

The constant for the first hydrolysis ($|\log K_{OH}|$) was correlated also with the covalent index $X_m^2 r$ in QSAR studies (Newman and McCloskey 1996; Tatara et al. 1997, 1998).

3.6 SOLUBILITY OF METAL COMPOUNDS AND METAL ION HYDRATION

3.6.1 THERMODYNAMIC ASPECTS OF SOLUBILITY

To describe the phenomena that occur during the solubility process, the relationship between the solubility equilibrium constant and the free enthalpy variation can be defined first:

$$\Delta G^\circ = -RT \ln K_{eq}. \tag{3.69}$$

As for other reactions, K_{eq}, and thus solubility, increases as the ΔG° becomes increasingly negative. Free enthalpy variation is related to the variation in both enthalpy and entropy:

$$\Delta G^\circ = \Delta H^\circ - T\Delta S^\circ, \tag{3.70}$$

During the dissolution process, two opposite processes occur—release of ions from the crystal lattice (endothermic process) and hydration of the liberated ions (exothermic process).

The energy necessary to remove the ions from the position that they hold in a crystal to the infinite is equal, but of opposite sign, to the lattice energy (U). During the separation of ions, the energy ($-U$) is high and positive. Nevertheless, the ions in solution are not separated through an infinite distance as is the case in a gaseous phase, but they are surrounded by, and strongly associated with, water dipoles. This association occurs with energy liberation, called the *free enthalpy of hydration* (ΔG_h°).

The dissolution capacity of salts depends on the energies of opposite signs, $-U$ and G_h°, which make the solubility characteristics irregular. The free energy of dissolution Q° can be considered as the sum of lattice energy $-U$ of the salt and the free energy of hydration ΔG_h° of metal ions (Figure 3.14). The free enthalpy of hydration (ΔG_h°), is the favoring factor of the dissolution process and is largely determined by the enthalpy of hydration of ions.

The hydration process is an exothermic one. In the case of cations, the enthalpy of hydration is dependent on the size and charge of metal ions. The smaller and higher charged a metal ion is, the more energy is released at its hydration. In Table 3.12, the enthalpies of hydration of some metal ions are presented. The enthalpy of hydration $\Delta_h H$ was used in QSAR studies for the analysis of metal ion toxicity (Enache et al. 1999; Enache et al. 2003).

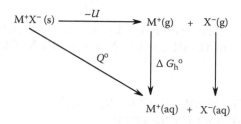

FIGURE 3.14   Born-Haber cycle for the dissolution process of a salt.

TABLE 3.12

Enthalpies of Hydration (kJ mol⁻¹) of Selected Metal Ions

H⁺	−1091			Cr²⁺	−1904	Cd²⁺	−1807
Li⁺	−519	Be²⁺	−2494	Mn²⁺	−1841	Hg²⁺	−1824
Na⁺	−406	Mg²⁺	−1921	Fe²⁺	−1946	Sn²⁺	−1552
K⁺	−322	Ca²⁺	−1577	Co²⁺	−1996	Pb²⁺	−1481
Rb⁺	−293	Sr²⁺	−1443	Ni²⁺	−2105	Al³⁺	−4665
Cs⁺	−264	Ba²⁺	−1305	Cu²⁺	−2100	Fe³⁺	−4430
Ag⁺	−473			Zn²⁺	−2046		
Tl⁺	−326						

Source: Data from W.L. Jolly, *Modern Inorganic Chemistry,* 2nd edition
(New York: McGraw-Hill, 1991).

3.6.2 The Solubility-Product Constant

A saturated solution of a slightly soluble salt in contact with undissolved salt involves an equilibrium like this:

$$M_xA_y(s) \rightleftharpoons xM^{y+}(aq) + yA^{x-}(aq). \tag{3.71}$$

In writing the equilibrium constant expression for a heterogeneous equilibrium, the concentrations of pure liquids and solids are considered to be one. So the equilibrium constant expression for the equilibrium above is the following:

$$K_{sp} = [M^{y+}]^x[A^{x-}]^y. \tag{3.72}$$

This equilibrium constant is called a *solubility-product constant*. In general, the solubility-product constant (K_{sp}) is the equilibrium constant for the equilibrium that exists between a solid ionic solute and its ions in a saturated aqueous solution.

Application of the solubility-product constant in QSAR studies started with the research of Shaw (Shaw 1954a, 1954b) in which he demonstrated that the affinity of the cations for the sulfhydryl group is directly proportional to the insolubility of the corresponding metal sulfide. In a similar deduction, the affinity of the cations for oxygen donor bioligands can be correlated with the solubility of the corresponding metal hydroxide. Thus, the most used solubility-related parameters used in SAR and QSAR studies are the solubility of sulfide (pK_{sp} sulfide) and hydroxide log $-K_{SO}$(MOH) or pK_{sp} hydroxide (Shaw 1954a, 1954b; Shaw and Grushkin 1957; Biesinger and Christensen 1972; Khangarot and Ray 1989). These and additional studies are discussed in Chapter 5. The values of the solubility-product constant for some hydroxide and sulfide of metal ions are presented in Table 3.13.

3.6.3 LSER for Inorganic Compounds

The linear solvation energy relationship (LSER) can be a useful predictive tool for environmental property estimations (Hickey 1999). In LSER, the solution behavior of a substance (e.g., solubility, bioaccumulation, and toxicity) is directly related to several aspects of its chemical structure. For example, a LSER equation is depicted in Equation (3.73):

$$Log(Property) = mV_i/100 + s\pi^* + b\beta_m + a\alpha_m \tag{3.73}$$

where V_i is the intrinsic (Van der Waals) molecular volume, π^* is the solute's ability to stabilize a neighboring charge or dipole by nonspecific dielectric interactions, and β_m and α_m are the solute's ability to accept or donate a hydrogen in a hydrogen bond. The equation coefficients m, s, b, and a are constants for a particular set of conditions, determined by multiple linear regression of the LSER variable values for a series of inorganic chemicals with the measured value for a particular chemical property (Hickey 1999).

TABLE 3.13

Solubility Product Constants at 298 K

Hydroxide	K_{sp}	Sulphide	K_{sp}
$Mg(OH)_2$	$1.5 \cdot 10^{-11}$	Tl_2S	$1 \cdot 10^{-22}$
$Ca(OH)_2$	$7.9 \cdot 10^{-6}$	Bi_2S_3	$2 \cdot 10^{-72}$
$Al(OH)_3$	$1.9 \cdot 10^{-33}$	Sb_2S_3	$2 \cdot 10^{-93}$
$Sn(OH)_2$	$2 \cdot 10^{-26}$	SnS	$8 \cdot 10^{-29}$
$Pb(OH)_2$	$2.8 \cdot 10^{-16}$	PbS	$8 \cdot 10^{-28}$
$Cr(OH)_3$	$7 \cdot 10^{-31}$		
$Mn(OH)_2$	$4.5 \cdot 10^{-14}$	MnS	$5 \cdot 10^{-15}$
$Fe(OH)_2$	$8 \cdot 10^{-15}$	FeS	$5 \cdot 10^{-18}$
$Fe(OH)_3$	$6.3 \cdot 10^{-38}$	Fe_2S_3	$1 \cdot 10^{-88}$
$Co(OH)_2$	$2 \cdot 10^{-16}$	$CoS(\alpha)$	$6 \cdot 10^{-21}$
$Co(OH)_3$	$4 \cdot 10^{-45}$	$CoS(\beta)$	$8 \cdot 10^{-23}$
$Ni(OH)_2$	$2 \cdot 10^{-15}$	$NiS(\alpha)$	$3 \cdot 10^{-21}$
$Zn(OH)_2$	$4.5 \cdot 10^{-17}$	ZnS	$1.1 \cdot 10^{-21}$
$Cu(OH)_2$	$2 \cdot 10^{-19}$	CuS	$8.7 \cdot 10^{-36}$
$Cd(OH)_2$	$1.2 \cdot 10^{-14}$	CdS	$4 \cdot 10^{-29}$
		Cu_2S	$1.6 \cdot 10^{-48}$
		Ag_2S	$7 \cdot 10^{-50}$
		HgS	$3 \cdot 10^{-53}$

Source: Data from B.G. Segal, *Chemistry: Experiments and Theory* (New York: Wiley, 1989).

Values for the steric parameter $V_i/100$ were calculated from atomic parachors, through a characteristic molecular volume (Vx) to a corresponding intrinsic molecular volume (V_i).

For π^*, dipole moments, polarizabilities and electronegativity were used, whereas for β and α ionization potentials (basicity), electron affinities (acidity) and pK values were the used data.

ACKNOWLEDGMENTS

The author gratefully thanks Dr. Monica Enache for the generous help given in writing this paper.

REFERENCES

Ahrland, S. 1968. Scales of softness for acceptors and donors. *Chem. Phys. Lett.* 2(5):303–306.

Ahrland, S., J. Chatt, and N.R. Davies. 1958. The relative affinities of ligand atoms for acceptor molecules and ions. *Quart. Rev.* (London) 12:265–276.

Allen, L.C. 1989. Electronegativity is the average one-electron energy of the valence-shell electrons in ground-state free atoms. *J. Am. Chem. Soc.* 111(25):9003–9014.

Allred, A.L., and E.G. Rochow. 1958. A scale of electronegativity based on electrostatic force. *J. Inorg. Nucl. Chem.* 5(4):264–268.

Babich, H., J.A. Puerner, and E. Borenfreund. 1986. In vitro cytotoxicity of metal to bluegill (BF-2) cells. *Arch. Environ. Contam. Toxicol.* 15(1):31–37.

Bailar, J.C. 1984. *Chemistry,* 2nd edition. New York: Academic Press.

Barrett, J. 2003. *Inorganic Chemistry in Aqueous Solution.* Cambridge, UK: The Royal Society of Chemistry.

Basolo, F., and E.G. Pearson. 1967. *Mechanisms of Inorganic Reactions: A Study of Metal Complexes in Solution,* 2nd edition. New York: John Wiley.

Bethe, H. 1929. Termaufspaltung in Kristallen [Term splitting in crystals]. *Ann. Phys.* 3(2):133–206.

Biesinger, K.L., and G.M. Christensen. 1972. Effects of various metals on survival growth reproduction and metabolism of *Daphnia magna. J. Fish. Res. Board Can.* 29(12):1691–1700.

Bratsch, S.G. 1988. Revised Mulliken electronegativities: I. Calculation and conversion to Pauling units. *J. Chem. Educ.* 65(1):34–41.

Can, C., and W. Jianlong. 2007. Correlating metal ionic characteristics with biosorption capacity using QSAR model. *Chemosphere,* 69(10):1610–1616.

Clare, B.W. 1994. Frontier orbital energies in quantitative structure-activity relationships: A comparison of quantum chemical methods. *Theor. Chim. Acta.* 87(6):415–430.

Cowan, J.A. 1996. *Inorganic Biochemistry: An Introduction.* New York: Wiley-VCH Inc.

De Proft, F., and P. Geerlings. 1997. Calculation of ionization energies, electron affinities, electronegativities and hardnesses using density functional methods. *J. Chem. Phys.* 106(8):3270–3280.

Enache, M., J.C. Dearden, and J.D. Walker. 2003. QSAR analysis of metal ion toxicity data in sunflower callus cultures (*Helianthus annuus,* Sunspot). *QSAR Comb. Sci.* 22(2):239–240.

Enache, M., P. Palit, J.C. Dearden, and N.W. Lepp. 1999. Evaluation of cation toxicities to sunflower: A study related to the assessment of environmental hazards. In *Extended Abstracts, 5th International Conference on the Biogeochemistry of Trace Elements, Vol.* 2, 1120–1121. Vienna, Austria, July 11–15.

Feig, A.L., and O.C. Uhlenbeckv. 1999. The role of metal ions in RNA biochemistry. In *The RNA World,* 2nd edition, edited by Raymond F. Gesteland, Thomas R. Cech, and John F. Atkins, 287–319. New York: Cold Spring Harbor Laboratory Press.

Geerlings, P., and F. De Proft. 2002. Chemical reactivity as described by quantum chemical methods. *Int. J. Mol. Sci.* 3(4):276–309.

Geerlings, P., F. De Proft, and J.M.L. Martin. 1996a. Density functional theory: Concepts and techniques for studying molecular charge distributions and related properties. In *Recent Developments in Density Functional Theory,* Vol. 4 of *Theoretical and Computational Chemistry,* edited by J. Seminario, 773–809. New York: Elsevier.

Geerlings, P., W. Langenaeker, F. De Proft, and A. Baeten. 1996b. Molecular electrostatic potentials vs. DFT descriptors of reactivity. In *Molecular Electrostatic Potentials: Concepts and Applications,* Vol. 3 of *Theoretical and Computational Chemistry,* edited by P. Politzer and Z.B. Maksic, 587–617. New York: Elsevier.

Gray, B., and G.P. Haight. 1988. *Principes de chimie.* Paris InterEditions.

Greenwood, N.N., and A. Earnshaw. 2006. *Chemistry of the Elements,* 2nd edition. New York: Elsevier Butterworth Heinemann.

Hickey, J.P., 1999. Estimation of environmental properties for inorganic compounds using LSER. Symposia papers presented before the Division of Environmental Chemistry American Chemical Society, Anaheim, CA March 21–25. *Preprints of Extended Abstracts* 39(1):154–157.

Hoeschele, J.D., J.E. Turner, and M.W. England. 1991. Inorganic concepts relevant to metal binding, activity, and toxicity in a biological system. *Sci. Tot. Environ.* 109/110:477–492.

Huang, Q.-G., L. Kong, and L.S. Wang. 1996. Application of frontier molecular orbital energies in QSAR studies. *Bull. Environ. Contam. Toxicol.* 56(5):758–765.

Huheey, J.E., E.A. Keiter, and R.L. Keiter. 1993. *Inorganic Chemistry: Principles of Structure and Reactivity,* 4th edition. New York: Harper Collins.

Iczkowski, R.P., and J.L. Margrave. 1961. Electronegativity. *J. Am. Chem. Soc.* 83(17):3547–3551.

Irving, H., and R.J.P. Williams. 1953. The stability of transition-metal complexes. *J. Chem. Soc.*:3192–3210.

Jolly, W.L. 1991. *Modern Inorganic Chemistry,* 2nd edition. New York: McGraw-Hill.

Jones, M.M., and W. Vaughn. 1978. HSAB theory and acute metal ion toxicity and detoxification processes. *J. Inorg. Nucl. Chem.* 40(12):2081–2088.

Kaiser, K.L.E. 1980. Correlation and prediction of metal toxicity to aquatic biota. *Can. J. Fish. Aquat. Sci.* 37(2):211–218.

Kaiser, K.L.E. 1985. Correlation of metal ion toxicities to mice. *Sci. Total Environ.* 46: 113–119.

Khangarot, B.S., and P.K. Ray. 1989. Investigation of correlation between physico-chemical properties of metals and their toxicity to the water flea *Daphnia magna* Straus. *Ecotoxicol. Environ. Saf.* 18(2):109–120.

Klopman, G. 1968. Chemical reactivity and the concept of charge- and frontier-controlled reactions. *J. Am. Chem. Soc.* 90(2):223–234.

Lepădatu, C., M. Enache, and J.D. Walker. 2009. Toward a more realistic QSAR approach to predicting metal toxicity. *QSAR Comb. Sci.* 28(5):520–525.

Lewis, D.F.V., M. Dobrota, M.G. Taylor, and D.V. Parke. 1999. Metal toxicity in two rodent species and redox potential: Evaluation of quantitative structure-activity relationships. *Environ. Toxicol. Chem.* 18(10):2199–2204.

Magwood, S., and S. George. 1996. *In-vitro* alternatives to whole animal testing. Comparative cytotoxicity studies of divalent metals in established cell-lines derived from tropical and temperate water fish species in a neutral red assay. *Mar. Environ. Res.* 42(1–4):37–40.

McCloskey, J.T., M.C. Newman, and S.B. Clark. 1996. Predicting the relative toxicity of metal-ions using ion characteristics: Microtox® bioluminescence assay. *Environ. Toxicol. Chem.* 15(10):1730–1737.

McKinney, J.D., A. Richard, C. Waller, M.C. Newman, and F. Gerberick. 2000. The practice of structure activity relationships (SAR) in toxicology. *Toxicol. Sci.* 56(1):8–17.

Miso, M., and Y. Saito. 1970. Evaluation of softness from the stability constants of metal-ion complexes. *Bull. Chem. Soc. Japan* 43:3680–3684.

Mulliken, R.S. 1934. A new electroaffinity scale: Together with data on valence states and on valence ionization potentials and electron affinities. *J. Chem. Phys.* 2(11):782–793.

Newman, M.C., and J.T. McCloskey. 1996. Predicting relative toxicity and interactions of divalent metal ions: Microtox® bioluminescence assay. *Environ. Toxicol. Chem.* 15(3):275–281.

Newman, M.C., J.T. McCloskey, and C.P. Tatara. 1998. Using metal-ligand binding characteristics to predict metal toxicity: Quantitative ion character-activity relationships (QICARs). *Environ. Health Perspect.* 106(Suppl. 6):1419–1425.

Nieboer, E., and D.H.S. Richardson. 1980. The replacement of the nondescript term "heavy metals" by a biologically and chemically significant classification of metal ions. *Environ. Poll. Series B—Chem. Phys.* 1(1):3–26.

Orgel, L.E. 1952. The effects of crystal fields on the properties of transition metal ions. *J. Chem. Soc.*:4756–4761.

Ownby, D.R., and M.C. Newman. 2003. Advances in quantitative ion character-activity relationships (QICARs): Using metal-ligand binding characteristics to predict metal toxicity. *QSAR Comb. Sci.* 22(2):241–246.

Parr, R.G., R.A. Donelly, M. Levy, and W.E. Pulke. 1978. Electronegativity: The density functional viewpoint. *J. Chem. Phys.* 68(8):3801–3807.

Parr, R.G., and R.G. Pearson. 1983. Absolute hardness: Companion parameter to absolute electronegativity. *J. Am. Chem. Soc.* 105(26):7512–7516.

Parr, R.G., and W. Yang. 1989. *Density-Functional Theory of Atoms and Molecules.* New York: Oxford University Press.

Pauling, L. 1932. The nature of the chemical bond. IV. The energy of single bonds and the relative electronegativity of atoms. *J. Am. Chem Soc.* 54(9):3570–3582.

Pearson, R.G. 1963. Hard and soft acids and bases. *J. Am. Chem. Soc.* 85(22):3533–3543.

Pearson, R.G. 1966. Acids and bases. *Science* 151:172–177.

Pearson, R.G. 1988. Absolute electronegativity and hardness: Application to inorganic chemistry. *Inorg. Chem.* 27(4):734–740.

Pearson, R.G., and R.J. Mawby. 1967. The nature of metal-halogen bonds. In *Halogen Chemistry*, Vol. 3, edited by V. Gutmann, 55–84. London: Academic.

Roat-Malone, R.M. 2003. *Bioinorganic Chemistry: A Short Course.* Hoboken, NJ: Wiley.

Sanderson, R.T. 1955. Relation of stability ratios to Pauling electronegativities. *J. Chem. Phys.* 23:2467–2468.

Segal, B.G. 1989. *Chemistry: Experiments and Theory.* New York: Wiley.

Segner, H., D. Lenz, W. Hanke, and G. Schüürmann. 1994. Cytotoxicity of metals toward rainbow trout R1 cell-line. *Environ. Toxicol. Water Qual.* 9(4):273–279.

Shaw, W.H.R. 1954a. The inhibition of urease by various metal ions. *J. Am. Chem. Soc.* 76(8):2160–2163.

Shaw, W.H.R. 1954b. Toxicity of cations toward living systems. *Science* 120(3114):361–363.

Shaw, W.H.R., and B. Grushkin. 1957. The toxicity of metal ions to aquatic organisms. *Arch. Biochem. Biophys* 67(2):447–452.

Somers, E. 1959. Fungitoxicity of metal ions. *Nature* 184(4684):475–476.

Tan, E-L., M.W. Williams, R.L. Schenley, S.W. Perdue, T.L. Hayden, J.E. Turner, and A.W. Hsie. 1984. The toxicity of sixteen metallic compounds in Chinese hamster ovary cells. *Toxicol. Appl. Pharmacol.* 74(3):330–336.

Tatara, C.P., M.C. Newman, J.T. McCloskey, and P.L. Williams. 1997. Predicting relative metal toxicity with ion characteristics: *Caenorhabditis elegans* LC50. *Aquat. Toxicol.* 393(3–4):279–290.

Tatara, C.P., M.C. Newman, J.T. McCloskey, and P.L. Williams. 1998. Use of ion characteristics to predict relative toxicity of mono-, di-, and trivalent metal ions: *Caenorhabditis elegans* LC50. *Aquat. Toxicol.* 42(4):255–269.

Taube, H. 1953. The mechanism of electron transfer in solution. *J. Am. Chem. Soc.* 75(16):4118–4119.

Todeschini, R., and V. Consonni. 2008. *Handbook of Molecular Descriptors.* New York: Wiley.

Torrent-Sucarrat, M, F. De Proft, P.W. Ayers, and P. Geerlings. 2010. On the applicability of local softness and hardness. *Phys. Chem. Chem. Phys.* 12(5):1072–1080.

Turner, D.R., W. Whitfield, and A. Dickson. 1981. The equilibrium speciation of dissolved components of freshwater and seawater at 25°C and 1 atm pressure. *Geochim. Cosmochim. Acta* 45(6):855–881.

Turner, J.E., E.H. Lee, K.B. Jacobson, N.T. Christie, M.W. Williams, and J.D. Hoeschele. 1983. Investigation of correlations between chemical parameters of metal ions and acute toxicity in mice and *Drosophila*. *Sci. Total Environ.* 28(1–3):343–354.

Vouk, V. 1979. General chemistry of metals. In *Handbook on the Toxicology of Metals*, edited by L. Friberg, G.F. Nordberg, and V.B. Vouk, 15–30. Amsterdam: Elsevier.

Walker, J.D., M. Enache, and J.C. Dearden. 2003. Quantitative cationic-activity relationships for predicting toxicity of metals. *Environ. Toxicol. Chem.* 22(8):1916–1935.

Walton, H.F. 1952. *Principles and Methods of Chemical Analysis.* New York: Prentice Hall.

Williams, M.W., J.D. Hoeschele, J.E. Turner, K.B. Jacobson, N.T. Christie, C.I. Paton, L.H. Smith, H.R. Witschi, and E.H. Lee. 1982. Chemical softness and acute metal toxicity in mice and *Drosophila*. *Toxicol. Appl. Pharmacol.* 63(3):461–469.

Williams, M.W., and J. Turner. 1981. Comments on softness parameters and metal ion toxicity. *J. Inorg. Nucl. Chem.* 43(7):1689–1691.

Williams, R.J.P., and J.D. Hale. 1966. The classification of acceptors and donors in inorganic reactions. *Struct. Bonding* 1:249–281.

Wulfsberg, G. 1991. *Principles of Descriptive Inorganic Chemistry.* Monterey, CA: Brooks/Cole.

Yatsimirskii, K.B. 1994. Stabilization of unstable d-metal oxidation states by complex formation, in *Coordination Chemistry*, ACS Symposium Series, 565:207–212.

4 Descriptors for Organometallic Complexes

4.1 DEFINITIONS

The term *molecule* means here a species that possesses, on the highest energy occupied orbital, two electrons, in contrast to a *radical*. The charge on system is null in molecules and different from zero in molecular ions. Radicals are not analyzed here, but some molecular ions are analyzed.

Here, an *organic* molecule or ion means a molecule or ion that includes at least one carbon atom. According to this definition, silicones, silazanes, phosphazines, borazine, hydrazine, and sulfur dioxide are not organic molecules, but carbon monoxide, ferrocene, and lead acetate are.

In any ensemble of atomic nuclei and electrons, such as a molecule, ion, radical, or molecular ion, the electronic cloud has different densities in different spatial zones. The *density* is the probability of existence of electrons in that zone. In chemistry, the electronic cloud between two neighboring atomic nuclei is called a *chemical bond* and it possesses a certain density. The electron cloud of different chemical bonds, for example, coordinative, single, and triple bonds, have different characteristics. The density can be evaluated by a calculable feature of the chemical bond, called the *bond order* (Coulson 1939; Chirgwin and Coulson 1950; Jug 1977). The bond order cannot be measured directly by experimental methods.

We are not going to include here the definition for metals and nonmetals because such definitions are discussed elsewhere in this book. We will define *organometallic species* as organic species that comprise at least one metal atom. These organometallic molecules and ions can be further classified by the type of the metal-to-nonmetal (M–E) chemical bond.

The type of the M–E chemical bond can be identified using the calculated value of the bond order. The M–E bond can be ionic, coordinative, single, aromatic, double, or triple.

If an organic compound includes an ionic M–E bond, we can talk about an organic salt. If a coordinative M–E bond is present, we can talk about an organometallic complex. If there are metal atoms, but there are no ionic or coordinative M–E bonds, we can talk about an organometallic compound. Consequently, organometallic complexes are nonradical species that include at least one carbon atom, at least one metal atom, and at least one coordinative chemical bond.

The term *descriptor* means any computable characteristic of the analyzed species, at the molecular level. Macroscopic properties such as density, hardness, melting point, biochemical activity, or viscosity are not descriptors in this sense. The statistical methodology quantitative structure-property relationship (QSPR) allows mathematical formulae to be obtained for macroscopic properties based on certain descriptors. In the QSPR abbreviation *property* (P) is often *biochemical activity* (A). The vast majority of QSPR programs, calculations, and equations are quantitative structure-activity relationship (QSAR) programs, calculations, and equations.

Structure is the term applied here to specify the position in 3D space of the atomic nuclei in the minimum energy conformer of the analyzed species. The computer-assisted searching approach of the structure is called *geometry optimization.* The correctness of geometry optimization is important because the value of some descriptors depends on the structure.

4.2   METHODS AND COMPUTER PROGRAMS

Within the software that optimizes geometry and calculates descriptors, organometallic complexes are treated as only a special type of organic species, that is, a set of atomic nuclei incorporated in a common molecular electronic cloud like "raisins in a cake". The results of geometry optimization and descriptor calculations are sufficiently accurate if these computer programs are properly parameterized for all elements and chemical bonds in the organometallic compound of interest.

The first aim of the molecular/quantum mechanics programs (e.g., Personal Computer MODELling, PCModel®; Molecular Orbital PACkage, MOPAC®; HyperChem®) is geometry optimization. The minimum energy conformer is then characterized by calculating a small number of descriptors characteristic of any particular program. Only descriptors that reflect the presence of the metal atom and the coordinative bonds could be considered specific descriptors for organometallic complexes.

Calculation of descriptors is an essential step in the operation of programs for QSPR calculations. Such programs are diverse (Mekenyan et al. 1990; SciQSAR®, Murugan et al. 1994; Cerius²®; TSAR®; Tarko 2005; Tarko 2008a) and include specialized modules for the calculation of a large number of descriptors. Some programs are used only for the calculation of descriptors (DRAGON®) that are later used for QSPR calculations. It should be noted that most of the programs used for calculation of descriptors and QSPR calculations are not parameterized for metal atoms (M) and M–E chemical bonds.

The algorithm of the statistical software used for QSAR calculations selects the significant descriptors by correlation with the biochemical activity and other descriptors. They identify significant molecular fragments and select the QSAR equations according to their quality. They also identify intramolecular synergetic effects and assess quantitatively the representative sample feature of the training set. They identify atypical molecules of what are called the *outliers for lead hopping* category (Cramer et al. 2004; Saeh et al. 2005) and they estimate the value of the biochemical activity for new, yet to be synthesized, molecules. Although relevant to the topic of this chapter, presentation of QSAR methodology is not the subject of this chapter.

To obtain the results that are illustrated here, the MMX method of the PCModel program was used to virtually build the complexes and execute a coarse geometry optimization. More accurate geometry optimization was then carried out using the Parameterization Model 6 (PM6) method of the MOPAC program. PM6 is a semiempirical quantum mechanics method (Stewart 2007) parameterized for 70 elements. Lacking more specific information, the species were analyzed with the environment dielectric constant set equal to unity. Certain net atomic charges were calculated for the analyzed complex. The total charge on system is the algebraic sum of the calculated net charges. For analysis by MOPAC, the total charge was set so that the analyzed organometallic complex is not going to be a radical.

How accurately does the obtained structure reflect that of the real one? We suppose that the analyzed complex interacts with the active site of a receptor molecule by a *key–lock* mechanism. The organometallic complex is flexible, due to the presence of coordinative bonds, even if the ligand is rigid. The geometry of the two systems of atoms starts to change gradually when the "key" approaches the "lock". The final complementary shape of the two components in the key–lock system is the complement of two *modified* geometries, not the geometry calculated prior to interactions. Accordingly, in QSAR calculations, molecular descriptors that have values poorly influenced by the 3D final shape of the analyzed complexes are very useful.

In cases in which the active site structure is known, an alternative approach would be geometry optimization and calculation of descriptors for the effector molecule-active site ensemble. Docking of the effector molecule in the active site can be done manually using computer-generated structures (Murcia et al. 2006; Huey et al. 2007; Weber et al. 2006; Tucinardi et al. 2007).

The descriptors were calculated here, for each complex, using the computer programs MOPAC/PM6, PRoperty Evaluation by CLAss Variables (PRECLAV) (Tarko 2005), and DESCRIPT (Tarko 2008a). The LogP descriptor was calculated using the KowWin algorithm of EPISuite software (EPISuite®; Meylan and Howard 1995). The type of chemical bonds was defined according to Table 4.1.

In order to explain the utilized methods, some structures that are not organometallic complexes are presented next.

Figure 4.1 presents the structure of auranofin, a compound considered to have anti-HIV potential (ChemIDplus/a). The type of chemical bonds is presented in a

TABLE 4.1
Conventional Type of Chemical Bonds According to Value of Bond Order

Computed Bond Order	Type of Chemical Bond
<0.10	Ionic
$0.10<B<0.76$	Coordinative
$0.76<B<1.05$	Single
$1.05<B<1.85$	Aromatic
$1.85<B<2.50$	Double
>2.50	Triple

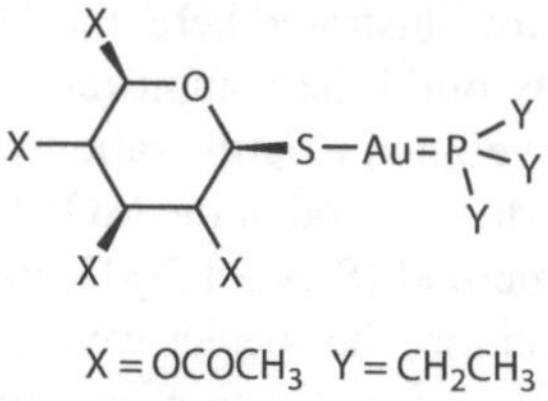

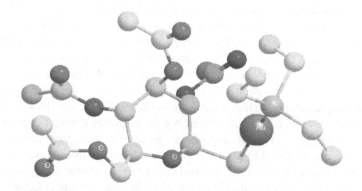

X = OCOCH₃ Y = CH₂CH₃

FIGURE 4.1A 2D structure of auranofin.

FIGURE 4.1B 3D structure of auranofin without hydrogen atoms.

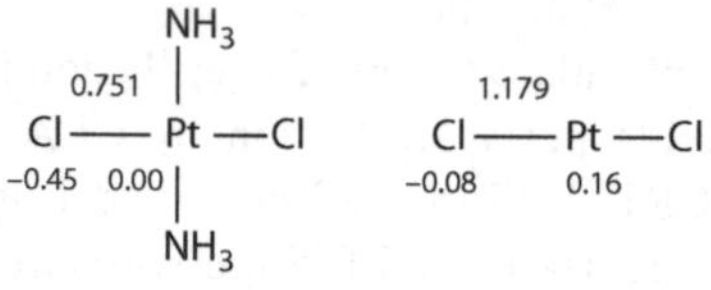

$$\underset{-0.45 \quad 0.00}{\underset{\begin{array}{c}NH_3\\ |\\ 0.751\end{array}}{Cl - Pt - Cl}} \qquad \underset{-0.08 \quad 0.16}{\overset{1.179}{Cl - Pt - Cl}}$$

FIGURE 4.2 Cisplatin and platinum chloride.

conventional manner. After geometry optimization, one calculated a bond order B = 1.476 (typical aromatic) for the S–Au bond, and a bond order B = 1.085 for the Au–P bond (value situated at the limit between a single and aromatic bond). The other chemical bonds were calculated as single bonds. Since it does not include coordinative bonds (see Table 4.1), auranofin is considered here an organometallic compound, not an organometallic complex.

Tetraethyl lead (Et₄Pb) and Grignard compound Ph–Mg–Br are modeled as organometallic compounds also, because $B_{C-Pb}=0.876$, $B_{C-Mg}=0.833$, and $B_{Mg-Br}=0.779$.

Figure 4.2 presents the planar structure of cisplatin, a compound used in anticancer therapy (ChemIDplus/b). The calculated values of the bond orders (truncated at three decimal places) and the calculated values of the net charges (two decimal places) are noted in the figure. For comparison, the same values in PtCl₂ are shown in Figure 4.2. The reader can see the significant decrease of the values of the bond orders of the Pt–Cl bonds and of the net charges of the Pt and Cl atoms, because of presence of the NH₃ ligand. The value of B for the Pt–N bond was

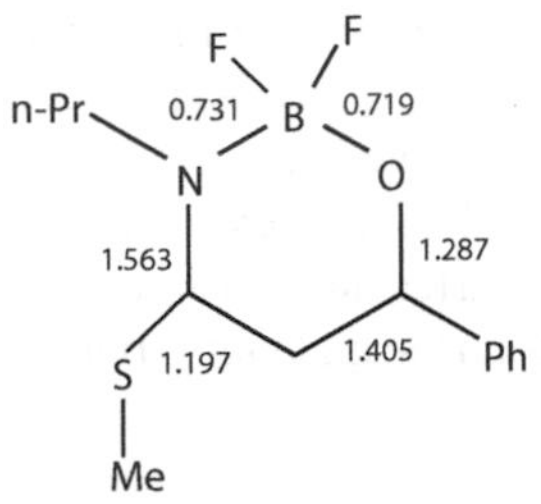

FIGURE 4.3 Nonmetallic organic complex.

calculated to be 0.657 and the calculated value of the net charges for the nitrogen atoms was −0.26. Lacking carbon atoms (see also Table 4.1), cisplatin is considered here an inorganic–metallic complex.

In (hypothetical?) double-hydrated $[(H_2O)_2 M]^{+n}$ cations, one computes B=0.258 for the Ca–O bond ($n=+2$), B=0.253 for the Cu–O bond ($n=+1$) and B=0.599 for the Fe–O bond ($n=+2$). According to Table 4.1, these species are also inorganic–metallic complexes.

Figure 4.3 shows a structure that was recently synthesized (Zhang et al. 2009). This figure presents the calculated values of the bond orders of the planar chemical bonds in cycle. We note the existence of both coordinative and aromatic bonds. The nitrogen, boron, oxygen, fluorine, and sulfur-containing compound in Figure 4.3 is lacking metal atoms; therefore, it is considered here an organic complex.

4.3 EXAMPLES OF DESCRIPTORS

The number of molecular characteristics that can be calculated (descriptors) is huge, and the number of formula and calculation procedures (methods) that are developed is growing fast. In the last 60 years, the number of descriptors in the "whole molecule" category has grown by 1 per month on average, and the number of descriptors in the "topological indices" category has grown by 1 per week on average. Even the number of categories (classes) of descriptors is large; there are tens of descriptor classes today.

There are many works that present a large number of descriptor categories (Rekker and Mannhold 1992; Karelson 2000; Tarko 2008b; Verma et al. 2010). The Sections 4.3.5, 4.3.6, 4.3.7, 4.3.8, 4.3.11, 4.3.13, 4.3.14 and 4.3.16 include formulas and paragraphs reproduced from Tarko 2008b. A very useful source of information is the documentation of programs for QSPR/QSAR computations. The appendix of this chapter lists almost 400 whole molecule descriptors computed by the PRECLAV software (Tarko 2005). COmprehensive DEscriptors for Structural and Statistical Analysis (CODESSA) computes about 1000 whole molecule descriptors (Murugan et al. 1994). DRAGON® is one of the best-known applications for the calculation of molecular descriptors. The latest version, 6.0, calculates 4885 descriptors and can work both in graphical and command-line interfaces (for batch processing of large datasets). The extension for the KNIME platform (http://www.knime.org) is now available, so that DRAGON can be easily used inside complex workflows. DRAGON is available for Windows and Linux platforms. Some programs also calculate non-linear functions of the descriptors (e.g., ratios, products, parabolic, exponential, logarithmic functions). Comparative molecular field analysis (CoMFA) provides 3D grid descriptors (Cramer et al. 1988). Descriptors are calculated according to certain characteristics of the atoms (e.g., Cartesian coordinates, atomic number, mass, volume, or net charge) and chemical bonds (e.g., length, bond order, or type of bound atoms).

4.3.1 BOND ORDERS

The sum of the kinetic energy and the potential energy for a specific system of atomic nuclei and electrons (such as an organometallic complex) can be estimated.

$$H = p^2/2m + V(r) = -(\hbar^2/2m)\nabla^2 + V(r) \tag{4.1}$$

The equation of vectors and eigenvalues for energy is:

$$[-(\hbar^2/2m)\nabla^2 + V(r)] \, |n\rangle = E_n \, |n\rangle \tag{4.2}$$

where $|n\rangle$ are eigenvectors corresponding to the E_n eigenvalues.

Equation (4.2), called the *static Schrödinger equation* (Schrödinger 1926), is a special case of the equation of vectors and eigenvalues, and is written in a particular representation (the representation of position). The E_n energies are solutions of Equation (4.2) and reflect *possible* energies of the system. On a case-by-case basis, each system is characterized by a different form of potential energy $V(r)$.

Equation (4.2) provides analytical solutions for mono-electronic systems only. Such solutions have been calculated for the hydrogen atom and for H_2^+, He^+, and Li^{+2} ions. For common organic species comprising many atomic nuclei and electrons, the expression of the potential energy is complicated because the electrostatic repulsion between nuclei, the electrostatic repulsion between electrons, and the electrostatic attraction between nuclei and electrons should be considered. Solving Equation (4.2) is very difficult for many common organic species. For systems including many electrons, the solutions of Equation (4.2) are obtained by the variational method. According to this method, every molecular orbital $|m\rangle$ is considered a linear combination of atomic orbitals $|\varphi_i\rangle$.

$$|m\rangle = \Sigma c_i \, |\varphi_i\rangle \tag{4.3}$$

The c_i coefficients must have values that give a minimum value to the energy E_m of that specific molecular orbital. In order to have other solutions besides the common (ordinary) solution, $c_1 = c_2 = c_3 = \dots = 0$, these must be solutions of the system of Equation (4.4).

$$\Sigma c_{mi} \cdot (F_{ij} - E_m \cdot S_{ij}) = 0 \tag{4.4}$$

and E_m is a solution of Equation (4.5)

$$|F_{ij} - E_{ij} \cdot S_{ij}| = 0 \tag{4.5}$$

where

$$F_{ij} = \langle i|H|j \rangle, \quad \text{and} \quad S_{ij} = \langle i|j \rangle. \tag{4.6}$$

If $i=j$, then F_{ij} is called the *Coulomb integral* and it represents the energy of an electron from an atomic orbital. If $i \neq j$, then F_{ij} is called the *bonding integral* and represents the energy of interaction of two atomic orbitals. The term S_{ij} is called the *covering integral* and is a measure of the interpenetration of two atomic orbitals. The presentation of the methods of solving Equations (4.4) and (4.5) and of calculating the values of the coefficients is too complex to cover in sufficient detail in this chapter.

The term c_{mi}^2 is considered the charge induced by the orbital i in atom m. The charge on a certain atom is also induced by orbitals of the neighboring atoms. These orbitals might or might not participate in the bond between the two atoms (Roothaan 1951). The total charge q_m of the atomic electronic cloud for atom m is calculated with Formula (4.7).

$$q_m = \Sigma(c_i^2 + c_i \cdot c_j \cdot S_{ij}) \tag{4.7}$$

The net charge (S) of atom m is given by the difference between the nuclear charge Z and the charge, q_m.

$$S = Z - q_m \tag{4.8}$$

The bond order (B) between any two atoms depends on the value of the coefficients in Equation (4.3) as calculated using Formula (4.9).

$$B = 2 \cdot \Sigma c_{mi} \cdot c_{mj} \tag{4.9}$$

The bond order is calculated for any pair of atoms, but has high values only if the atoms are neighbors. The smaller the value of B for two neighboring atoms, the more ionic is the character of the chemical bond. On the other hand, the larger the value of B, the more multiple is the character (aromatic, double, triple) of the chemical bond. The calculation methods used by the quantum mechanics programs are parameterized so that they calculate the bond order $B \sim 1$ for the single chemical bond, $B \sim 1.5$ for the aromatic chemical bond, $B \sim 2$ for the double chemical bond, and $B \sim 3$ for the triple chemical bond.

Figure 4.4 presents the values of the bond orders calculated for the bonds in the cycle of 6 atoms of the two isomers in the guanine molecule. In isomer A, the C–N bond in the amide group is single, and the other bonds between the heavy atoms (other than hydrogen) are calculated as aromatic, even the $C=O$ or $C–NH_2$ bonds. In the B isomer, all bonds between the heavy atoms are calculated as aromatic.

Knowledge of the bond order allows the grouping of that particular chemical bond in a specific category (see Table 4.1), the counting of the chemical bonds, the counting of the chemical bonds of a certain type, and the calculation of the Shannon entropy of the chemical bonds. Furthermore, the value of the bond order allows the identification of the bound and unbound atoms, after empirically setting a minimum

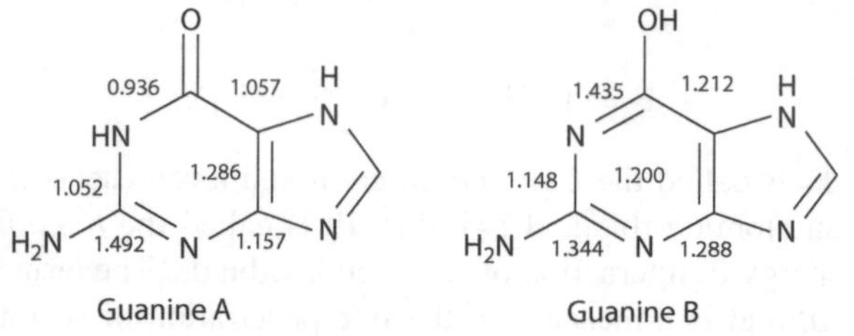

FIGURE 4.4   Some bond orders in two isomers of guanine.

limit for bound atoms. An alternative solution for the identification of the bound atoms is the use of a formula that considers the type of atoms and the distance between the atoms. The parameterization of such a formula is difficult, especially in the case of organometallic complexes because of the diversity of the metal atoms and associated chemical bonds.

Many descriptors are calculated for *all pairs of bonded atoms* and/or for *all pairs of atoms*. Matrices of adjacencies and distances contain information about the binding (connection) mode of the heavy (nonhydrogen) atoms and the topological distances between heavy atoms that can be applied to the calculation of topological indices.

4.3.2   FREE VALENCE

In theory, the free valence of a specific atom is defined and computed as the difference FV (Sannigrahi 1992).

$$FV = V_{max} - S \tag{4.10}$$

where V_{max} is the maximum valence of atom and S is the valence of atom, that is, the sum of bond orders ($S = \Sigma B_i$).

As a rule, the value of V_{max} is the number of atomic orbitals occupied by one electron. Setting the value of V_{max} for metal atoms is frequently difficult because some atomic orbitals *s, p,* and *d,* in the high-energy electronic shells, are hybridized.

When calculating the sum S, all bonds of the analyzed atom must be considered, including the very weak ionic bonds with very small values of B (see Table 4.1), that is, bonds to atoms that are not bound directly. The effect of these weak bonds is important because of their large number.

A positive value of the free valence suggests that particular atom is available for interaction with atoms other than those to which it is bound in the molecule. A negative value of the free valence suggests that the chemical bonds involving that particular atom were relatively established with the participation of the neighboring atomic orbitals. Some authors (Coulson and Lonquet-Higgins 1947) consider that radical substitutions will take place at the heavy atoms that possess high values of the free valence.

The reader can notice in Table 4.2 the differences between the values of the free valence calculated for the heavy atoms in the molecules of the two isomers of

TABLE 4.2
Free Valence of Heavy Atoms in Guanine Isomers

Heavy Atom in Figure 4.5	Type	Free Valence Guanine A	Guanine B	Difference
1	C	+0.192	+0.133	+0.059
2	N	−0.164	+0.127	−0.291
3	C	+0.126	+0.122	+0.004
4	N	+0.071	+0.072	−0.001
5	C	+0.072	+0.079	−0.007
6	C	+0.173	+0.156	+0.017
7	N	−0.526	−0.480	−0.046
8	C	+0.091	+0.094	−0.003
9	N	+0.006	+0.010	−0.004
10	O	−0.025	−0.103	+0.078
11	N	−0.134	−0.204	+0.070

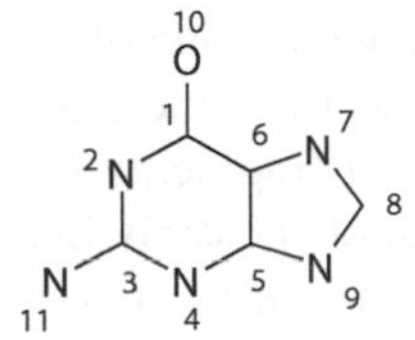

FIGURE 4.5   Indices of heavy atoms in guanine isomers.

guanine from Figure 4.4. The heavy atoms are indexed as in Figure 4.5. The biggest differences are calculated for the heavy atoms in isomers A, 1, 2, and 10, which are part of the amide group.

4.3.3 DESCRIPTORS OF MOLECULAR FRAGMENTS

In order to calculate descriptors for the molecular fragments, the term *fragment* must be defined mathematically so as to be consistent with the computer program algorithm. Fragments can be defined by comparing groups of atoms that are identified on the molecular map with groups included in previous lists, created empirically (Satoh et al. 1997; Japertas et al. 2002). Another method is based on the idea that each fragment contains only nonrotatable bonds and the fragments are connected by rotatable bonds (Terwillinger et al. 2006; Choi 2006; Zhu and Agrafiotis 2007). However, mathematical definition of a rotatable bond is itself difficult (Tarko 2011a). The identification of fragments can be done considering the value of the bond order and the types of atoms that are connected, in an attempt to define fragments that are as similar to classical chemical groups as possible (Tarko 2004a). A much simpler method takes into account only the value of the bond order for the chemical bonds of the heavy atoms (Tarko 2004b). Each fragment contains only multiple bonds

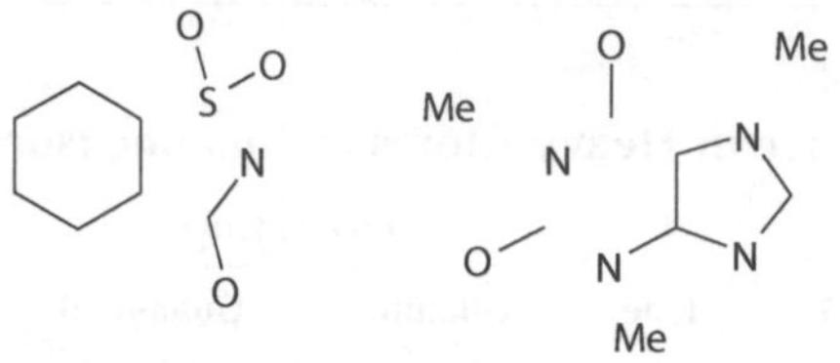

FIGURE 4.6 The identified molecular fragments in saccharin CAS 81-07-2 and caffeine CAS 58-08-2.

($B_{PM6} > 1.051$), and the fragments are connected by single or coordinative bonds. Hydrogen atoms are included in the same fragment as the heavy atoms to which they are bound. They can be identified, and fragments that contain certain types of atoms, for example, OH, NH_2, CO, can be noted. Applying this method to the saccharin molecule (CAS 81-07-2), three molecular fragments are identified. Seven fragments are identified for the caffeine molecule (CAS 58-08-2) (Figure 4.6).

After the identification of the molecular fragments, various descriptors can be calculated (e.g., number of fragments, maximum and average mass of the fragments, weight percentage of fragments, variance coefficient of mass fragments). If fragments have been defined only from the value of the bond order, then the value of the fragmentation descriptors is a measure of the size of the molecule and the continuity of conjugation. A small number of fragments in an analyzed complex reflects the existence of an extended conjugation, possibly ligandwide.

In QSAR calculations, the values of the mass percentage of a certain fragment that is present in the training set molecules are found to be correlated statistically to varying degrees with the values of the biochemical activity. The value and the algebraic sign of the Pearson linear correlation coefficient will indicate if the presence of a certain molecular fragment is influencing the value of the biochemical activity strongly or weakly, or positively or negatively.

4.3.4 NET CHARGES DESCRIPTORS

The net charges are calculated by Formula (4.8). The formation of hydrogen bonds or coordinative bonds substantially changes the value of the net charge of the involved atoms. Table 4.3 presents the net charges of the heavy atoms in guanine (Figure 4.5). We note that the biggest difference involves the carbon atom that is bound to the oxygen atom.

The knowledge of the values of the net charges allows, for example, the calculation of the standard deviation σ for the positive and negative charges, for the negative charges and then calculation of the electrostatic balance parameter descriptor (Murray et al. 1993).

$$\nu = \sigma_+^2 \cdot \sigma_-^2 / \sigma_{all}^2 \qquad (4.11)$$

The minimal, average, and maximum value of the attraction–repulsion force for different types of atoms can be calculated based on the values of the net charges and interatomic distances.

TABLE 4.3

Net Charges of Heavy Atoms in Guanine Isomers

Heavy Atom in Figure 4.5	Type	Net Charge		Difference
		Guanine A	Guanine B	
1	C	+0.68	+0.53	+0.15
2	N	−0.56	−0.60	+0.04
3	C	+0.52	+0.51	+0.01
4	N	−0.47	−0.48	+0.01
5	C	+0.37	+0.38	−0.01
6	C	−0.43	−0.39	−0.04
7	N	−0.14	−0.18	+0.04
8	C	+0.06	+0.10	−0.04
9	N	−0.33	−0.35	+0.02
10	O	−0.51	−0.44	−0.07
11	N	−0.52	−0.47	−0.05

The dipole moment (μ) of a neutral ensemble of electrical charges is defined as the product of the length D of a vector that joins the geometric center of the negative charges to the geometric center of the positive charges and the sum of the positive charges.

$$\mu = D \cdot \Sigma q_i \qquad (4.12)$$

The position of atomic nuclei (position of the geometric center of the positive charges) is the result of geometry optimization. To calculate as accurately as possible the coordinates of the geometric center of negative charges, quantum mechanics calculates the density of the molecular electronic cloud in different zones of the space.

The value of the dipole moment is influenced substantially by the molecular geometry. A small calculated value of the dipole moment—for example, for hexa-iodine-benzene, or dodecane—is the result of a high symmetry of the molecule and/or the effect of a small value of the net charges (Formula [4.12]).

A value of $\mu = 2.674$ is calculated for the A isomer of guanine in Figure 4.4 and a much bigger value is calculated for isomer B ($\mu = 4.249$).

The term *polarity* causes some confusion because it is defined in several ways. Sometimes the dipole moment is used as a measure of polarity. In this case, the *para*-dinitro-benzene appears as a nonpolar molecule due to its high symmetry. Other times, polarity means the difference between the maximum and the minimum value of the net charges in the molecule that is analyzed. In this case, the *para*-dinitro-benzene appears as a molecule with high polarity because the difference is large between the net (positive) charge of the nitrogen atoms and the net (negative) charge of the oxygen atoms.

A *polarity parameter* is also calculated with the ratio of polarity and the distance between the atoms that possess the maximum and minimum charge. In this case, the *para*-dinitro-benzene appears as a very polar molecule because the distance between the atoms of nitrogen and oxygen is small.

As another measure of polarity, the dielectric constant of the pure substance made only of the analyzed molecules can also be calculated.

No matter how it is defined, measured, or calculated, the polarity of molecules correlates with macroscopic properties such as density, miscibility/solubility, boiling point, melting point, critical point, retention time in chromatography, dilation capacity, and viscosity.

The maximum net charge X of hydrogen atoms (X>0) can be considered the measure of H atom donor capacity for the analyzed molecule. The minimum net charge Y of heteroatoms (Y<0) can be considered the measure of H atom acceptor capacity of analyzed molecule. Here, the descriptor H Donor–H Acceptor capacity gap is defined as the sum of $2 \cdot X + Y$. In the calculation of this sum, all hydrogen atoms and all heteroatoms of the complex of interest are considered. If this sum is positive, the analyzed species as a whole is considered an H donor. For water, the following values are calculated: $X = 0.309$, $Y = -0.619$ and $2 \cdot X + Y \sim 0$. We can say that water has equal hydrogen donor and acceptor capacities.

The pK_a descriptor is defined by the usual Formula (4.13), based on the acidity constant, K_a.

$$pK_a = -Log_{10}K_a \qquad (4.13)$$

This acidity constant is a measure of the strength of an acid, that is, a measure of the capacity of the OH, COOH, C(S)OH, and other groups to give away hydrogen atoms as H^+ ions.

$$K_a = [A^-][H^+]/[AH] \qquad (4.14)$$

The PM6 method calculates pK_a for the hydrogen atoms that are bound to oxygen by taking into account the (calculated) net charge of the hydrogen atom and the (calculated) length of the O–H bond. For the hydrogen atoms involved in hydrogen bonds, the net charge is sometimes overestimated, and the value of pK_a is underestimated. For example, the following values are determined experimentally for trifluoroacetic acid $pK_a = 0.00$ (calculated -0.18), tyrosine $pK_a = 2.20$ (calculated 2.43), 2,2-dimethylpropionic acid $pK_a = 5.03$ (calculated 5.23), *meta*-cresol $pK_a = 10.08$ (calculated 9.63), water $pK_a = 15.74$ (calculated 15.75), and *tert*-butanol $pK_a = 17.00$ (calculated 16.25). The pK_a is calculated to be -2.86 for the hydrogen atom of the OH group present in the B isomer of guanine (Figure 4.4).

4.3.5 Energy of Molecular Orbitals

In species that are not radicals or ion-radicals, each low-energy orbital is occupied by two electrons. There are also higher-energy orbitals unoccupied by electrons. The occupied orbital with the highest energy is referred to as the highest occupied molecular orbital (HOMO), while the unoccupied orbital with the lowest energy is referred to as the lowest unoccupied molecular orbital (LUMO). The orbitals' energies are solutions of Equation (4.2). The sign for the energies of the occupied orbitals energies is negative by convention. The ionization potential of a molecule is the energy of the HOMO orbital with the sign changed.

An electron that receives energy equal to the energy of the HOMO orbital leaves the molecule via ionization. An electron that receives energy equal to the difference in energy between two molecular orbitals can occupy a molecular orbital that was unoccupied initially (molecular excitation).

The number of occupied orbitals, the orbitals' energies, and the differences between the energy of orbitals are descriptors. For example, expression of the ionization potential and the electron affinity through the energies of the HOMO and LUMO orbitals (Delchev et al. 2006) allows definition of the chemical hardness as $(E_{LUMO} - E_{HOMO})/2$. The value of these descriptors often correlates with certain macroscopic properties such as reactivity, magnetic, electrical, and optical properties.

For the guanine A isomer in Figure 4.4, $E_{LUMO} = -0.59$ and $E_{HOMO} = -9.04$ are calculated, and for the guanine B isomer, $E_{LUMO} = -0.67$ and $E_{HOMO} = -8.88$ are calculated. The value of the chemical hardness for isomer B is smaller than that of isomer A.

4.3.6 STATIC INDICES OF REACTIVITY

Some authors (Fukui et al. 1954; Klopman 1974; Fukui 1975) believe that the HOMO and LUMO orbitals have a particular importance for the reactivity of the molecules.

In the case of a donor–acceptor type of interaction, an electron transfer takes place between molecule A and B. In the case of a change type of interaction, an electron transfer takes place between molecule A and B. As a rule, when two molecules A and B interact, the electron transfer that requires the smallest energy is the transfer from the HOMO orbital of a molecule to a LUMO orbital of the other molecule. Which molecule participates in the interaction with the HOMO orbital and which participates with the LUMO orbital depends on the relative value of the differences Δ_1 and Δ_2 between the energies of those particular orbitals.

$$\Delta_1 = E_{LUMO}^A - E_{HOMO}^B \quad \Delta_2 = E_{LUMO}^B - E_{HOMO}^A \tag{4.15}$$

The literature in the field does not specify the value for which the differences Δ_1 and Δ_2 should be considered large or small. When the differences Δ_1 and Δ_2 are large, the donor–acceptor interactions are charge-controlled reactions. In this case, the interaction between the molecules A and B is made through the atoms that possess the largest net charges (of opposite sign). When the differences Δ_1 and Δ_2 are small, the donor–acceptor interactions are orbital-controlled reactions. In this case, the interaction between the molecules A and B is made through the atoms that possess the highest absolute values of the orbitals' coefficients.

For each heavy atom in the analyzed molecule, one can calculate (Fukui 1975) the reactivity indices I_N (nucleophilic), I_E (electrophilic), and I_R (one-electron).

$$I_N = \Sigma\, c^2_{HOMO}/(1 - E_{HOMO}) \tag{4.16}$$

$$I_E = \Sigma\, c^2_{LUMO}/(10 + E_{LUMO}) \tag{4.17}$$

$$I_R = \Sigma\Sigma\, c_{HOMO}\, c_{LUMO}/(E_{LUMO} - E_{HOMO}) \tag{4.18}$$

TABLE 4.4

Reactivity Fukui Indices for Heavy Atoms in Guanine Isomers

| Heavy Atom in Figure 4.5 | Type | Isomer | | | | | |
| | | Guanine A | | | Guanine B | | |
		I_N	I_E	I_R	I_N	I_E	I_R
1	C	1.27	11.63	4.42	3.32	27.31	11.12
2	N	0.83	7.29	2.82	1.84	12.09	5.52
3	C	4.52	9.36	7.48	3.93	0.16	0.94
4	N	21.54	0.22	2.47	19.95	9.51	16.09
5	C	7.76	21.98	15.01	4.64	13.95	9.40
6	C	20.99	13.38	19.27	19.24	1.07	5.30
7	N	12.63	16.15	16.42	14.23	10.29	14.14
8	C	5.93	18.46	12.03	3.50	25.81	11.11
9	N	5.46	0.05	0.61	6.79	2.95	5.23
10	O	5.64	5.23	6.25	2.31	3.70	3.42
11	N	12.84	2.02	5.86	21.23	0.06	1.34

The Fukui Equations (4.16), (4.17), and (4.18) reactivity indices are considered a measure of the reactivity of the molecule (as nucleophil, electrophil, or radical) when the reaction center is the atom for which the indices have been calculated. The values of the Fukui indices, calculated for different atoms, allow the comparison of the atoms and the identification, as a rule, of the reaction center in different types of chemical reactions. The reactivity estimated in this way does not consider steric factors. Table 4.4 presents the values of the Fukui reactivity indices for the isomers of guanine that are analyzed here. The indices I_N, I_E, and I_R do not reflect the deformation of the electron clouds of two molecules that are approaching in order to react; 1 therefore they are called *static*. The reactivity indices that consider this factor also are called *dynamic*. Their calculation requires much more complicated formulas (Roothaan 1951) and involves the comparison of the energy of the molecular orbitals of the isolated molecules to the energy of the molecular orbitals of the supermolecule (i.e., the ensemble of the A and B molecules, situated at very short distances).

4.3.7 HEAT OF FORMATION

The heat of formation is the energy released as heat when atoms situated at theoretically infinite distance approach, bind, and form the molecule of interest. The core includes, by definition, the atomic nucleus and the electrons that do not participate in chemical bonds, that is, the nonvalence electrons. The semiempirical method PM6 estimates the heat of formation as the sum of the total repulsion energy of the cores and the total heat of formation of the atoms. Each semiempirical quantum mechanics method calculates, in its own manner, the energy of repulsion of the cores and utilizes a different set of values for the atomic heat of formation. Consequently, the values of the heat of formation determined for the same molecule by different semiempirical methods are quite different. It seems that the accuracy of the PM6 method in the calculation of this descriptor is higher than the accuracy of some *ab-initio* quantum mechanics methods (Stewart 2007).

For isomer A of guanine in Figure 4.4, $\Delta H_f = +76.3$ KJ is calculated, while for isomer B of guanine $\Delta H_f = +120.9$ KJ. The value of this descriptor suggests that the isomer A of guanine is much more stable and is present in a much larger quantity in the mixture of the two isomers. A considerable heat of formation is also calculated for other isomers of guanine when the hydrogen atoms are bound to the nitrogen atoms in a different manner.

The value of this descriptor depends on the number and type of atoms, and on the number and type of chemical bonds in the analyzed species. Accordingly, the heat of formation weighted by the number of atoms and the heat of formation weighted by the number of chemical bonds are also used as descriptors.

4.3.8 GEOMETRICAL DESCRIPTORS

The values of the diverse geometrical descriptors depend only on the type of atoms and on the Cartesian coordinates of these atoms in the minimum energy conformer. The type of a particular atom can be identified by the tabulated values of the atomic number, mass, or van der Waals radius. Often included in this category are descriptors whose value depends on other characteristics of atoms. For instance, topological indices are sometimes considered geometrical descriptors. However, topological indices can only be calculated after the identification of the way in which atoms are connected.

The principal moments of inertia are simple geometrical descriptors defined as the sum of products of the atomic masses with the distance to the main rotation axis x.

$$I_x = \sum m_i [r^2_i]_x \qquad (4.19)$$

The three principal moments of inertia characterize the mass distribution in the analyzed molecule. The maximum and minimum values of the inertia moments, I_M and I_m, allow the calculation of eccentricity (Arteca 1991), which is a measure of the deviation from the spherical shape.

$$\varepsilon = [(I_M)^2 - (I_m)^2]^{1/2}/I_M \qquad (4.20)$$

Other very simple geometrical descriptors are the average distance to geometric center, the maximum distance to geometric center, the standard deviation (SD) of distances to geometric center, and the variance coefficient of distances to geometric center. The variance coefficient of distances is SD/average ratio. If the variance coefficient is calculated considering only the peripheral atoms, this descriptor can be considered a measure of the unevenness of the molecular surface, a measure of the deviation of the form of the molecule from the spherical shape, or even a symmetry index.

Other geometrical descriptors are the volume and the area of the parallelepiped, sphere, and the ellipsoid circumscribed to the analyzed molecule. The area of the ellipsoid can be calculated only by numerical methods because there is no analytical formula for its calculation.

The molecular volume and area of molecular surface are two geometrical descriptors used by all programs for QSAR calculations.

Molecules could be considered ensembles of rigid spheres that overlap, and in this conceptualization, the calculation of the volume of the molecules using geometrical methods would be simple. The atomic radius that should be considered is the van der Waals atomic radius. After geometry optimization, the position of the center of the spheres is known, and the van der Waals radii of the atoms can be determined using physical methods (Bondi 1966). Consequently, the volume of the region of interpenetration of the atomic spheres could be calculated. This volume would be subtracted later from the total volume of the spheres. But the atomic spheres are elastic and deformable, and their overlapping is negligible. Therefore, the values of molecular volumes, calculated through geometrical methods, are significantly different from those obtained by physical measurements. This being the case, they can be considered wrong.

To calculate the molecular volume, statistical methods that involve the following steps are usually employed:

- An easily calculated virtual body (such as parallelepiped, sphere, ellipsoid) circumscribed around the analyzed molecule is generated, and a body with volume V is estimated.
- Inside the circumscribed body, a large number of N points are generated randomly.
- n points within the molecule are identified and counted (within a certain atomic sphere).
- The volume v of the molecule is calculated with formula (4.21).

$$v = V \cdot n / N \qquad (4.21)$$

The molecular surface is calculated through similar methods. Points are randomly generated on the surface of each atomic sphere. A certain percentage of the points generated on the surface of a certain atomic sphere is situated outside of any other atomic spheres. Thereby, for each atomic sphere, the "exposed" van de Waals surface is determined. The total van der Waals surface will be the sum of the exposed surfaces. The PCModel program computes for the guanine isomer A in Figure 4.4, Surface Area = 175.4 $Å^2$ and Volume = 156.1 $Å^3$, and for guanine isomer B, Surface Area = 173.9 $Å^2$, Volume = 155.2 $Å^3$.

The shadow areas of molecules (Rohrbaugh and Jurs 1987) are the projections of the molecules on the three planes obtained after the rotation of the molecule in a standard position. Because of the irregular shape of these projections, their surface is calculated using numerical methods. This descriptor is a measure of the size of the molecule as seen from a certain direction. The comparison of the surfaces of the projections is another way of assessing the deviation from the spherical shape.

Weighted holistic invariant molecular (WHIM) descriptors (Todeschini et al. 1994) are sophisticated geometrical descriptors based on statistical indices calculated using the elements s_{jk} of matrix S.

$$s_{jk} = [\Sigma w_i (q_{ij} - Q_j)(q_{ik} - Q_k)]/\Sigma w_i \qquad (4.22)$$

where

w_i are various weighting factors (such as atomic masses, van der Waals volumes, electronegativities, polarizabilities, and electrotopological indices);

q_{ij} and q_{ik} are the j, kth Cartesian co-ordinates of atom i (j,k=1, 2, 3);

Q_j and Q_k are the averages of j, kth Cartesian coordinates.

As a rule, WHIM descriptors offer some information about size, shape, and atomic 3D distribution. The program DRAGON computes almost 100 WHIM descriptors.

4.3.9 SOLVENT-ACCESSIBLE SURFACE

Only a part of the van der Waals surface is accessible to solvent molecules because, as a rule, the solvent doesn't access the cavities of the analyzed molecule. The mathematical definition of a cavity, and then the identification and calculation of the surface of molecular cavities are not simple tasks (Pascual-Ahuir and Silla 1990; Rinaldi et al. 1992; Ulmscheider and Penigault 1999a; Ulmscheider and Penigault 1999b). As a rule, the solvent-accessible surface is the locus of the center of a probe sphere rolling over the van der Waals surface. It was found that some biochemical properties correlate acceptably with the surface accessible to the solvent (Lee and Richards 1971; Shrake and Rupley 1973) and with the surface not accessible to the solvent (Connolly 1992).

The COnductor-like SCreening MOdel (COSMO) is a method that computes the electrostatic interaction of the analyzed molecule with a certain solvent by considering the dielectric continuum surrounding the solute molecule outside of molecular cavities (Klamt and Schüürmann 1993). The COSMO method can be used by all methods that compute the net atomic charges in analyzed molecules, for example, the semiempirical quantum mechanics method PM6.

For the guanine isomer A in Figure 4.4, $COSMO_{area} = 165.0$ Å^2 and $COSMO_{volume} = 159.0$ Å^3 are calculated, and for isomer B, $COSMO_{area} = 165.6$ Å^2 and $COSMO_{volume} = 159.3$ Å^3.

The approximated surface calculation (ASC) procedure calculates partial atomic van der Waals surface areas through an analytical method (Ulmscheider and Penigault 1999a) and then the Gibbs free energy of hydration is calculated by considering it to be an additive property. The ASC procedure considers the hybridization state of the atoms.

4.3.10 CHARGED PARTIAL SURFACE AREA (CPSA) DESCRIPTORS

The CPSA descriptors (Stanton and Jurs 1990; Stanton et al. 1992) are calculated according to the size and electrical charge of the atomic spheres and are considered a measure of the polar interaction between molecules. Total surface, total positive surface, and total negative surface are calculated, as are the numerous differences, ratios, and products of these types of surfaces. Moreover, some CPSA descriptors are calculated using only the donor–acceptor hydrogen atoms, while other descriptors use only the surface that is accessible or inaccessible to the solvent. There are almost 40 CPSA descriptors.

4.3.11 MOLECULAR SHAPE AND SIZE DESCRIPTORS

The value of these descriptors depends on the spatial distribution of atoms; therefore, they are a measure of the shape and size of a molecule. The descriptors that do not consider the connectivity of the atoms could be considered geometrical descriptors. The descriptors that do consider the connectivity of the atoms are called *topological descriptors*.

One of the simplest descriptors of molecular shape is the radius of gyration (R_G).

$$R^2_G = 1/M \cdot \Sigma \, m_i r^2_i. \qquad (4.23)$$

In Formula (4.23), M is the molecular mass, m is the mass of the atom i, and r is the distance between the atom i and the center of mass. Using the tabulated values of the atomic mass m and interatomic distances r (distances that are a result of geometry optimization), a gravitational index I_{grav} is calculated.

$$I_{grav} = \Sigma \, (m_1 \cdot m_2 / r_{12}^2). \qquad (4.24)$$

The summation in Formula (4.24) can be calculated for all atom pairs or only for all pairs of bonded atoms. The gravitational index is a measure of the spatial distribution of masses in the molecule and is similar to the geometrical principal moments of the inertia descriptor. An *electronic index* I_{elec} is calculated with a similar formula

$$I_{elec} = \Sigma \, (|\, q_1 - q_2 \,| / r_{12}^2). \qquad (4.25)$$

where q are net atomic charges and r is the distance between the atoms with the charges q_1 and q_2.

The I_{elec} index is a measure of the spatial distribution of the electrical charges in the molecule. Similar to the I_{grav} index, the electronic index I_{elec} can be calculated for all pairs of atoms or only for all pairs of bonded atoms.

A wide range of topological indices are calculated for molecular graphs. In these calculations, kenographs are usually used, that is, the graphs in which the vertices are the heavy atoms (any type) and the edges are chemical bonds (any type also). The g degree of the vertex is the number of vertices linked directly to the considered vertex, and the vertices having g = 1 are the peripheral atoms.

The values in the adjacency matrix and the values in the distances matrix are used for calculations. In the adjacency matrix, there is the value 1 if vertices on the line i and column j are connected, and the value 0 if the vertices are not connected. The distances matrix contains the topological distances between vertices, that is, the least number of steps in the graph of the molecule to reach the vertex on column j from the vertex on line i of the matrix. In fact, topological indices are mathematical functions where the variables are values in the adjacency and distances matrices.

As an example, we present the calculation of the topological indices W (Hosoya 1971), χ (Randic 1975), J (Balaban 1982), and MSD (Balaban 1983), chosen here because they are very simple formulas/procedures of calculation.

The Wiener index (W) is calculated as the half-sum of all topological distances collected in the distances matrix.

$$W = \frac{1}{2} \cdot \sum_{i=1}^{N} \sum_{j=1}^{N} d_{ij} \qquad (4.26)$$

The Randic index (χ) is the sum of the μ terms. Let δ be the degree of a vertex of the graph and then define δ_i and δ_j as the degrees of two vertices that are connected. The link (edge) between these vertices is characterized by μ (the inverse of the geometrical mean of the degrees). The sum of all μ values is the index Randic χ.

$$\mu = (\delta_i \cdot \delta_j)^{-1/2} \qquad (4.27)$$

$$\chi = \Sigma \mu_i \qquad (4.28)$$

For the guanine molecule, $\chi = 5.2709$ is calculated.

The Balaban index J is computed using the following procedure. Let N be the number of vertices and L the number of edges of a certain kenograph. A square matrix **M** that has N lines is built. Each element of the matrix is the topological distance between the vertex i and vertex j. Consequently, any line k and column k of the matrix will be identical, and the diagonal will include null values.

Atom	M_{ij}	Σd_j
1C	0 1 2 3 3 3 2 1 2 1 3	21
2N	1 0 1 2 4 4 3 2 3 2 2	24
3C	2 1 0 1 3 4 4 3 2 3 1	24
4N	3 2 1 0 2 3 3 2 1 4 2	23
5C	2 3 2 1 1 2 2 1 0 3 3	20
6C	1 2 3 2 2 2 1 0 1 2 4	20
7N	2 3 4 3 2 1 0 1 2 3 5	26
8C	3 4 4 3 1 0 1 2 2 4 5	29
9N	3 4 3 2 0 1 2 2 1 4 4	26
10O	1 2 3 4 4 4 3 2 3 0 4	30
11N	3 2 1 2 4 5 5 4 3 4 0	33

For the guanine molecule indexed as in Figure 4.5 (N = 11 and L = 12), the matrix **M** is obtained. The vertex with index j has a value on the line j of the matrix. For each vertex, the measure Σd_j is calculated as a sum of the elements on the corresponding line j of the matrix. For each link connecting two vertices i and j, μ is calculated, and then the Balaban index (J) is calculated. For guanine, J = 2.0404 is calculated.

$$d_j = \Sigma\, M_{ij} \tag{4.29}$$

$$\mu = (d_i \cdot d_j)^{-1/2} \tag{4.30}$$

$$J = L/(L - N + 2)\, \Sigma\, \mu_k \tag{4.31}$$

The mean square distance (MSD) index is a weighted version of the W index, where the weighting factor Z is the product $N \cdot (N - 1)$.

$$MSD = 1/Z \cdot \left[\sum_{i=1}^{N} \sum_{j=1}^{N} d_{ij}^2 \right]^{1/2} \tag{4.32}$$

The cyclomatic number (C) descriptor is simply defined to be $L - N + 1$, where L is the number of chemical bonds and N is the number of atoms. In calculation of the cyclomatic number, the molecular graph can be used with or without hydrogen atoms.

In QSAR calculations, the number and size of circuits are descriptors. Circuits are cyclic topological paths that, with the exception of the starting node, visit the vertices only once. There are procedures that allow the identification, on the molecular graph, of all topological paths, and therefore of all circuits (Balaban et al. 1985). The minimum length of the circuits is 3 (cycle of 3 atoms). The identified circuits can be ordered by length, starting with the shortest one. The first C circuits on the list (C is the cyclomatic number) are the chemical cycles. For acenaphthene in Figure 4.7, $C = 14 - 12 + 1 = 3$ is calculated and there are identified one circuit of length 5, two circuits of length 6, two circuits of length 9, one circuit of length 10, and one circuit of length 11. Formula (4.33), where C is cyclomatic number $(C > 1)$ and T is number of circuits, calculates, according to PRECLAV documentation, a *crowding index* CIC of circuits. This descriptor has values within the [0, 1] range.

$$CIC = (T - C)/(2^C - C - 1) \tag{4.33}$$

The CIC = 1 for acenaphthene; therefore, the degree of condensation of the cycles in this molecule is maximum.

A broad category of topological indices are calculated based on the values of the Detour (*Maximal Topological Distance*) Matrix (Ivanciuc and Balaban 1994). Recently, new descriptors have been proposed in this category, which are useful for evaluating solubility in water (Talevi et al. 2006).

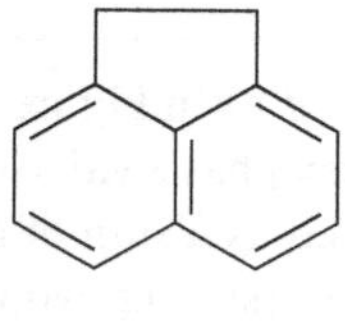

FIGURE 4.7   Acenaphthene.

The value of the topological indices calculated using the kenograph depends neither on the type of heavy atoms nor the type of chemical bonds. For this reason, the information that they offer regarding the shape and size of the molecules is relatively poor. There are currently used mainly topological indices weighted with other descriptors, including atomic masses, atomic volumes, atomic surfaces, atomic electrical charges, atomic polarizabilities, and bond orders.

A small error in the computation of the values of the descriptors, compared to the error of the experimental (observed) values of the dependent property, is an important condition for correct functioning of statistical computations. Compared to other descriptors used in QSAR calculations, the topological indices have the advantage that their value is determined with a much smaller error than the value of the biochemical activity.

4.3.12 SHANNON ENTROPY (INFORMATION CONTENT)

There are many papers in QSAR literature that describe the use of the discontinuous form of the Shannon entropy (Shannon 1948) as a molecular descriptor. This descriptor is computed using the values of certain other descriptors and Formula (4.34). In QSAR calculations, the Shannon entropy (SE) is considered a measure of diversity of descriptor values and it is usually called *information content*.

$$SE = - \sum_{i=1}^{k} n_i / N \cdot Log(n_i / N) \tag{4.34}$$

In Formula (4.34), N is the total number of the descriptor values and n_i is the number of the values included in category i. Therefore, to apply Formula (4.34), one must use an adequate criterion for putting a certain value into a particular category (Tarko 2011b). In this chapter, the base of the logarithm is the Euler's number e (i.e. $Log \equiv Ln$). The number k_{def} of defined categories can be large, but Formula (4.34) uses only nonempty categories $(n_i > 0$ and $k \le k_{def})$. If $k_{def} < N$, the value of SE is within the $[0, Log\, k_{def}]$ range. Otherwise, the value of SE is within the $[0, Log\, N]$ range. Consequently, the value of SE is within the range marked by the inequalities in Equation (4.35).

$$0 \le SE \le \min(Log\, k_{def}, Log\, N) \tag{4.35}$$

If $k = k_{def}$ and each category includes the same number of values, SE in Formula (4.34) has the maximum value. If $k = 1$ (all values are included within the same category), $SE = 0$ because $n_k = N$. The ratio between SE and its maximum value is another descriptor. A very simple example is the computation of SE for guanine ($C_5H_5N_5O$) when the descriptor is the atomic number. The total number of atoms is 16 ($N = 16$). There are four nonempty categories of atoms ($k = 4$). These categories include 5, 5, 5, and 1 atoms, respectively. Consequently, the value of SE_{atoms} is

$$SE_{atoms} = -[5/16 \cdot Ln(5/16) + 5/16 \cdot Ln(5/16) + 5/16 \cdot Ln(5/16) + 1/16 \cdot Ln(1/16)] \sim 1.2637.$$

In the guanine molecule (number of chemical bonds $N = 12$), according to Table 4.1, there are only two nonempty categories (single and aromatic) of chemical bonds ($k = 2$). These categories include 2 and 10 bonds, respectively. Consequently, the value of SE_{bonds} is

$$SE_{bonds} = -[2/12 \cdot Ln(2/12) + 10/12 \cdot Ln(10/12)] \sim 0.4506.$$

In guanine, the diversity of chemical bonds (from the point of view of bond orders) is much smaller than the diversity of atoms (from the point of view of atomic numbers). In addition, one can say that the amide chemical group in guanine seems to be nonexistent because the bond order of the C–N bond is too small (Figure 4.4).

Putting the values of many descriptors (net charges, geometric distances, features of molecular fragments, etc.) into categories is a difficult task. In contrast, it is easy to put the features of kenographs into categories.

4.3.13 PARTITION COEFFICIENT IN OCTANOL–WATER SYSTEM

A pure substance, dissolved in a heterogeneous system consisting of water and a nonmiscible solvent, is characterized, at equilibrium, by the concentration c_1 in water and the concentration c_2 in solvent. When the concentrations c_1 and c_2 are much lower than the solubilities (the maximum concentrations at working temperatures), the partition coefficient is defined as the ratio $P = c_2/c_1$.

There are numerous experimental methods for determining the P value for organic substances (de Bruijn et al. 1989; Tayar et al. 1991). Since experimentally P values are within a very broad range, the logarithm of P is used instead. The LogP descriptor is, by definition, a measure of the lipophilic nature of the molecules that make up the dissolved organic compound. Sometimes there is a high statistical correlation between biochemical activity and the LogP values and/or of the values of the parabola $a + b \cdot LogP + c \cdot (LogP)^2$. It was found that the best results are obtained when the organic solvent is n-octanol, perhaps because the octanol–water system is a better model for the characteristics of the biophase. Therefore, calculated or experimentally derived LogP in the octanol–water system is the most utilized descriptor.

It is believed that a strong lipophilic character (high LogP value) reflects a high capacity of the molecule to pass through cell membranes. On the other hand, the finding that a pronounced lipophilic character is not an obstacle in the transport of molecules to the target cells is explained by the assumption that, in the aqueous blood environment, the bioactive molecules form heterogeneous microemulsions or microsuspensions, not homogeneous solutions.

The development of an algorithm for calculating the LogP value is a difficult task, because terms such as *functional group* (Tarko 2004a), *conjugation, position in the carbon chain*, and *proximity effect* are difficult to express mathematically. Some software considers LogP an additive property (Hansch et al. 1968; Hansch and Leo, 1979; Rekker and Nys 1974; Rekker and de Kort 1979), but others calculate LogP as a function of descriptors correlated strongly with LogP (Gaillard et al. 1994).

The ACD/LogP algorithm (ACD/LogP®) considers numerous types of carbon atoms (relative to hybridization, number of hydrogen atoms attached, the position in the carbon chain, presence in cycles, and proximity of aromatic cycles), hundreds of functional groups and thousands of pairs of neighboring functional groups. In addition, the ACD/LogP algorithm proposes a formula for calculating the solubility in water, depending on the melting point and calculated LogP value of the molecule of interest.

The calculation of LogP is inaccurate when the value of the denominator and/or the numerator in the c_2/c_1 ratio is very large or very small. Examples include octanol ($c_2 \sim \infty, c_1 \sim 0$), macromolecules ($c_2 \sim 0, c_1 \sim 0$), perfluorinated compounds ($c_2 \sim 0$, $c_1 \sim 0$), and amino acids ($c_2 \sim 0, c_1 \sim \infty$). For isomer A of guanine in Figure 4.4, one calculates LogP$=-1.05$, while for isomer B, one calculates LogP$=-0.96$.

4.3.14 FLEXIBILITY DESCRIPTORS

The possibility of rotation of some groups of atoms around some chemical bonds is intuitively associated with molecular flexibility. Consequently, the number of rotatable bonds and the percentage of rotatable bonds are the simplest descriptors that measure the molecular flexibility. More complicated formulas calculate flexibility descriptors by weighting some topological descriptors; therefore, they consider the shape and size of the molecules, the number of atoms, and the sum of bond orders (Tarko 2004c).

$$\text{flex}_1 = W/(N \cdot S) \tag{4.36}$$

$$\text{flex}_2 = \text{flex}_1/(1+p) \tag{4.37}$$

where
   W = Wiener topological index
   N = number of heavy atoms
   S = bond order sum (all bonds which bond heavy atoms)
   p = percentage of aromatic bonds (using only bonds that bond two heavy atoms)

Much utilized is the Kier flexibility index (Kier 1990), defined as a weighted product of some shape indices of the molecules.

$$K_k = (N+\alpha-k+1) \cdot (N+\alpha-k)^2 \cdot (P_k+\alpha)^2 \tag{4.38}$$

where
   k is the order of the shape index and has the value 1 or 2; for k > 2, Kier proposes
       formulas different from (4.38)
   N is the number of atoms in the molecule
   P_k is the number of topological routes, of length equal to k steps, that can be taken
       on the graph of the molecule; P_1 (k = 1) is the number of chemical bonds in
       the molecule
   α is the sum after $N\alpha = 1/r_C \Sigma r_i$

r_C is a constant (the radius of the carbon atom in the sp^3 hybridization state)

r_i are constants (tabulated values for atomic radii in different hybridization states)

Using the shape indices K_1 and K_2, the Kier algorithm calculates the flexibility index, Φ.

$$\Phi = K_1 \cdot K_2 / N \tag{4.39}$$

According to the value of the flex$_1$ index, propane is very rigid; however, according to the Φ index, this hydrocarbon has an average flexibility. The flex$_1$ descriptor estimates the order of flexibility 2,2,3,3 – tetramethyl-butane < *trans*-decaline < cycloheptane < *n*-butane, yet, for the same four hydrocarbons, the flexibility order according to the index Φ is exactly the reverse. The inverse of the descriptors of flexibility flex$_1$, flex$_2$, and Φ can be considered rigidity descriptors.

4.3.15 AROMATICITY DESCRIPTORS

The application of quantitative methods for the evaluation of the molecular aromaticity is common (Bird 1992; Schleyer et al. 1996; Fores et al. 1999; Cyranski et al. 2002; Bultinck et al. 2005; Espinosa et al. 2005; Mitchell et al. 2005; Zborowski et al. 2005; Shishkin et al. 2006; Tarko 2008a; Tarko 2010); however, the listing of aromaticity descriptors has emerged recently (Tarko 2008a; Tarko 2010).

The aromaticity A for (just) one chemical bond can be computed using the empirical formula based on the B_{PM6} bond order of analyzed bond,

$$A = 1000 - 6250 \cdot (1.4406 - B_{PM6})^2. \tag{4.40}$$

The range of A values for aromatic bonds is [0, 1000]. The range of A values for anti-aromatic bonds is [−1000, −350]. For groups of adjacent chemical bonds, that is, various topological routes on the kenograph, the formula for the calculation of aromaticity is much more complicated because it measures the alternating character of chemical bonds (Tarko 2008a).

The number of aromatic bonds; minimum, average, and maximum value of the aromaticity of the aromatic bonds; percentage of aromatic bonds; number of aromatic molecular zones; or aromaticity of the peripheral topological path can be calculated as descriptors. Some descriptors simultaneously reflect the aromaticity and size of the molecule. Other descriptors reflect the concentration of aromaticity in small areas of the molecule.

When the values of aromaticity descriptors have been obtained, the application of QSPR/QSAR methodology makes possible the verification of the dependence of some macroscopic properties (magnetic, electrical, optical, biochemical) on aromaticity (Tarko 2010).

Figure 4.8 shows the aromaticity map for Black Indanthrone (Tarko 2010), a molecule having semiconductor properties (Inokuchi 1952). The presence of single bonds that are invisible on the map induces the existence of some nonaromatic cycles and the fragmentation of the molecule in five aromatic areas having low, medium,

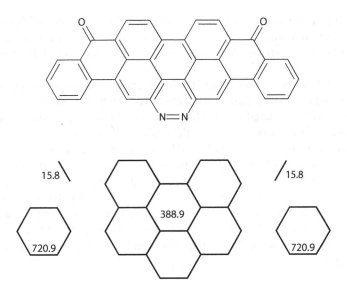

FIGURE 4.8 Aromaticity map of Black Indanthrone.

and high aromaticity (Figure 4.8). For Black Indanthrone one calculates, for instance, the following descriptors:

- percentage of aromatic bonds 87.2%
- average aromaticity of aromatic bonds 415.3
- number of aromatic zones 5
- variation coefficient of masses in aromatic zones 106%
- average aromaticity of aromatic zones 372.4 ± 352.7
- variation coefficient of aromaticity in aromatic zones 94.7%
- length of LATP (Longest Aromatic Topological Path) 22
- aromaticity of LATP 395.4

4.3.16 3D DESCRIPTORS

In order to calculate 3D descriptors, the analyzed molecule is placed into a regular, virtual, three-dimensional network of points. The 3D descriptors are basically characteristics of some physical quantities, calculated in the points of the network. The value of the 3D descriptors depends on the position in space of the considered point and on the characteristics of the atoms in certain areas of the molecule. When a group of molecules is analyzed, the network of points includes the package of overlapped molecules. The methods used to overlap molecules are highly diverse. For example, molecules can be overlapped on the common skeleton if present. Other methods overlap the vectors of the dipole moments of the molecules in the package. The value of the 3D descriptors is influenced by the method used to overlap the molecules and by the distance between the network points. A different approach is to use the molecules in the package in their position within the effector–active site ensemble if such position can be determined.

The CoMFA method (Cramer et al. 1988) considers the points of the virtual network to be virtual protons and calculates, for each point, steric and electric potentials that depend on the electrical charge and the van der Waals radii of the atoms of each molecule.

The PRECLAV program (Tarko 2005) calculates electrostatic forces and parallaxes for each point. The parallax is the angle under which a certain pair of atoms is viewed from the considered point. The maximum and average values of parallaxes are a measure of the shape and size of the analyzed molecule.

The hypothetical active site lattice (HASL) method (Doweyko 1988) identifies the points of the network associated to the atoms of the molecule of interest. Then it gives a fraction of the value of the biochemical property to these points. This fraction is characteristic of the analyzed molecule. By repeating the procedure for all the molecules in a given set, some points of the network acquire the summing of the assigned values that are different from null values. These points describe a structure as a map of the active site of the receptor macromolecule that interacts with the effector molecules.

4.4 EXAMPLES OF DESCRIPTORS CALCULATION FOR ORGANOMETALLIC COMPLEXES

When a metal atom ion enters a living organism, it often forms organometallic complexes in which the ligand is a protein. In many studies, the 3D structure of these complexes was obtained by X-ray spectroscopy and entered into various databases such as the Protein Data Bank (n.d.; see, for instance, many ferrodoxins, ferritins, myelin transcription factors, etc.).

Following are some examples of calculation of descriptors for organometallic complexes in which the ligand has low molecular weight. Descriptors were chosen whose values seem to be more sensitive to the presence and type of metal atom, as well as to the number, shape, and size of ligands. Some of the analyzed complexes have been synthesized and have various practical applications.

In Figures 4.9–4.25, obtained using PCModel (after geometry optimization by MOPAC), the coordinative chemical bonds are visible only if the calculated value of the bond order is high enough. To calculate the free valence, we used for V_{max} in Formula (4.10) the value of maximum oxidation states, that is, 2 (for Zn, Mg and Cd), 3 (for Fe), 4 (for Cu, Ni and Ge), 5 (for Au, Co and V), 6 (for Cr and Pd), and 7 (for Mn).

4.4.1 EXAMPLE 1

Figure 4.9 presents the optimized geometry of a hypothetical complex, Cu–glycine (charge on system = +1). The amino acid has been analyzed in zwitterion state. All of the heavy atoms are (calculated to be) in the same plane. According to Table 4.1, the following bonds are identified in the kenograph of the complex: two Cu–O coordinative bonds, one O–O coordinative bond, two C–O aromatic bonds, one C–C single bond, and one C–N single bond.

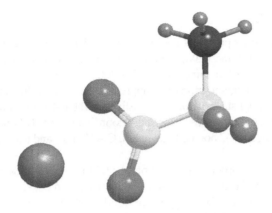

FIGURE 4.9 Cu-glycine complex.

The values of bond orders for Cu–O bonds: 0.270 and 0.253
Sum of bond orders for Cu (all types of bonds): 0.591
Free valence for Cu: +3.409
Net charge of Cu: +0.67 (in c. e. u. charge electrostatic units)
Values of net charges for O atoms: −0.65 and −0.50 (in c. e. u.)
Donor H–Acceptor H capacity gap: −0.15 (in c. e. u.)
Topological indices:

- Gravitation index (all atoms): 1001.7
- Gravitation index (bonded atoms): 712.7
- Electronic index (all atoms): 6.83
- Electronic index (bonded atoms): 4.49
- Wiener index: 26.0
- Randic index: 2.87
- Balaban index: 2.10

$E_{LUMO} = -4.44$ $E_{HOMO} = -13.64$ $E_{LUMO} - E_{HOMO} = 9.20$ $(E_{LUMO} - E_{HOMO})/2 = 4.60$ (in eV)

Fukui indices for O atoms (multiplied by 1000):

- I_N = 14.87 and 16.80
- I_E = 1.44 and 1.43
- I_R = 10.51 and 10.79

Heat of formation: 383.7 (in KJ)
COSMO area: 117.66 (in Å^2)
COSMO volume: 101.25 (in Å^3)
Empirical formula: $C_2H_5NO_2Cu$
Shannon entropy of atomic numbers: 1.4143
Shannon entropy of bond orders in kenograph: 1.0790

4.4.2   EXAMPLE 2

Figure 4.10 shows the optimized geometry of the hypothetical complex of Zn with 2,2`-Bipyridin and chlorine (charge on system=0). The angle between the planes of the aromatic cycles (in the same ligand) is ~20°. The structures of the two ligands are nearly perpendicular, and moreover, approximately perpendicular to the plane that includes the Cl–Zn–Cl substructure. According to Table 4.1, the following bonds are identified in the kenograph of the complex: 4 coordinative Zn–N bonds, 2 coordinative Zn–Cl bonds, 2 single C–C bonds, 16 aromatic C–C bonds, and 8 aromatic C–N bonds.

The values of bond orders for Zn–N bonds: 0.300, 0.271, 0.300 and 0.271
The values of bond orders for Zn–Cl bonds: 0.432 and 0.432
Sum of bond orders for Zn (all types of bonds): 2.075
Free valence for Zn: −0.075
Net charge of Zn: +0.51
Values of net charges for N atoms: −0.14, −0.17, −0.14 and −0.17
Values of net charges for Cl atoms: −0.66 and −0.66
Donor H–Acceptor H capacity gap: −0.26
Topological indices:

- Gravitation index (all atoms): 7108.0
- Gravitation index (bonded atoms): 2511.1

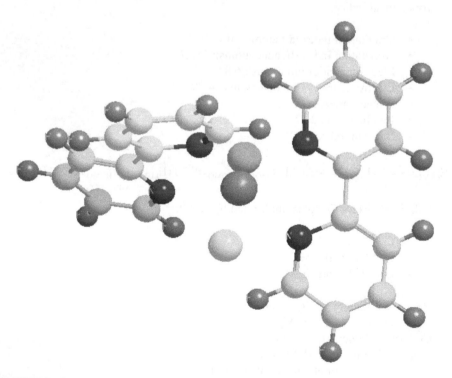

FIGURE 4.10   [2,2`-Bipyridin]$_2$–ZnCl$_2$ complex.

- Electronic index (all atoms): 19.06
- Electronic index (bonded atoms): 7.79
- Wiener index: 1158.0
- Randic index: 12.42
- Balaban index: 0.93

$$E_{LUMO} = -1.05 \quad E_{HOMO} = -8.84 \quad E_{LUMO} - E_{HOMO} = 7.79 \quad (E_{LUMO} - E_{HOMO})/2 = 3.90$$

Fukui indices for N atoms:

- $I_N = 0.34, 0.26, 0.27,$ and 0.31
- $I_E = 4.53, 4.86, 1.64,$ and 1.78
- $I_R = 2.22, 3.59, 2.21,$ and 2.12

Heat of formation: 196.8
COSMO area: 360.4
COSMO volume: 453.0
Empirical formula: $C_{20}H_{16}N_4Cl_2Zn$
Shannon entropy of atomic numbers: 1.1750
Shannon entropy of bond orders in kenograph: 0.7029

4.4.3 EXAMPLE 3

Figure 4.11 presents the optimized geometry of Ferrocene (charge on system = 0).

Ferrocene and its derivatives are used as fuel additives, anticancer drugs, and catalysts. The (calculated) angle between the planes of the aromatic cyclic ions

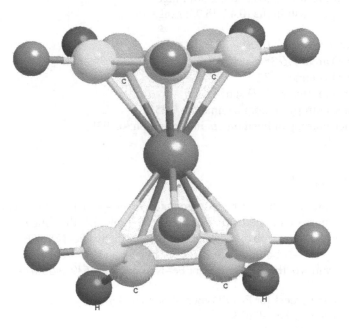

FIGURE 4.11   Ferrocene.

is ~ 15°. There is a calculated asymmetry between the characteristics of the carbon atoms (bond orders and length of the Fe–C bonds, net charges, Fukui indices). According to Table 4.1, the following bonds are identified in the kenograph of the complex: 10 coordinative Fe–C bonds and 10 aromatic C–C bonds.

The values of bond orders for Fe–C bonds are within the [0.267, 0.316] range.
Sum of bond orders for Fe (all types of bonds): 2.980
Free valence for Fe: +0.020
Net charge of Fe: +0.62
Values of net charges for C atoms are within the [−0.36, −0.18] range.
Topological indices:

- Gravitation index (all atoms): 2276.1
- Gravitation index (bonded atoms): 1595.7
- Electronic index (all atoms): 11.01
- Electronic index (bonded atoms): 6.23
- Wiener index: 90.0
- Randic index: 5.16
- Balaban index: 2.46

$$E_{LUMO} = +0.72 \quad E_{HOMO} = -8.79 \quad E_{LUMO} - E_{HOMO} = 9.51 \quad (E_{LUMO} - E_{HOMO})/2 = 4.76$$

Fukui indices for N atoms:

- I_N are within the [0.04, 16.27] range
- I_E are within the [1.06, 2.36] range
- I_R are within the [0.47, 15.98] range

Heat of formation: 202.8
COSMO area: 176.3
COSMO volume: 187.1
Empirical formula: $C_{10}H_{10}Fe$
Shannon entropy of atomic numbers: 0.8516
Shannon entropy of bond orders in kenograph: 0.6931
LogP: 3.28

4.4.4 EXAMPLE 4

Figure 4.12 presents the optimized geometry of Fe–phthalocyanine, calculated as nonplanar (charge on system = 0). According to Table 4.1, the following bonds are identified in the kenograph of the complex: 2 coordinative Fe–N bonds, 4 single C–C bonds, 4 single C–N bonds, 2 single Fe–N bonds, and 40 aromatic C–C and C–N bonds. The values of the bond orders for Fe–N bonds are 0.539 and 1.025, respectively.

Sum of bond orders for Fe (all types of bonds): 3.673
Free valence for Fe: −0.673
Net charge of Fe: +1.28

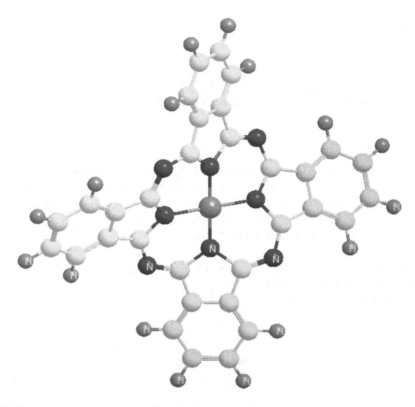

FIGURE 4.12 Iron phthalocyanine complex.

Values of net charges for N atoms in Fe–N bonds are –0.39 and –0.19, respectively.
Donor H–Acceptor H capacity gap: –0.04
Topological indices:

- Gravitation index (all atoms): 10734.6
- Gravitation index (bonded atoms): 4343.4
- Electronic index (all atoms): 22.28
- Electronic index (bonded atoms): 8.83
- Wiener index: 4044.0
- Randic index: 20.35
- Balaban index: 0.83

$E_{LUMO} = -2.00$ $E_{HOMO} = -7.85$ $E_{LUMO} - E_{HOMO} = 5.85$ $(E_{LUMO} - E_{HOMO})/2 = 2.93$

Fukui indices for N atoms in Fe–N bonds:

- I_N are 0.02 and 0.15, respectively
- I_E are 0.03 and 11.40, respectively
- I_R are 0.15 and 1.87, respectively

Heat of formation: 1080.6
COSMO area: 480.2
COSMO volume: 575.5
Empirical formula: $C_{32}H_{16}N_8Fe$
Shannon entropy of atomic numbers: 1.0273
Shannon entropy of bond orders in kenograph: 0.6442

4.4.5 Example 5

Figure 4.13 shows the optimized geometry of Mg–phthalocyanine, calculated as planar (charge on system = 0). According to Table 4.1, the following bonds are identified in the kenograph of the complex: 4 coordinative Mg–N bonds, 8 single C–C and C–N bonds, and 40 aromatic C–C and C–N bonds. The values of bond orders for Mg–N bonds are within the range [0.372, 0.426].

Sum of bond orders for Mg (all types of bonds): 1.687
Free valence for Mg: +0.313
Net charge of Mg: +0.89

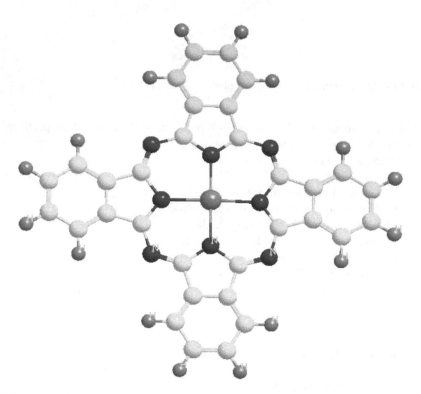

FIGURE 4.13   Magnesium phthalocyanine complex.

Values of net charges for N atoms in Mg–N bonds are within the range [−0.31, −0.58].

Donor H–Acceptor H capacity gap: −0.23

Topological indices:

- Gravitation index (all atoms): 10420.2
- Gravitation index (bonded atoms): 4265.5
- Electronic index (all atoms): 29.55
- Electronic index (bonded atoms): 12.69
- Wiener index: 4044.0
- Randic index: 20.35
- Balaban index: 0.83

The reader can compare the value, naturally close, of the topological indices calculated for the two phthalocyanines in Examples 4 and 5.

$$E_{LUMO} = -2.14 \; E_{HOMO} = -7.14 \; E_{LUMO} - E_{HOMO} = 5.00 \; (E_{LUMO} - E_{HOMO})/2 = 2.50$$

Fukui indices for N atoms in Mg–N bonds:

- I_N are 0.00 and 0.18, respectively
- I_E are 0.00 and 8.01, respectively
- I_R are 0.00 and 1.90, respectively

Heat of formation: 877.2

COSMO area: 486.3

COSMO volume: 579.6

Empirical formula: $C_{32}H_{16}N_8Mg$

Shannon entropy of atomic numbers: 1.0273

Shannon entropy of bond orders in kenograph: 0.6871

4.4.6 EXAMPLE 6

Complexes Cu–Hem and V–Hem are computed as planar (Figure 4.14a), the complex Fe–Hem is computed as almost planar and the complex Co–Hem is computed as rather nonplanar (Figure 4.14b). The vinyl groups and the aromatic system are not in the same plane ($\alpha \sim 80°$). There is an intramolecular hydrogen bond between the carboxyl groups.

Some features of these complexes are presented in Table 4.5. According to the average value of the bond orders of metal–nitrogen bonds, the ion having vanadium seems to be an organometallic compound, not an organometallic complex. However, two of these bonds are coordinative and two are single. The reader can compare the value, naturally close, of the topological indices calculated for the four complexes in Example 6 (Table 4.6).

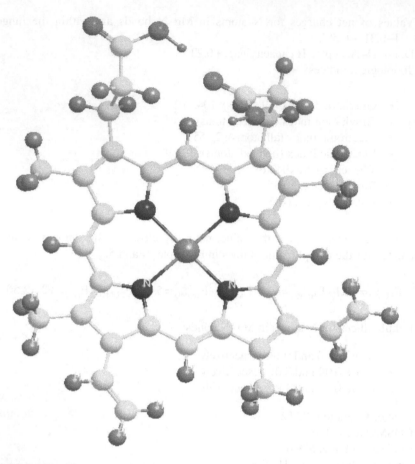

FIGURE 4.14A   Planar complex M-Hem.

Other features of analyzed complexes are presented in Table 4.7 and Table 4.8. The symbol Δ in Table 4.7 is the difference $E_{LUMO} - E_{HOMO}$. The value of pK_a is computed for H atoms in carboxyl groups.

Empirical formula: $C_{34}H_{32}N_4O_4$ Metal
Shannon entropy of atomic numbers: 1.0923

Shannon entropy SE_{bonds} of the bond orders in the kenograph was computed using the values in Table 4.9. Some C–O bonds in ester groups are computed as aromatic.

The LogP value was calculated for some metal–Hem complexes:

- 6.04 for V–Hem
- 3.48 for Fe–Hem
- 4.48 for Co–Hem

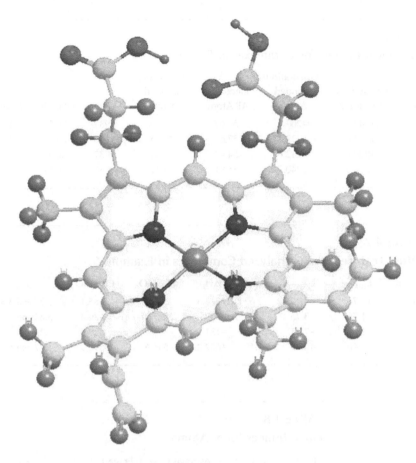

FIGURE 4.14B Nonplanar complex M-Hem.

TABLE 4.5
Various Features of Analyzed Complexes

Metal	Charge on System	Average B_{M-N}	ΣB_M	Free Valence for M	Charge on Metal	Average Charge on Nitrogen	Donor–Acceptor Gap
Cu	−1	0.199	0.980	+0.020	0.66	− 0.32	+0.26
V	−1	0.826	3.881	+1.119	0.64	− 0.25	+0.42
Fe	0	0.485	2.530	+0.470	1.04	− 0.23	+0.36
Co	+1	0.444	2.387	+2.613	0.12	− 0.24	+0.48

TABLE 4.6
Topological Indices for Complexes in Example 6

Metal	Gravitation All Atoms	Gravitation Bonded Atoms	Electronic All Atoms	Electronic Bonded Atoms	Wiener	Randic	Balaban
Cu	11149.0	4267.4	51.67	24.05	4937	20.57	1.14
V	11183.0	4238.7	57.64	25.64	4937	20.57	1.14
Fe	11280.8	4282.6	54.43	23.69	4937	20.57	1.14
Co	11200.4	4240.7	55.91	25.21	4937	20.57	1.14

TABLE 4.7
Other Features of the Analyzed Complexes in Example 6

Metal	E_{LUMO}	E_{HOMO}	$\Delta/2$	ΔH_f	$COSMO_A$	$COSMO_V$	pK_a
Cu	+1.74	−3.49	2.62	−214.3	537.4	685.1	3.77 and 3.39
V	+1.76	−3.73	2.75	−559.4	527.9	689.4	3.80 and 2.99
Fe	−1.61	−7.71	3.05	−181.5	536.7	688.5	3.26 and 3.50
Co	−6.46	−10.83	2.19	+602.2	545.6	690.7	1.77 and 5.12

TABLE 4.8
Fukui Indices for N Atoms

Metal	Average I_N	Average I_E	Average I_R
Cu	13.37	2.71	16.54
V	5.06	0.85	0.82
Fe	9.07	5.76	9.45
Co	0.19	12.26	5.95

TABLE 4.9
Shannon Entropy of Bond Orders

Metal	Number of Bonds Coordinative	Single	Aromatic	Double	SE_{bonds}
Cu	4	16	28	2	1.0201
V	2	18	28	2	0.9500
Fe	4	16	28	2	1.0201
Co	4	25	18	3	1.0852

4.4.7 EXAMPLE 7

The complex [L]$_2$Cu (charge on system = −1) in Figure 4.15 was the subject of a study of the chromatic dependence of the Cu complex dyes on the type of the coupling component (Emandi et al. 2004). According to Table 4.1, the following bonds are identified in the kenograph of the complex: 4 coordinative bonds, 16 single bonds, and 24 aromatic bonds. The C = O and C = N bonds, conventionally represented as double, are calculated to be aromatic. The N = O bonds in the NO$_2$ groups are also calculated as aromatic. The N–N bond in the cycle is calculated as single, but the N–N bond outside the cycle is calculated as aromatic.

The values of bond orders for Cu–N bonds: 0.238
The values of bond orders for Cu–O bonds: 0.113
Sum of bond orders for Cu (all types of bonds): 0.856
Free valence for Cu: +3.144
Net charge of Cu: +0.63
Values of net charges for N atoms in Cu–N bonds: −0.32
Values of net charges for O atoms in Cu–O bonds: −0.48
Donor H–Acceptor H capacity gap: −0.08

The value of pK$_a$ (12.09 and 12.20) is computed for H atoms in hydroxyl groups.

Topological indices:

- Gravitation index (all atoms): 10297.0
- Gravitation index (bonded atoms): 4216.75
- Electronic index (all atoms): 43.63
- Electronic index (bonded atoms): 21.84
- Wiener index: 4336.0
- Randic index: 18.59
- Balaban index: 0.94

E_{LUMO} = +0.63 E_{HOMO} = −5.82 E_{LUMO} − E_{HOMO} = 6.45 (E_{LUMO} − E_{HOMO})/2 = 3.23

Heat of formation: 43.5
COSMO area: 493.7
COSMO volume: 568.8

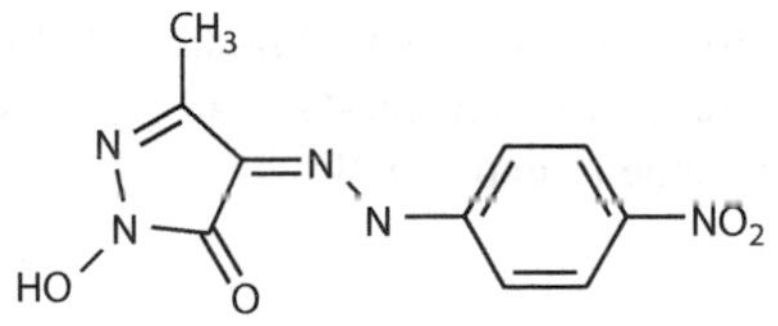

FIGURE 4.15A The ligand in example 7.

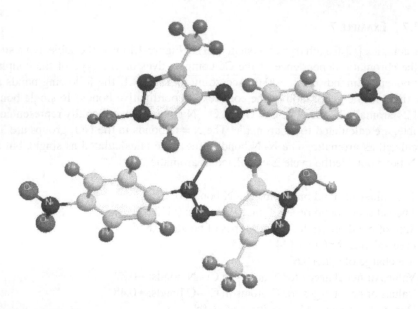

FIGURE 4.15B   The complex in example 7.

Empirical formula: $C_{20}H_{16}N_{10}O_8Cu$
Shannon entropy of atomic numbers: 1.3903
Shannon entropy of bond orders: 0.9165

4.4.8   EXAMPLE 8

The complex $[L]_2[H_2O]Zn$ (L = isoniazid, charge on system = +2) in Figure 4.16 was recently synthesized (Kriza et al. 2009). According to Table 4.1, the following bonds are identified: 6 coordinative bonds, 4 single bonds, and 16 aromatic bonds. The C = O and C–N bonds in the amide group are calculated as aromatic. The C–C bond between the amide group and the aromatic cycle and the N–N bond are calculated as single.

The values of bond orders for Zn–N bonds: 0.393
The values of bond orders for Zn–O bonds: 0.322
The values of bond orders for Zn–O (water) bonds: 0.293
Sum of bond orders for Zn (all types of bonds): 2.126
Free valence for Zn: −0.126
Net charge of Zn: +0.51
Values of net charges for N atoms in Zn–N bonds: −0.16
Values of net charges for O atoms in Zn–O bonds: −0.49
Values of net charges for O atoms in Zn–O (water) bonds: −0.52
Donor H–Acceptor H capacity gap: +0.21
Topological indices:

- Gravitation index (all atoms): 5134.1
- Gravitation index (bonded atoms): 2466.9

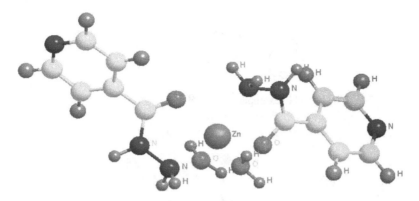

FIGURE 4.16   The complex in example 8.

- Electronic index (all atoms): 28.52
- Electronic index (bonded atoms): 14.15
- Wiener index: 1142.0
- Randic index: 10.90
- Balaban index: 0.82

$$E_{LUMO} = -8.23 \quad E_{HOMO} = -15.26 \quad E_{LUMO} - E_{HOMO} = 7.03 \quad (E_{LUMO} - E_{HOMO})/2 = 3.52$$

Heat of formation: 1010.0
COSMO area: 348.9
COSMO volume: 383.6
Empirical formula: $C_{12}H_{18}N_6O_4Zn$
Shannon entropy of atomic numbers: 1.3199
Shannon entropy of bond orders: 0.9251

If the water molecules are replaced by Cl^- anions, the charge on analyzed complex becomes null. In this case, the values of some descriptors are changed. For instance:

The values of bond orders for Zn–N bonds: 0.294
The values of bond orders for Zn–O bonds: 0.235
The values of bond orders for Zn–Cl bonds : 0.489
Sum of bond orders for Zn (all types of bonds): 2.092
Free valence for Zn: −0.092
Net charge of Zn: + 0.48

4.4.9   EXAMPLE 9

The complex $[L]_2Zn$ (charge on system = +2) in Figure 4.17 was recently synthesized in order to study the bactericidal activity of the ligand and the complex (Reiss et al. 2009). According to Table 4.1, the following bonds are identified: 4 coordinative

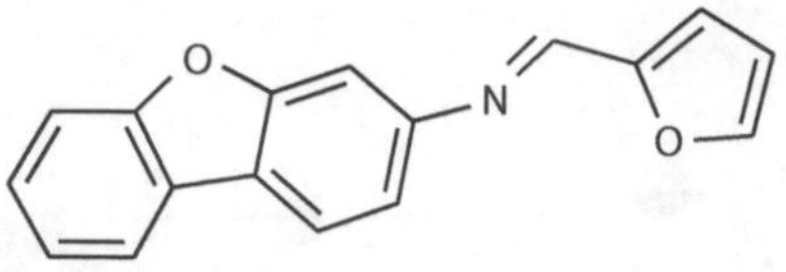

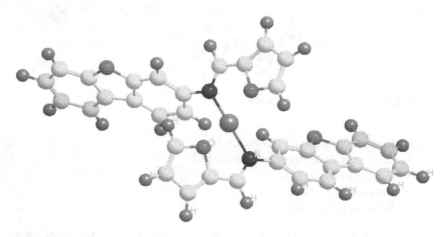

FIGURE 4.17A   The ligand in example 9.

FIGURE 4.17B   The complex in example 9.

bonds, 8 single bonds, and 38 aromatic bonds. Most of the C–O bonds are calculated as single.

The values of bond orders for Zn–N bonds: 0.671
The values of bond orders for Zn–O bonds: 0.319
Sum of bond orders for Zn (all types of bonds): 2.166
Free valence for Zn: −0.166
Net charge of Zn: +0.45
Values of net charges for N atoms in Zn–N bonds: −0.06
Values of net charges for O atoms in Zn–O bonds: −0.28
Donor H–Acceptor H capacity gap: +0.18
Topological indices:

- Gravitation index (all atoms): 9947.2
- Gravitation index (bonded atoms): 4232.3
- Electronic index (all atoms): 26.88
- Electronic index (bonded atoms): 12.03
- Wiener index: 5124.0
- Randic index: 20.29
- Balaban index: 0.77

$E_{LUMO} = -8.00 \ E_{HOMO} = -13.43 \ E_{LUMO} - E_{HOMO} = 5.43 \ (E_{LUMO} - E_{HOMO})/2 = 2.72$

Fukui indices for N atoms in Zn–N bonds:

- I_N are 0.00 and 0.05, respectively
- I_E are 50.42 and 49.85, respectively
- I_R are 0.49 and 4.46, respectively

Fukui indices for O atoms in Zn–O bonds:

- I_N are 0.00 and 0.00, respectively
- I_E are 24.77 and 24.62, respectively
- I_R are 0.31 and 0.78, respectively

Heat of formation: 1742.0
COSMO area: 523.6
COSMO volume: 629.0
Empirical formula: $C_{34}H_{22}N_2O_4Zn$
Shannon entropy of atomic numbers: 1.0506
Shannon entropy of bond orders: 0.7038

4.4.10 EXAMPLE 10

The complex $[L]_2Cl_2Ni$ (charge on system = 0) in Figure 4.18 was recently synthesized in order to study its antimicrobial activity (Mitu et al. 2009). According to Table 4.1, the following bonds are identified: 4 coordinative bonds, 8 single bonds, and 48 aromatic bonds. In the amide groups the C–O bonds are calculated as aromatic, but the C–N bonds are calculated as single. The N–N bonds are calculated as single. The CH = N are calculated as aromatic. There are no coordinative Ni–O bonds.

The values of bond orders for Ni–N bonds: 0.563
The values of bond orders for Ni–Cl bonds: 0.522
Sum of bond orders for Ni (all types of bonds): 2.475
Free valence for Ni: +1.525
Net charge of Ni: +0.53
Values of net charges for N atoms in Ni–N bonds: +0.19
Values of net charges for Cl atoms in Ni–Cl bonds: −0.61

The heteroatoms having the lowest (negative) net charge are the chlorine atoms.
Donor H–Acceptor H capacity gap: −0.14

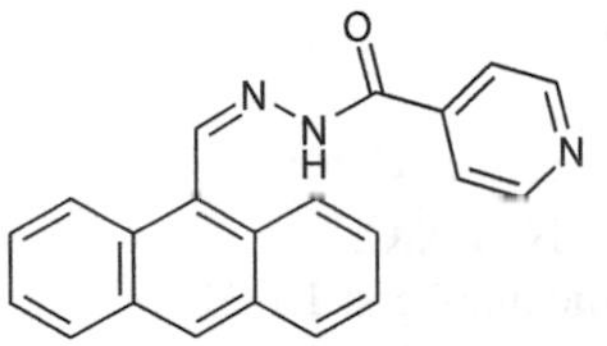

FIGURE 4.18A   The ligand in example 10.

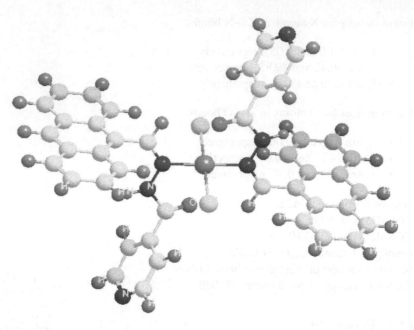

FIGURE 4.18B   The complex in example 10.

Topological indices:

- Gravitation index (all atoms): 15233.4
- Gravitation index (bonded atoms): 5310.0
- Electronic index (all atoms): 46.20
- Electronic index (bonded atoms): 16.26
- Wiener index: 9948.0
- Randic index: 25.86
- Balaban index: 0.68

$E_{LUMO} = -1.34$ $E_{HOMO} = -7.98$ $E_{LUMO} - E_{HOMO} = 6.64$ $(E_{LUMO} - E_{HOMO})/2 = 3.32$

Fukui indices for N atoms in Ni–N bonds:

- I_N are 0.34 and 0.35, respectively
- I_E are 0.38 and 1.02, respectively
- I_R are 1.14 and 1.97, respectively

Heat of formation: 401.2
COSMO area: 655.1
COSMO volume: 834.4
Empirical formula: $C_{42}H_{30}N_6O_2NiCl_2$
Shannon entropy of atomic numbers: 1.1352
Shannon entropy of bond orders: 0.6277
LogP: 5.22

4.4.11 EXAMPLE 11

The complex $L_1[L_2]_2Co$ (charge on system = +1) in Figure 4.19, where the ligand L_1 is the isoniazid (see Example 8) and the ligand L_2 is the acetate ion, was synthesized recently in order to study its tuberculostatic activity (Kriza et al. 2010). According to Table 4.1, the following bonds are identified in this complex: 5 coordinative Co–O bonds, 1 Co–N coordinative bond, 4 single bonds, and 12 aromatic bonds. The C–O bonds in the amide and ester groups are calculated as aromatic, and the C–N bond in the amide group is also calculated as aromatic.

The values of bond order for Co–N bond: 0.423
The values of bond orders for Co–O bonds are within the [0.348, 0.418] range
Sum of bond orders for Co (all types of bonds): 2.802
Free valence for Co: +2.198
Net charge of Co: +1.66
Values of net charge for N atom in Co–N bond: −0.18
Values of net charges for O atoms in Co–O bonds are within the [−0.62, −0.39] range
Donor H–Acceptor H capacity gap: −0.03
Topological indices:

- Gravitation index (all atoms): 4538.6
- Gravitation index (bonded atoms): 2380.9
- Electronic index (all atoms): 28.13
- Electronic index (bonded atoms): 15.99
- Wiener index: 585.0
- Randic index: 8.73
- Balaban index: 1.10

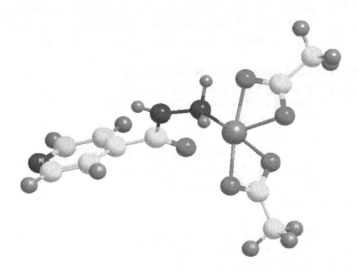

FIGURE 4.19 The complex in example 11.

$E_{LUMO} = -4.99$ $E_{HOMO} = -9.82$ $E_{LUMO} - E_{HOMO} = 4.83$ $(E_{LUMO} - E_{HOMO})/2 = 2.42$

Fukui indices for N atom in Co–N bond:

- I_N is 3.36
- I_E is 26.75
- I_R is 39.44

Fukui indices for O atom in Co–O bonds:

- I_N is within the [0.11, 1.07] range
- I_E is within the [0.07, 8.85] range
- I_R is within the [0.37, 11.62] range

Heat of formation: −261.3
COSMO area: 285.4
COSMO volume: 301.8
Empirical formula: $C_{10}H_{13}N_3O_5Co$
Shannon entropy of atomic numbers: 1.3497
Shannon entropy of bond orders: 0.9949

4.4.12 EXAMPLE 12

The $[L]_2Cd$ complex (charge on the system = 0) in Figure 4.20 was synthesized recently in order to study its antifungal activity (Kumar and Kumar 2007). During geometry optimization, the PM6 method adjusts the value of the bond orders so that the structure in Figure 4.20a becomes the structure in Figures 4.20b. Many of the chemical bonds described conventionally as single and double are computed as aromatic. According to Table 4.1, the following bonds are identified in this complex: 2 coordinative Cd–O bonds, 2 coordinative Cd–N bonds, 6 single bonds, and 40 aromatic bonds. All C–O and C–N bonds are calculated as aromatic.

The values of bond orders for Cd–N bonds: 0.495
The values of bond orders for Cd–O bonds: 0.223
Sum of bond orders for Cd (all types of bonds): 1.644
Free valence for Cd: +0.356
Net charge of Cd: +1.00
Values of net charges for N atoms in Cd–N bonds: −0.41
Values of net charges for O atoms in Cd–O bonds: −0.65
Donor H–Acceptor H capacity gap: −0.25
Topological indices:

- Gravitation index (all atoms): 10237.6
- Gravitation index (bonded atoms): 4002.2
- Electronic index (all atoms): 41.83
- Electronic index (bonded atoms): 17.37

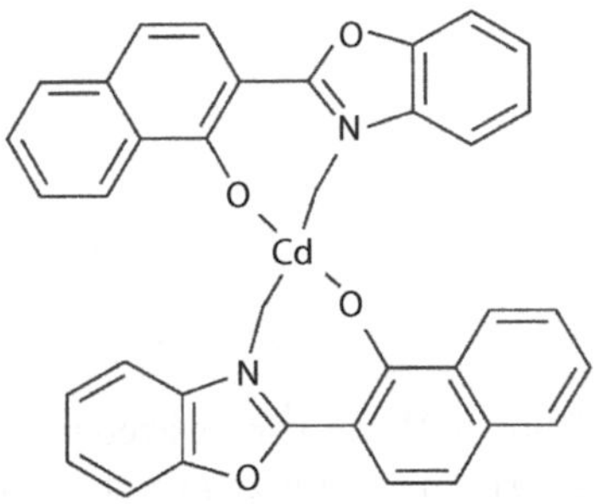

12a 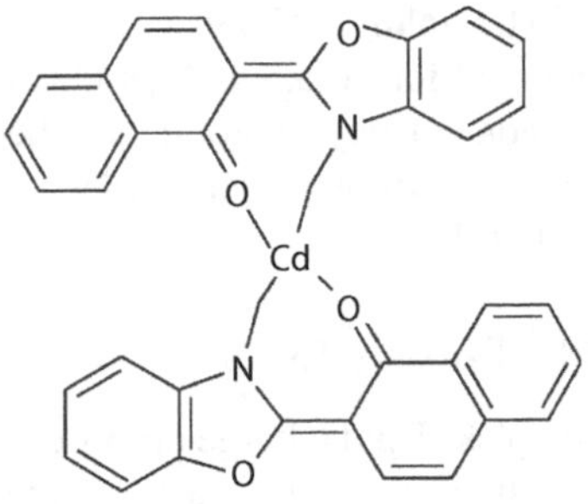   12b

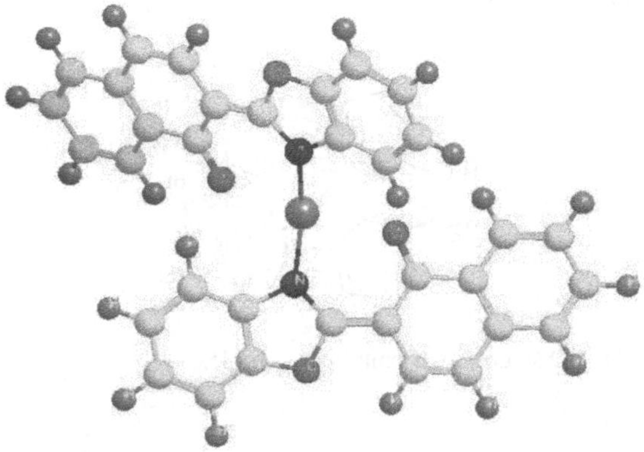

FIGURE 4.20 The complex in example 12.

- Wiener index: 5216.0
- Randic index: 20.17
- Balaban index: 0.72

$E_{LUMO} = -0.86$ $E_{HOMO} = -8.22$ $E_{LUMO} - E_{HOMO} = 7.36$ $(E_{LUMO} - E_{HOMO})/2 = 3.68$

Fukui indices for N atoms in Cd–N bonds:

- I_N is 4.27 and 4.12
- I_E is 1.73 and 3.12
- I_R is 4.15 and 3.09

Fukui indices for O atom in Co–O bonds:

- I_N is 3.99 and 3.82
- I_E is 1.22 and 2.17
- I_R is 3.51 and 10.71

Heat of formation: +24.1
COSMO area: 502.0

COSMO volume: 598.9
Empirical formula: $C_{34}H_{20}N_2O_4Cd$
Shannon entropy of atomic numbers: 1.0495
Shannon entropy of bond orders: 0.6350
Log P: 6.19

4.4.13 EXAMPLE 13

The complex [L]Pd (charge on system = 0) in Figure 4.21 was synthesized recently in order to study its catalytic activity (Nan et al. 2011). The optimized geometry of this complex is nonplanar. According to Table 4.1, the following bonds can be identified in this complex: 2 Pd–C coordinative bonds, 2 Pd–Br coordinative bonds, 14 single bonds, and 15 aromatic bonds.

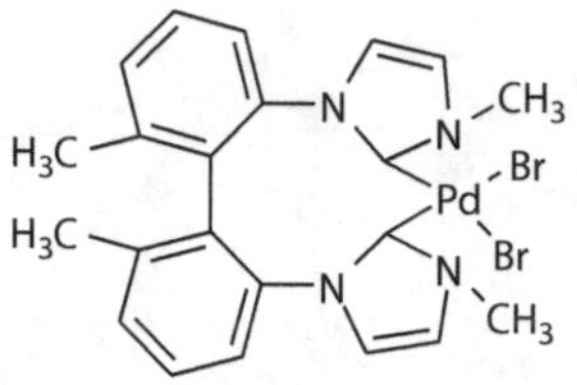

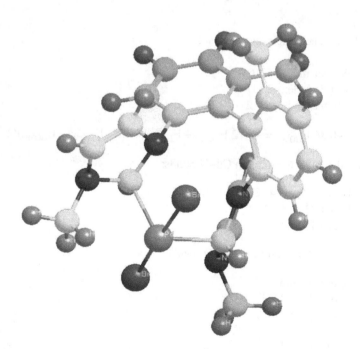

FIGURE 4.21A   The 2D structure of paladium complex.

FIGURE 4.21B   The 3D structure of paladium complex.

The values of bond orders for Pd–C bonds: 0.687 and 0.665
The values of bond orders for Pd–Br bonds: 0.369 and 0.417
Sum of bond orders for Pd (all types of bonds): 2.551
Free valence for Pd: +3.449
Net charge of Pd: +0.50
The Br atoms are the atoms having the smallest (negative) net charge.
Values of net charges for C atoms in Pd–C bonds: +0.33 and +0.36
Values of net charges for Br atoms in Pd–Br bonds: −0.60 and −0.65
Donor H–Acceptor H capacity gap: −0.28
Topological indices:

- Gravitation index (all atoms): 10634.2
- Gravitation index (bonded atoms): 2959.8
- Electronic index (all atoms): 33.68
- Electronic index (bonded atoms): 13.65
- Wiener index: 1568.0
- Randic index: 13.49
- Balaban index: 0.94

$E_{LUMO} = -0.18\ E_{HOMO} = -6.94\ E_{LUMO} - E_{HOMO} = 6.76\ (E_{LUMO} - E_{HOMO})/2 = 3.38$

Fukui indices for C atoms in Pd–C bonds:

- I_N is 0.81 and 1.65
- I_E is 9.08 and 1.70
- I_R is 10.96 and 6.48

Heat of formation: +355.1
COSMO area: 366.1
COSMO volume: 506.2
Empirical formula: $C_{22}H_{24}N_4PdBr_2$
Shannon entropy of atomic numbers: 1.1173
Shannon entropy of bond orders: 0.9779
Log P: 3.50

4.4.14 EXAMPLE 14

The bischelate complex $[L_1][L_2]GeCl_2$ in Figure 4.22 was prepared by double trans-metallation (Baukov et al. 2008). The charge on the system is null. According to Table 4.1, the following bonds are identified in this complex: 3 coordinative Ge–O bonds, 2 coordinative Ge–Cl bonds, 12 single bonds, and 19 aromatic bonds. The Ge–C bond seems to be at the border between the single and coordinative bonds (B = 0.761). The C–O and C–N bonds in the substituted amide groups are calculated as aromatic. The O atom next to the aromatic cycle is connected to it through a C–O aromatic bond.

The value of bond order for Ge–C bonds 0.761
The values of bond orders for Ge–O bonds are within the [0.353, 0.681] range

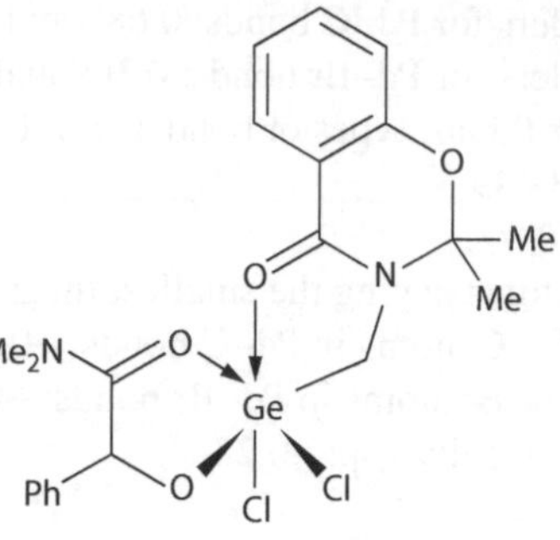

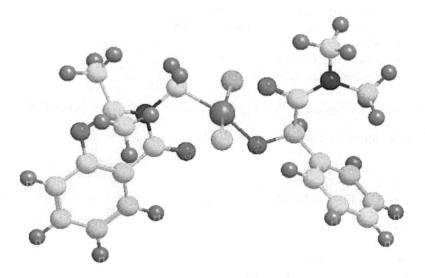

FIGURE 4.22A   The 2D structure of germanium complex.

FIGURE 4.22B   The 3D structure of germanium complex.

The values of bond orders for Ge–Cl bonds are 0.558 and 0.585

Sum of bond orders for Ge (all types of bonds): 3.392

Free valence for Ge: +0.608

Net charge of Ge: +1.25

Value of net charge for C atoms in Ge–C bond: −0.37

Values of net charges for O atoms in Ge–O bonds are within the [−0.60, −0.47]
   range

Value of net charges for Cl atoms in Ge–Cl bonds: −0.53 and − 0.55

Donor H–Acceptor H capacity gap: −0.19

Topological indices:

- Gravitation index (all atoms): 8513.5
- Gravitation index (bonded atoms): 3286.3
- Electronic index (all atoms): 43.93

- Electronic index (bonded atoms): 19.46
- Wiener index: 2086.0
- Randic index: 13.94
- Balaban index: 0.94

$E_{LUMO} = -0.81$ $E_{HOMO} = -9.17$ $E_{LUMO} - E_{HOMO} = 8.36$ $(E_{LUMO} - E_{HOMO})/2 = 4.18$

Fukui indices for O atoms in Ge–O bonds:

- I_N are 3.22, 0.38 and 1.46
- I_E are 0.04, 0.05 and 6.03
- I_R are 1.03, 0.40 and 5.47

Heat of formation: −827.8
COSMO area: 417.5
COSMO volume: 513.5
Empirical formula: $C_{21}H_{24}N_2O_4GeCl_2$
Shannon entropy of atomic numbers: 1.2385
Shannon entropy of bond orders: 0.9777
Log P: 3.39

4.4.15 EXAMPLE 15

The catecholase activity of the manganese complex in Figure 4.23 has been evaluated (Wellington et al. 2008). The charge on the analyzed system is +1. The geometry of the bonds involving Mn is asymmetrical. Thus, Mn–N and Mn–Cl bonds are calculated as single. The Mn–O bond (with the carbonyl group) is calculated as coordinative. One of the Mn–O bonds (with water) is calculated as coordinative, and the other is nonexistent (B = 0.093). Overall, the following bonds are identified in this complex: 2 Mn–O coordinative bonds, 7 single bonds, and 14 aromatic bonds. The C–O and C–N bonds in the amide group are calculated as aromatic.

The value of bond order for Mn–O bond (in carbonyl group): 0.402
The value of bond order for Mn–O bond (in water): 0.264
The values of bond orders for Mn–Cl and Mn–N: ~ 0.9
Sum of bond orders for Mn (all types of bonds): 3.913
Free valence for Mn: +3.087

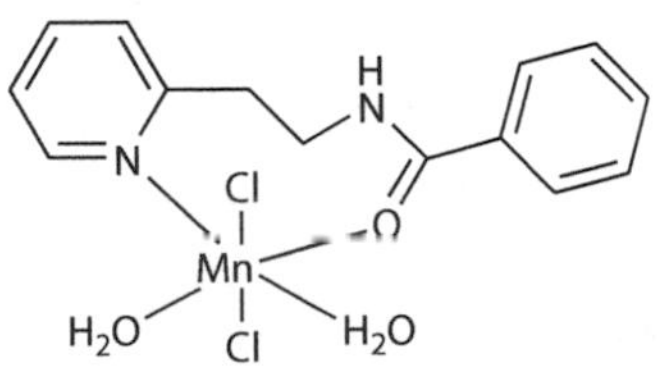

FIGURE 4.23A   The 2D structure of manganese complex.

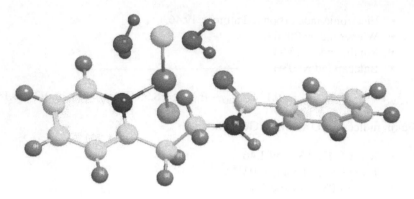

FIGURE 4.23B The 3D structure of manganese complex.

Net charge of Mn: +1.06
Value of net charge for N atom in Mn–N bond: −0.07
Values of net charge for O atom in Mn–O bond (carbonyl group): −0.53
Values of net charge for O atom in Mn–O bond (water): −0.53
Value of net charges for Cl atoms in Mn–Cl bonds: −0.37 and −0.41
Donor H–Acceptor H capacity gap: +0.15
Topological indices:

- Gravitation index (all atoms): 6038.1
- Gravitation index (bonded atoms): 2396.5
- Electronic index (all atoms): 27.67
- Electronic index (bonded atoms): 12.22
- Wiener index: 834.0
- Randic index: 9.94
- Balaban index: 0.87

$E_{LUMO} = -3.84$ $E_{HOMO} = -9.65$ $E_{LUMO} - E_{HOMO} = 5.81$ $(E_{LUMO} - E_{HOMO})/2 = 2.91$

Fukui indices for O atom in Mn–O bond (carbonyl group):

- I_N are 0.19
- I_E are 13.06
- I_R are 4.32

Fukui indices for N atom in Mn–N bond:

- I_N are 0.25
- I_E are 9.09
- I_R are 5.82

Heat of formation: −122.7
COSMO area: 323.1
COSMO volume: 387.5

Empirical formula: $C_{14}H_{18}N_2O_3MnCl_2$
Shannon entropy of atomic numbers: 1.3128
Shannon entropy of bond orders: 0.8766

4.4.16 Example 16

The bimetallic complex in Figure 4.24 was obtained recently by a synthesis involving a small number of steps (Mendoza-Espinosa et al. 2011). The charge on the analyzed system is +1.

The value of bond order for Pd–Cl: 0.789
The value of bond order for Pd–C: 0.802
Sum of bond orders for Pd (all types of bonds): 1.898
Free valence for Pd: +4.102

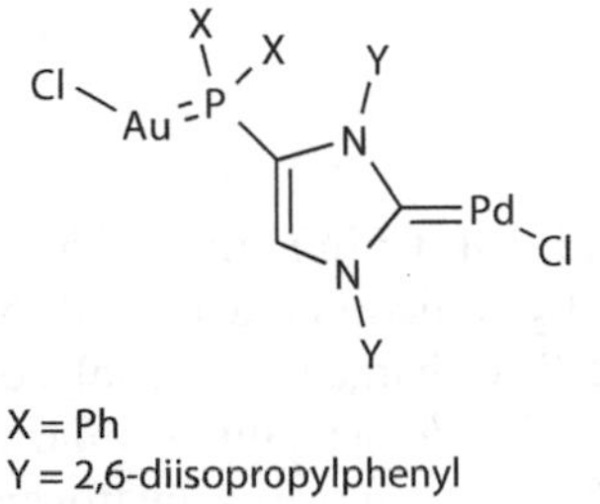

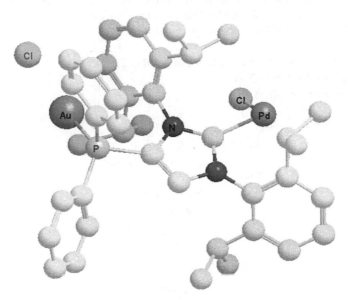

X
X
Y
Cl
Au
P
N
Pd
Cl
N
Y
X = Ph
Y = 2,6-diisopropylphenyl

FIGURE 4.24A The 2D structure of bimetallic complex in example 16.

Cl
Cl
Au
Pd
N
P

FIGURE 4.24B The 3D structure of bimetallic complex in example 16 without hydrogen atoms.

Net charge of Pd: +0.65
Value of net charge for Cl atom in Pd–Cl bond: −0.35
Values of net charge for C atom in Pd–C bond: −0.01
The value of bond order for Au–Cl: 0.277
The value of bond order for Au–P: 0.477
Sum of bond orders for Au (all types of bonds): 1.050
Free valence for Au: +3.950
Net charge of Au: +0.39
Value of net charge for Cl atom in Au–Cl bond: −0.78
Values of net charge for P atom in Au–P bond: +0.58

$E_{LUMO} = -4.48\ E_{HOMO} = -11.06\ E_{LUMO} - E_{HOMO} = 6.58\ (E_{LUMO} - E_{HOMO})/2 = 3.29$

Heat of formation: +756.2
COSMO area: 549.9
COSMO volume: 850.0

4.4.17 EXAMPLE 17

In the hypothetical complex $L_1L_2L_3Cr$ in Figure 4.25, the ligand L_1 is oximethionine CAS 4385-91-5, the ligand L_2 is nicotinic acid CAS 59-67-6, and the ligand L_3 is phenylalanine CAS 63-91-2. The charge on the analyzed system is null. The ligands were built virtually as zwitterions/inner salts, similar to Example 1. During geometry optimization, the PM6 method moves a hydrogen atom of the NH_3^+ group of oximethionine to the COO^- group of nicotinic acid. In the L_1 and L_3 ligands, the C–O bonds are calculated as aromatic. In ligand L_2, one of the C–O bonds is calculated as aromatic and the other as single. The Cr atom is involved in 5 Cr–O coordinative bonds $0.250 < B < 0.374$ and one single bond $B = 0.847$.

Sum of bond orders for Cr (all types of bonds): 2.660
Free valence for Cr: +3.340
Net charge of Cr: +0.58

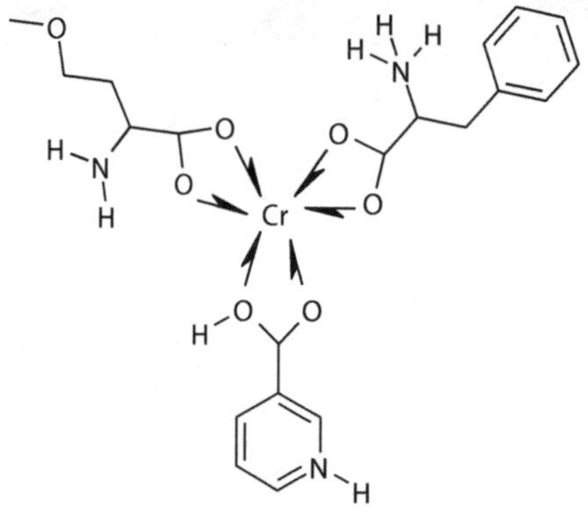

FIGURE 4.25A   The 2D structure of Cr complex with three ligands.

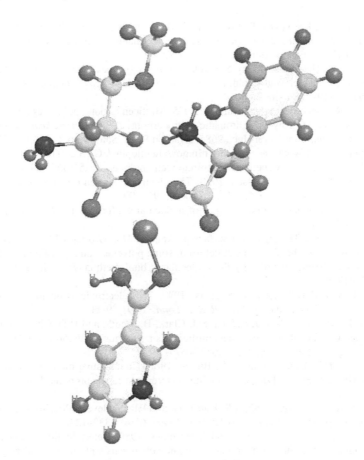

FIGURE 4.25B The 3D structure of Cr complex with three ligands.

$E_{LUMO} = -0.38$ $E_{HOMO} = -5.89$ $E_{LUMO} - E_{HOMO} = 5.51$ $(E_{LUMO} - E_{HOMO})/2 = 2.76$

Heat of formation: -1091.1
COSMO area: 439.2
COSMO volume: 520.3

REFERENCES

ACD/LogP®. Advanced Chemistry Development Inc., 133 Richmond Street West, Toronto, http://www.acdlabs.com/products/phys_chem_lab/logp/.

Arteca, G.A. 1991. Molecular shape descriptors. In *Reviews in Computational Chemistry*, Vol. 9, edited by K. B. Lipkowitz and D. B. Boyd, 191–253. New York: VCH Publishers.

Balaban, A.T. 1982. Distance connectivity index. *Chem. Phys. Lett.* 89:399–404.

Balaban, A.T. 1983. Topological indices based topological distances in molecular graphs. *Pure & Appl.Chem.* 55:199–206.

Balaban, A.T., P. Filip, and T.S. Balaban. 1985. Computer program for finding all possible cycles in graphs. *J. Comput. Chem.* 6:316–326.

Baukov, Y. I., A.A. Korlyukov, E.P. Kramarova, A.G. Shipov, S.Y. Bylikin, V.V. Negrebetsky, and M.Y. Antipin. 2008. Hexacoordinate germanium mixed bischelates with the $GeCO_3Cl_2$ ligand environment. *Arkivoc* 4:80–89.

Bird, C. W., 1992. Heteroaromaticity, 5, a unified aromaticity index. *Tetrahedron* 48 (2):335–340.

Bondi, A., 1966. Vander Waals volumes and radii of metals in covalent compounds. *J. Phys. Chem.* 70 (9):3006–3007.

Bultinck, P., R. Ponec, and S. van Damme. 2005. Multicenter bond indices as a new measure of aromaticity in polycyclic aromatic hydrocarbons. *J. Phys. Org. Chem.*18:706–718.

Cerius[2]®. Molecular Simulations Inc., 9685 Scranton Road, San Diego, CA 92121-3752.

ChemIDplus/a. http://www.chem.sis.nlm.nih.gov/chemidplus/, CAS 34031-32-8.

ChemIDplus/b; http://www.chem.sis.nlm.nih.gov/chemidplus/, CAS 15663-27-1.

Chirgwin, B. H., and C.A. Coulson. 1950. The electronic structure of conjugated systems. VI. *Proc. R. Soc. London Ser. A.* 201(1065):196–209.

Choi, V. 2006. On updating torsion angles of molecular conformations. *J. Chem. Inf. Model.* 46(1):438–444.

Connolly, M.L. 1992. The molecular surface package, *J. Mol. Graphics* 11:139–141.

Coulson, C.A. 1939. The electronic structure of some polyenes and aromatic molecules. VII. Bonds of fractional order by the molecular orbital method. *Proc. R. Soc. London A* 169:413–428.

Coulson, C.A. and H.C. Lonquet-Higgins. 1947. The electronic structure of conjugated systems. I. General theory. *Proc. R. Soc. Lond. A* 191:39–60.

Cramer, R.D., R.J. Jilek, S. Guessregen, S.J. Clark, B. Wendt, and R.D. Clark. 2004. "Lead Hopping": Validation of topomer similarity as a superior predictor of similar biological activities. *J. Med. Chem.,* 47(27):6777–6791.

Cramer, R.D., D.E. Patterson, and J.D. Bunce. 1988. Comparative molecular field analysis (CoMFA). 1. Effect of shape on binding of steroids to carrier proteins. *J. Am. Chem. Soc.,* 110(18):5959–5967.

Cyranski, M.K., T.M. Krygowski, A.R. Katritzky, and P.R. Schleyer. 2002. To what extent can aromaticity be defined uniquely? *J. Org. Chem.* 67(4):1333–1338.

De Bruijn, J., F. Busser, W. Seiner, and J. Hermens. 1989. Determination of octanol/water partition coefficients for hydrophobic organic chemicals with the slow-stirring method. *Environ. Toxicol. Chem.* 8:499–512.

Delchev, Y.I., A.I. Kuleff, J. Maruani, T. Mineva, and F. Zahariev. 2006. Strutinsky's shell-correction method in the extended Kohn-Sham scheme: Application to the ionization potential, electron affinity, electroneg ativity and chemical hardness of atoms. In *Recent Advances in the Theory of Chemical and Physical Systems*, edited by Jean-Pierre Julien, Jean Maruani, and Didier Mayou, 159–177. New York: Springer-Verlag.

Doweyko, A.M. 1988. The hypothetical active site lattice. An approach to modeling active sites from data on inhibitor molecules. *J. Med. Chem.* 31(7):1396–1406.

DRAGON®, Talete srl., via V. Pisani, 13-20124, Milano, Italy, http://www.talete.mi.it.

Emandi, A., M. Calinescu, and C. Ciomaga. 2004. Electronic studies of the new 2:1 Cu(II): Complexes of 4-arylazo-pyrazol-5-one dyes. *J. Mol. Cryst. & Liq. Cryst.* 416:13–30.

EPISuite®, Exposure Assessment Tools and Models, October 21, 2011, http://www.epa.gov/oppt/exposure/pubs/episuitedl.htm.

Espinosa, A., A. Frontera, R. Garcia, M.A. Soler, and A. Tarraga. 2005. Electrophilic behaviour of 3-methyl-2-methylthio-1,3,4-thiadiazolium salts: A multimodal theoretical approach. *Arkivoc* 9:415–437.

Fores M., M. Duran, M. Sola, and L. Adamowicz. 1999. Excited-state intramolecular proton transfer and rotamerism of 2-(2'-hydroxyvinyl) benzimidazole and 2-(2'-hydroxyphenyl) imidazole. *J. Phys. Chem A*, 103(22):4413–4420.

Fukui, K. 1975. *Theory of Orientation and Stereoselection*. Berlin: Springer-Verlag.

Fukui, K., T. Yonezawa, and C. Nagata. 1954. Theory of substitution in conjugated molecules. *Bull. Chem. Soc. Japan* 27:423–427.

Gaillard, P., P.A. Carrupt, B. Testa, and A. Boudon. 1994. Molecular lipophilicity potential, a tool in 3D-QSAR: Method and applications. *J. Comput. Aided Mol. Design* 8:83–96.

Hansch, C., and A. Leo. 1979. *Substituent Constants for Correlation Analysis in Chemistry and Biology.* New York: John Wiley & Sons.

Hansch, C., J.E. Quinlan, and G.L. Lawrence. 1968. Linear free-energy relationship between partition coefficients and the aqueous solubility of organic liquids. *J. Org. Chem.* 33:347–350.

Hosoya, H. 1971. Topological index: A newly proposed quantity characterizing the topological nature of structural isomers of saturated hydrocarbons. *Bull. Chem. Soc. Jap.* 44(9):2332–2339.

Huey, R., G.M. Morris, A.J. Olson, and D.S. Goodsell. 2007. A semiempirical free energy force field with charge-based desolvation. *J. Comp. Chem.* 28(6):1145–1152.

HyperChem® Hypercube Inc., 1115 N.W., 4th-Street, Gainesville, FL 32601, USA, http://www.hyper.com/.

Inokuchi, H. 1952. The electrical conductivity of the condensed polynuclear aza-aromatic compounds. *Bull. Chem. Soc. Japan* 25:28–32.

Ivanciuc, O., and A.T. Balaban. 1994. Design of topological indices. Part 8. Path matrices and derived molecular graph invariants. *MATCH (Commun. Math. Chem.)* 30:141–152.

Japertas, P., R. Didziapetris, and A. Petrauskas A., 2002. Fragmental methods in the design of new compounds: Applications of the advanced algorithm builder. *Quant. Struct.-Act. Relat.* 21(1):23–37.

Jug, K., 1977. A maximum bond order principle. *J. Am. Chem. Soc.* 99(24):7800–7805.

Karelson, M. 2000. *Molecular Descriptors in QSAR/QSPR.* New York: John Wiley & Sons.

Kier, L.B. 1990. *Computational Chemical Graph Theory.* New York: Nova Science.

Klamt, A., and G. Schüürmann. 1993. COSMO: A new approach to dielectric screening in solvents with explicit expressions for the screening energy and its gradient. *J. Chem. Soc., Perkin Transaction* 2:799–805.

Klopman, G. (ed.). 1974. *Chemical Reactivity and Reaction Paths.* New York: John Wiley.

Kriza, A., L. Ababei, N. Stanica, and I. Rau. 2009. Complex combinations of some transitional metals with the isonicotinic acid hydrazide. *Rev. Chim. (Bucuresti)* 60(8):774–777.

Kriza, A., L.V. Ababei, N. Cioatera, I. Rau, and N. Stanica. 2010. Synthesis and structural studies of complexes of Cu, Co, Ni and Zn with isonicotinic acid hydrazide and isonicotinic acid (1- naphthylmethylene). *J. Serb. Chem. Soc.* 75(2):229–242.

Kumar, A., and D. Kumar. 2007. Synthesis and antimicrobial activity of metal complexes from 2-(1'/2'-hydroxynaphthyl)benzoxazoles. *Arkivoc* 14:117–125.

Lee, B., and F.M. Richards. 1971. The interpretation of protein structures: Estimation of static accessibility. *J. Mol. Biol.* 55(3):379–400.

Mekenyan, O., S. Karabunarliev, and D. Bonchev. 1990. The microcomputer OASIS system for predicting the biological activity of chemical compounds. *Comput. & Chem.* 14:193–200.

Mendoza-Espinosa, D., B. Donnadieu, and G. Bertrand. 2011. Facile preparation of homo- and hetero-dimetallic complexes with a 4-phosphino-substituted NHC ligand: Toward the design of multifunctional catalysts. *Chem. Asian J.* 6(4):1099–1103.

Meylan, W.M., and P.H. Howard. 1995. Atom/fragment contribution method for estimating octanol-water partition coefficients. *J. Pharm. Sci.* 84:83–92.

Mitchell, R.H., R. Zhang, W. Fan, and D.J. Berg. 2005. Measuring antiaromaticity by an analysis of ring current and coupling constant changes in a cyclopentadienone-fused dihydropyrene. *J. Am. Chem. Soc.* 127(46):16251–16254.

Mitu, L., N. Raman, A. Kriza, N. Stanica, and M. Dianu. 2009. Template synthesis, characterization and antimicrobial activity of some new complexes with Isonicotinoyl hydrazone ligands. *J. Serb. Chem. Soc.*, 74(10):1075–1084.

MOPAC®, http://www.openmopac.net/.

MOPAC manual, http://www.openmopac.net/manual/index.html.

Murcia, M., A. Morreale, and A.R. Ortiz. 2006. Comparative binding energy analysis considering multiple receptors: A step toward 3D-QSAR models for multiple targets. *J. Med. Chem.* 49(21):6241–6253.

Murray, J. S., P. Lane, T. Brinck, and P. Politzer. 1993. Relationships between computed molecular properties and solute–solvent interactions in supercritical solutions. *J. Phys. Chem.* 97(19):5144–5148.

Murugan, R., M.P.P. Grendze, J.J.E. Toomey, A.R. Katritzky, M. Karelson, V.S. Lobanov, and P.P. Rachwal. 1994. Predicting physical properties from molecular structure. *CHEMTECH* 24:17–23.

Nan, G., B. Rao, and M. Luo. 2011. *Cis*-chelated palladium(II) complexes of biphenyl-linked bis(imidazolin-2-ylidene): Synthesis and catalytic activity in the Suzuki-Miyaura reaction. *Arkivoc* 2:29–40.

Pascual-Ahuir, J.L., and E. Silla. 1990. GEPOL: An improved description of molecular surfaces, I: building the spherical surface set. *J. Comput. Chem.* 11:1047–1060.

PCModel® Serena Software, Box 3076, Bloomington, Indiana, USA, http://www.serenasoft.com/gmmx.html; e-mail: gilbert@serenasoft.com.

Protein Data Bank. Biological Macromolecular Resource, http://www.pdb.org/pdb/home/home.do, for instance many Ferrodoxins, Ferritins, Myelin Transcription Factors etc.

Randic, M. 1975. On characterization of molecular branching. *J. Am. Chem. Soc.* 97:6609–6615.

Reiss, A., T. Caproiu, and N. Stanica. 2009. Synthesis, characterization and antibacterial studies of some metal complexes with the Schiff base of a heterocyclic aldehyde. *Bull. Chem. Soc. Ethiop.* 23(1):63–68.

Rekker, R.F., and H.M. de Kort. 1979. The hydrophobic fragmental constant: An extension to a 1000 data-point set. *Eur. J. Med. Chem.* 14:479–488.

Rekker, R.F., and R. Mannhold. 1992. *Calculation of Drug Lipophilicity: The Hydrophobic Fragmental Constant Approach.* Weinheim, Germany: VCH.

Rekker, R.F., and G.G. Nys. 1974. The concept of hydrophobic fragmental constants (f-values). 11. Extension of its applicability to the calculation of lipophilicities of aromatic and heteroaromatic structures. *Eur. J. Med. Chem.* 9:361–375.

Rinaldi, D., J.-L. Rivail, and N. Rguini. 1992. Fast geometry optimization in self-consistent reaction field computations on solvated molecules. *J. Comput. Chem.* 13:675–680.

Rohrbaugh, R.H., and P.C. Jurs. 1987. Descriptions of molecular shape applied in studies of structure/activity and structure/property relationships. *Anal. Chim. Acta* 199:99–109.

Roothaan, C.C.J. 1951. New developments in molecular orbital theory. *Rev. Mod. Phys.* 23:69–89.

Saeh, J.C., P.D. Lyne, B.K. Takasaki, and D.A. Cosgrove. 2005. Lead hopping using svm and 3D pharmacophore fingerprints. *J. Chem. Inf. Model.* 45(4):1122–1133.

Sannigrahi, A.B. 1992. Abinitio molecular orbital calculations of bond index and valency. *Adv. Quantum Chem.* 23:301–351.

Satoh, K., S. Azuma, H. Satoh, and K. Funatsu. 1997. Development of a program for construction of a starting material library for AIPHOS. *J. Chem. Software* 4:101–107.

Schleyer, P.R., C. Maerker, A. Dransfeld, H. Jiao, and N.J.R.E. Hommes. 1996. Nucleus-independent chemical shifts: A simple and efficient aromaticity probe. *J. Am. Chem. Soc.* 118:6317–6318.

Schrödinger, E. 1926. Quantisierung als Eigenwert problem/Quantization as a problem of proper values. Part I. *Ann. Physik* 384(4):361–376.

SciQSAR®, SciVision, Inc., 200 Wheeler Road, Burlington, MA, 01803, http://www. scivision.com.

Shannon, C.E. 1948. A mathematical theory of communication. *Bell System Tech. J.* 27:379–423, 623–656.

Shishkin, O.V., I.V. Omelchenko, M.V. Krasovska, R.I. Zubatyuk, L. Gorb, and J. Leszczynski. 2006. Aromaticity of monosubstituted derivatives of benzene: The application of out-of-plane ring deformation energy for a quantitative description of aromaticity. *J. Mol. Struct.* 791:158–164.

Shrake, A., and J.A. Rupley. 1973. Environment and exposure to solvent of protein atoms. Lysozyme and insulin. *J. Mol. Biol.* 79(2):351–371.

Stanton, D.T., L.M. Egolf, P.C. Jurs, and M.G. Hicks. 1992. Computer-assisted prediction of normal boiling points of pyrans and pyrroles. *J. Chem. Inf. Comp. Sci.* 32:306–316.

Stanton, D.T., and P.C. Jurs. 1990. Development and use of charged partial surface area structural descriptors in computer assisted quantitative structure property relationship studies. *Anal. Chem.* 62:2323–2329.

Stewart, J.J.P. 2007. Optimization of parameters for semiempirical methods V: Modification of NDDO approximations and application to 70 elements. *J. Mol. Model.* 13:1173–1213.

Talevi, A., E.A. Castro, and L.E. Bruno-Blanch. 2006. New solubility models based on descriptors derived from the Detour Matrix. *An. Asoc. Quím. Argent.* 94(1–3):129–141.

Tarko, L., 2004a. A procedure for virtual fragmentation of molecules into functional groups. *Arkivoc* 14:74–82.

Tarko, L., 2004b. Virtual fragmentation of molecules and similarity evaluation. *Rev. Chim. (Bucuresti)* 55:539–544.

Tarko, L., 2004c. Molecular flexibility descriptors for QSAR calculations. *Rev. Chim. (Bucuresti)* 55:169–173.

Tarko, L., 2005. QSPR/QSAR computations by PRECLAV software. *Rev. Chim. (Bucuresti)* 56(9):639–644.

Tarko, L., 2008a. Aromatic molecular zones and fragments. *Arkivoc* 11:24–45.

Tarko, L., 2008b. *Abordarea QSPR/QSAR in proiectarea moleculara.* Bucuresti: Ed. Universitara.

Tarko, L., 2010. *Aromatic, Non-aromatic, Anti-aromatic.* Bucuresti: Ed. Universitara.

Tarko, L., 2011a. Using the bond order calculated by quantum mechanics to identify the rotatable bonds. *Rev. Chim. (Bucuresti)* 62:135–138.

Tarko, L., 2011b. A new manner to use application of Shannon Entropy in similarity computation. *J. Math. Chem.* 49:2330–2344.

Tayar, N., A. Mark, R.-S. Tsai, P. Vallat, C. Altomare, and B. Testa. 1991. Measurement of partition coefficients by various centrifugal partition chromatographic techniques: A comparative evaluation. *J. Chromatog.* 556:181–194.

Terwillinger, T.C., H. Klei, P.D. Adams, N.W. Moriarty, and J.D. Cohn. 2006. Automated ligand fitting by core-fragment fitting and extension into density. *Acta Cryst.* 62:915–922.

Todeschini, R., M. Lasagni, and E. Marengo. 1994. New molecular descriptors for 2D and 3D structures. *Theory. J. Chemom.* 8:263–272.

TSAR®, Oxford Molecular Ltd., The Medawar Centre, Oxford Science Park, Oxford, OX4, 4GA, UK, http://www.oxmol.com.

Tuccinardi, T., E. Nuti, G. Ortore, C.T. Supuran, A. Rossello, and A. Martinelli. 2007. Analysis of human carbonic anhydrase II: Docking reliability and receptor-based 3D-QSAR study. *J. Chem. Inf. Model.* 47:515–525.

Ulmschelder, M., and E. Penigault. 1999a. Analytical model for the calculation of molecular van der Waals surfaces and solvent-accessible surfaces. *J. Chim. Phys.* 96:543–565.

Ulmscheider, M., and E. Penigault. 1999b. An approximate procedure for the calculation of van der Waals and solvent-accessible surface areas: Computing Gibbs free energies of hydration. *J. Chim. Phys.* 96:566–590.

Verma, J., V.M. Khedkar, E.C. Coutinho. 2010. 3D-QSAR in drug design: A review. *Curr. Top. Med. Chem.* 10(1):95–115.

Weber, A., M. Böhm, C.T. Supuran, A. Scozzafava, C.A. Sotriffer, and G. Klebe. 2006. 3D QSAR selectivity analyses of carbonic anhydrase inhibitors: Insights for the design of isozyme-selective inhibitors. *J. Chem. Inf. Model.* 6(46):2737–2760.

Wellington, K.W., P.T. Kaye, and G.M. Watkins. 2008. Designer ligands. Part 14. Novel Mn(II), Ni(II) and Zn(II) complexes of benzamide- and biphenyl-derived ligands. *Arkivoc* 17:248–264.

Zborowski, K., R. Grybos, and L.M. Proniewicz. 2005. Theoretical studies on the aromaticity of selected hydroxypyrones and their cations and anions. Part 1: Aromaticity of hetero-cyclic pyran rings. *J. Phys. Org. Chem.* 18:250–254.

Zhang, T., Y.-M. Jia, S.-J. Yan, C.-Y. Yu, and Z.-T. Huang. 2009. Synthesis of BF_2 complex of 3-methylthio enaminones. *Arkivoc* 14:156–170.

Zhu, F., and D.K. Agrafiotis. 2007. A self-organizing superposition (SOS) algorithm for con-formational sampling. *J. Comput. Chem.* 28:1234–1239.

APPENDIX: LIST OF PRECLAV WHOLE MOLECULE DESCRIPTORS

- Number of atoms
- Molecular mass
- Number of H, C, N, O, F, Cl, Br, I, S and P atoms
- Percentage of H, C, N, O, F, Cl, Br, I, S and P atoms
- Number of bonds
- Number of weak, single, single-conjugated, aromatic, double, and triple bonds
- Percentage of weak, single, single-conjugated, aromatic, double, and triple bonds
- Minimum, average, and maximum aromaticity of aromatic chemical bonds
- Number of O–H, N–H, S–H, C–O, C–N, and N–O single or weak bonds
- Number of C–C, C–O, C–N, and N–O aromatic bonds
- Number of C–C, C–O, C–N, and N–O double bonds
- Number of C–C and C–N triple bonds
- Sum of bond orders
- Average bond order (all bonds)
- Average bond order (bonds of heavy atoms)
- Average bond order (bonds of N, O, S, and P atoms)
- Maximum value of bond orders
- Maximum bond order in C–C bonds
- Maximum bond order in C–A or A–A bonds (A = heteroatom)
- Average bond order in C–C bonds
- Average bond order in C–A or A–A bonds (A = heteroatom)
- Maximum value of coordination (carbon atoms)
- Maximum value of coordination (heteroatoms)
- Minimum, average, and maximum distance between O atoms
- Minimum, average, and maximum distance between N atoms
- Minimum, average, and maximum distance between O or N atoms
- Minimum, average, and maximum distance between OH/NH chemical groups

- Cyclomatic number
- Molecular volume
- Area of molecular surface
- Volume of circumscribed parallelepiped, ellipsoid, and sphere
- Area of circumscribed ellipsoid
- Symmetry index
- Average, maximum, and root mean square (RMS) for distance(s) to geometric center
- RMS of distances to geometric center (H and halogen atoms)
- Variance of distances to geometric center
- Variance of distances to geometric center (H and halogen atoms)
- Roughness index
- Spherical shape index
- Moment of inertia A, B, and C
- Gravitation index (all atoms)
- Gravitation index (bonded atoms)
- Randic χ topologic index
- Balaban J topologic index
- Platt topologic index
- Zagreb topologic index
- Kier-Hall (order 2) topologic index
- Wiener W topologic index
- Harary topologic index
- Jh (Ivanciuc) topologic index
- Hyper-Wiener topologic index
- Shannon entropy (of atomic numbers and atomic volumes)
- Shannon entropy (of bond orders)
- Shannon entropy of topologic distances
- Shannon entropy for sum of topologic distances
- Minimum and maximum net charge (all atoms)
- Sum of absolute net charges
- Minimum, average, and maximum net charge of H, C, N, O, F, Cl, Br, I, S, and P
- Minimum net charge of heteroatoms
- Minimum and maximum net charge (heavy atoms)
- Maximum charge of H in H–C bonds
- Maximum charge of H in H–A bonds (A = heteroatom)
- Minimum, average and maximum force in C–H, O–H, N–H, S–H, C–F, C–Cl, C–Br, and C–I bonds
- Dipole moment (total and X, Y, and Z component)
- Molecular polarity
- Polarity parameter
- Topologic electronic index (all atoms)
- Topologic electronic index (bonded atoms)
- Area of positive, negative and neutral charged molecular surface
- Positive molecular area–negative molecular area gap

- Sum of [net charge · exposed atomic surface] products (charge > 0)
- Sum of [net charge · exposed atomic surface] products (weak charge)
- Sum of [net charge · exposed atomic surface] products (charge < 0)
- Sum of [net charge · exposed atomic surface] products (all atoms)
- COSMO area and volume
- Capability to form hydrogen bonds (function #1–#4)
- Donor H–Acceptor H capacity gap
- Heat of formation
- 1 – E(homo) gap
- 10 + E(lumo) sum
- E(lumo+1) energy
- E(homo−1) energy
- E(lumo) − E(homo) gap
- E(lumo) − E(homo−1) gap
- E(lumo+1) − E(homo) gap
- E(lumo+1) − E(homo−1) gap
- 1/[E(lumo) − E(homo)] ratio
- 1/[E(lumo) − E(homo−1)] ratio
- 1/[E(lumo+1) − E(homo)] ratio
- 1/[E(lumo+1) − E(homo−1)] ratio
- Number of double-occupied molecular orbitals
- Molecular orbital maximum bonding and antibonding contribution
- Minimum, average, and maximum free valence of H, C, N, O, F, Cl, Br, I, S, and P
- Maximum and average atomic electrophilic reaction index for C, N, and O atoms
- Maximum and average atomic nucleophilic reaction index for C, N, and O atoms
- Maximum and average atomic one-electron reaction index for C, N, and O atoms
- Total energy/1000
- Electronic energy/1000
- Core–Core energy/1000
- Flexibility and hardness indices #1 and #2
- Number of fragments
- Average and maximum mass of fragments
- Weight percent of largest molecular fragment
- Variance of mass fragments
- Variance of sum of net charges in fragments
- Number of circuits
- Number of circuits, cyclomatic number gap
- Number of 3–10 and > 10 atoms circuits
- Number of aromatic circuits
- Number of atoms in circuits
- Percentage of atoms in circuits
- Number of atoms in aromatic circuits

- Percentage of atoms in aromatic circuits
- Maximum aromaticity of aromatic circuits
- Number of bonds in circuits
- Percentage of bonds in circuits
- Number of Hamiltonian circuits
- Crowding index of circuits
- Number of rotatable bonds
- Percentage of rotatable bonds
- Number of rotatable bonds in cycles

5 QSARs for Predicting Cation Toxicity, Bioconcentration, Biosorption, and Binding

5.1 INTRODUCTION

This chapter discusses quantitative structure-activity relationships (QSARs) for predicting cation toxicity, bioconcentration, biosorption, and binding strength. Several approaches were used to identify these QSARs. First, the test systems, test substances, QSARs, and statistical analyses of each QSAR were extracted from the references cited by Walker et al. (2003). These efforts produced 21 references associated with 97 QSARs for predicting cation toxicities (Table 5.1). These QSARs are discussed in more detail in chapter Sections 5.2, 5.3, and 5.5.

Second, Table 3 of Walker et al. (2003) was analyzed to determine the number of times a metal ion was used as a test substance to determine if there was a correlation between a specific physical property and the metal ion's toxicity. The analysis revealed that 46 metal ions were used as test substances and 23 of these metal ions were used as test substances at least 10 times. The Chemical Abstracts Service Registry Numbers (CASRNs) of the 46 metals were incorporated into a table along with the CASRNs of the 23 metal ions that were used as test substances at least 10 times (Table 5.2).

Third, to identify QSAR metal ion papers that were published after 2003, the metal names and CASRNs of the 23 metal ions that were used as test substances at least 10 times were used to search a number of databases (e.g., PubMed, Toxline, RTECS, TSCATS, TSCATS2, TSCA8(e), TOXCENTER, BIOSIS, CAPLUS) and journals (e.g., *Environmental Toxicology and Chemistry*, *SAR & QSAR Environmental Research*, and *QSAR and Combinatorial Science*). The results of the search are provided in Table 5.3. Repeating the search with the 23 metal ions that were used as test substances less than 10 times did not add any new publications. The search identified 16 studies published from 2004 to 2011 that satisfied the search criteria (Table 5.3). These studies examined correlations between physical or chemical properties of metal ions with bioconcentration potential and toxicity (Van Kolck et al. 2008), biosorption capacity (Can and Jianlong 2007; Chen and Wang 2007; and Zamil et al. 2009), binding strength (Kinraide and Yermiyahu 2007; Zhou et al. 2011), and toxicity (Hickey 2005; Walker et al. 2007; Workentine et al. 2008; Khangarot and Das 2009; Kinraide 2009; Lepădatu et al. 2009;

TABLE 5.1
References from Walker et al. (2003) That Provided QSARs

Reference	QSARs
Biesinger and Christensen (1972)	$pM = -3 + 5 X_{AR}$
	$pM = -1.9 + 1.6 \log K_{eq}$
	$pM = 2.5 + 0.13 \, pK_{sp}$
Jones and Vaughn (1978)	$LD_{50} = -1.706 + 11.35\sigma_p$
	$LD_{50} = -2.999 + 18.79\sigma_p$
	$LogLD_{50} = -1.95 + 18.76\sigma_p$
	$LogLD_{50} = -1.64 + 9.94\sigma_p$
Kaiser (1980)	$pTm = 11.63(\pm0.09) + 4.87(\pm0.17)\log(AN/\Delta IP) - 4.29(\pm0.14)\Delta E_0$
	$pTm = 4.80(\pm0.32) + 6.15(\pm0.45)\log(AN/\Delta IP) - 3.11(\pm0.32)\Delta E_0$
Williams and Turner (1981)	$logLD_{50} = -2.30 + 14.1\sigma_p \; r^2 = 0.656$; no other statistics provided
Williams et al. (1982) Mouse	$logLD_{50} = -1.71 + 9.56\sigma_p \; r^2 = 0.360, \, sd = 0.574, \, p = 0.05, \, n = 14$
Williams et al. (1982) *Drosophila*	$logLC_{50} = 0.0993 + 9.77\sigma_p \; r^2 = 0.503, \, sd = 9.404, \, p = 0.02, \, n = 11$
Jacobson et al. (1983) *Drosophila*	$logLC_{50} = 0.109 + 9.74\sigma_p \; r^2 = 0.494, \, sd = 0.409, \, p = 0.02, \, n = 11$
Turner et al. (1983) Mouse	$logLD_{50} = -3.35 + 23.1\sigma_p \; r^2 = 0.879, \, sd = 0.228, \, p = 0.001, \, n = 8$
	$logLD_{50} = -2.56 + 15.0\sigma_p \; r^2 = 0.588, \, sd = 0.518, \, p = 0.01, \, n = 11$
	$logLD_{50} = -1.71 + 9.56\sigma_p \; r^2 = 0.360, \, sd = 0.574, \, p = 0.05, \, n = 14$
Tan et al. (1984) CHO cells	$logCE_{50} = -0.28 + 24.1\sigma_p \; r^2 = 0.449, \, sd = 1.1, \, p = 0.05, \, n = 11$
Turner et al. (1985) Mouse	$logLD_{50} = -3.35 + 23.1\sigma_p \; r^2 = 0.879, \, sd = 0.228, \, p = 0.001, \, n = 8$
Kaiser (1985)	$pT = -0.0004 + 2.47 \log(AN/\Delta IP) - 1.17\Delta E_0$
	$pT = 13.94 + 5.57 \log(AN/\Delta IP) - 6.65\Delta E_0$
	$pT = 6.38 + 2.39 \log(AN/\Delta IP) - 3.06\Delta E_0$
Babich et al. (1986)	$NR_{50} = 0.01 + 31.9\sigma_p \; r^2 = 0.826$
Magwood and George (1996)	$NR_{50} = -0.45 + 18.9\sigma_p \; r^2 = 0.865$
McCloskey et al. (1996)	$\text{Log EC50} = 0.14 + 1.62(\Delta E_0)$
	$\text{Log EC50} = 5.59 - 1.40\left(X_m^2 r\right)$
	$\text{Log EC50} = 2.51 + 0.51(|\log K_{OH}|)$
	$\text{Log EC50} = -3.25 + 39.95(\sigma_p)$
	$\text{Log EC50} = 0.95 - 0.19(AN/\Delta IP) + 1.73(\Delta E_0)$
	$\text{Log EC50} = 1.30 - 2.03(\log AN/\Delta IP) + 1.64(\Delta E_0)$
	$\text{Log EC50} = 1.48 - 1.00\left(X_m^2 r\right) + 0.32(|\log K_{OH}|)$
	$\text{Log EC50} = 6.36 - 1.39\left(X_m^2 r\right) - 0.16(Z^2/r)$
Newman and McCloskey (1996)	$\text{Log EC50} = -16.65 + 1.87(|\log K_{OH}|)$
	$\text{Log EC50} = -1.42 + 2.10(\Delta E_0)$
	$\text{Log EC50} = 5.71 - 1.56\left(X_m^2 r\right)$
	$\text{Log EC50} = -0.55 + 1.44(\Delta\beta)$
	$\text{Log EC50} = 12.59 - 0.78(\text{Log-}K_{so}MOH)$
	$\text{Log EC50} = 12.87 - 110.9(\sigma_p)$
	$\text{Log EC50} = 4.79 - 1.04(AN/\Delta IP) + 0.83(\Delta E_0)$
	$\text{Log EC50} = 7.03 - 11.99(\log AN/\Delta IP) + 0.78(\Delta E_0)$
	$\text{Log EC50} = 1.70 - 1.62\left(X_m^2 r\right) + 0.690(Z^2/r)$
	$\text{Log EC50} = -5.22 + 1.36(\Delta\beta) + 0.98(Z^2/r)$

TABLE 5.1 (*Continued*)
References from Walker et al. (2003) That Provided QSARs

Reference	QSARs
Sauvant et al. (1997)	RNA: $logIC_{50} = 2.012 - 0.226IP(eV)$
	MTT: $logIC_{50} = 7.024 - 0.796IP(eV)$
	NRI: $logIC_{50} = 7.641 - 0.847IP(eV)$
	CB: $logIC_{50} = 7.814 - 0.859IP(eV)$
	CGR: $logIC_{50} = 5.375 - 0.566IP(eV)$
	DTP: $logIC_{50} = 2.988 - 0.303IP(eV)$
Tatara et al. (1997)	$Log\ LC_{50} = -1.27 + 0.98(\Delta E_0)$
	$Log\ LC_{50} = 3.06 - 1.23(X_m^2 r)$
	$Log\ LC_{50} = -1.21 + 1.11(\Delta\beta)$
	$Log\ LC_{50} = 8.64 - 0.58(Log\text{-}K_{so}MOH)$
	$Log\ LC_{50} = -13.83 + 1.47(|log\ K_{OH}|)$
	$Log\ LC_{50} = -10.47 + 83.49(\sigma_p)$
	$Log\ LC_{50} = 3.93 - 0.87(AN/\Delta IP) + 0.08(\Delta E_0)$
	$Log\ LC_{50} = -3.13 - 0.86(X_m^2 r) + 1.06(Z^2/r)$
	$Log\ LC_{50} = -5.54 + 1.03(\Delta\beta) + 0.90(Z^2/r)$
Tatara et al. (1998)	$Log\ LC_{50} = 0.29 + 0.69(\Delta E_0)$
	$Log\ LC_{50} = 2.59 - 0.64(X_m^2 r)$
	$Log\ LC_{50} = -1.23 + 16.72(\sigma_p)$
	$Log\ LC_{50} = -1.43 + 0.27(|log\ K_{OH}|)$
	$Log\ LC_{50} = 0.53 - 0.06(AN/\Delta IP) + 0.72(\Delta E_0)$
	$Log\ LC_{50} = 3.22 - 0.60(X_m^2 r) - 0.14(Z^2/r)$
	$Log\ LC_{50} = 0.05 - 0.39(X_m^2 r) + 0.20(|log\ K_{OH}|)$
	$Log\ LC_{50} = 1.61 + 0.35(\Delta E_0) - 0.41(X_m^2 r)$
	$Log\ LC_{50} = -1.21 + 0.22(\Delta E_0) + 0.22(|log\ K_{OH}|)$
	$Log\ LC_{50} = -1.63 + 4.58(\sigma_p) + 0.22(|log\ K_{OH}|)$
Lewis et al. (1999)	$pT50^{mouse} = 0.41\ E^o + 3.72$
Enache et al. (1999)	$-(log\ EC_{50}) = 3.59 - 14.4(\sigma_p)\ r^2 = 0.83,\ sd = 0.21,\ F = 45.7,\ n = 10$
	$-(log\ EC_{50}) = 2.67 + 0.000012(\Delta H_o)$
	$-(log\ EC_{50}) = 2.07 + 0.00876(\Delta H_{aq\ ion})$
Enache et al. (2000)	$-(log\ Rrg_{50}) = 8.57 - 7.26\ X_{AR} - 2.51\ \Delta E_0 + 4.45\ X$
	$-(log\ Rrg_{50}) = -1.36 - 1.53\ \Delta E_0 + 1.20\ X^2 r + 0.239\ \Delta IP$
	$-(log\ Rrg_{50}) = -0.74 + 0.0195\ AW - 1.94\ \Delta E_0 + 0.295\ \Delta IP$
	$-(log\ Rrg_{50}) = -1.07 + 0.0519\ AN - 1.91\ \Delta E_0 + 0.299\Delta IP$
	$-(log\ Rrg_{50}) = -0.39 + 0.713\ Z - 2.47\ \Delta E_0 + 1.43\ X^2 r$
	$-(log\ Rrg_{50}) = 0.58 + 0.250\ Z/r - 2.41\ \Delta E_0 + 1.35\ X^2 r$
	$-(log\ Rrg_{50}) = 12.2 - 8.54\ X_{AR} - 1.59\ \Delta E_0 - 3.04\ X$
	$-(log\ Rrg_{50}) = -1.02 + 0.0187\ AW - 1.10\ \Delta E_0 + 0.264\ \Delta IP$
	$-(log\ Rrg_{50}) = -1.33 + 0.0498\ AN - 1.07\ \Delta E_0 + 0.268\ \Delta IP$
	$-(log\ Rrg_{50}) = -1.55 + 0.0442\ AN + 0.246\ \Delta IP$
	$-(log\ Rrg_{50}) = -1.28 + 0.0165\ AW + 0.242\ \Delta IP$

(Continued)

TABLE 5.1 (Continued)
References from Walker et al. (2003) That Provided QSARs

Reference	QSARs
Enache et al. (2003)	$-(\log EC_{50}) = 3.74 - 0.0167\ \Delta H_h$
	$-(\log EC_{50}) = 0.867 + 0.000296\ \rho$
	$-(\log EC_{50}) = -0.145 + 0.182\ IP$
	$-(\log EC_{50}) = 7.93 - 0.502\ |\log K_{OH}|$
	$-(\log EC_{50}) = 0.430 + 0.165\ \log K(EDTA)$
	$-(\log EC_{50}) = 0.212 + 1.37\ X_m^2 r$
	$-(\log EC_{50}) = -1.04 + 0.445\ \Delta IP$
	$-(\log EC_{50}) = -0.149 + 1.48\ X_m^2 r$
	$-(\log EC_{50}) = -1.74 + 2.91\ X$
	$-(\log EC_{50}) = 3.58 + 0.881\ E°$
	$-(\log EC_{50}) = 2.01 - 3.64\ IR$
	$-(\log EC_{50}) = 1.87 - 0.0222\ AR$
	$-(\log EC_{50}) = -3.83 + 0.552\ \log K(ATP)$
	$-(\log EC_{50}) = -3.30 + 2.33\ \log K(AMP)$
	$-(\log EC_{50}) = 1.23 + 1.47\ Y'$
	$-(\log EC_{50}) = -3.14 + 2.35\ \log K(AMP\text{-}2)$
	$-(\log EC_{50}) = -3.55 + 2.59\ \log K(AMP\text{-}3)$
	$-(\log EC_{50}) = -0.475 + 2.19\ \log K(Ac)$

Roy et al. 2009; Sacan et al. 2009; Mendes et al. 2010; Su et al. 2010). Analysis of these studies produced 14 references associated with 183 QSARs for predicting cation bioconcentration potential, biosorption capacity, binding strength, and toxicity (Table 5.4). These QSARs are discussed in more detail in Sections 5.2, 5.3, 5.5, 5.6, and 5.7. The acronyms used in Chapter 5 and associated tables are defined in the Chapter 5 appendix.

5.2 MOST COMMON PHYSICOCHEMICAL PROPERTIES USED TO PREDICT CATION TOXICITY

Walker et al. (2003) cited more than 100 studies that described the relationships among 24 properties of cations and their toxic actions. However, Walker et al. (2003) did not provide any of the QSARs used to describe those relationships. The purpose of this chapter section is to discuss the most commonly used physicochemical properties in QSARs used to predict cation toxicity.

5.2.1 Standard Electrode Potential

The standard electrode potential is a characteristic of bulk metal reflecting the capacity of a metal to generate positive ions in aqueous solution. Eight studies used standard electrode potential ($E°$) to predict cation toxicity (Table 5.5). All these studies were

TABLE 5.2
Chemical Abstract Service Registry Numbers (CASRNs) of Metals and Metal Ions That Appeared in Table 3 of Walker et al. (2003)

Metals	CASRNs	Ions	Ion CASRNs for Metals That Appeared More Than 10 Times in Table 3 of Walker et al. (2003)	Number of Times Appearing in Table 3 of Walker et al. (2003)
Copper	7440-50-8	Cuprous ion	17493-86-6	39
		Copper, ion (Cu2+)	15158-11-9	
Zinc	7440-66-6	Zinc cation	23713-49-7	38
Cadmium	7440-43-9	Cadmium, ion (Cd2+)	22537-48-0	37
Nickel	7440-02-0	Nickel, ion (Ni1+)	14903-34-5	36
Manganese	7439-96-5	Manganese, ion (Mn2+)	16397-91-4	35
		Manganese (3+), ion	14546-48-6	
		Manganese, ion (Mn1−)	22325-60-6	
		Manganese, ion (Mn(7+))	23317-90-0	
		Manganate (MnO4(2−))	14333-14-3	
		Permanganic acid	14333-13-2	
Cobalt	7440-48-4	Cobalt, ion (Co 2)	22541-53-3	35
		Cobalt, ion (Co3+)	22541-63-5	
Mercury	7439-97-6	Mercury, ion (Hg1+)	22542-11-6	33
		Mercury, ion (Hg2+)	14302-87-5	
Lead	7439-92-1	Lead, ion (Pb2+)	14280-50-3	33
		Lead, ion (Pb4+)	15158-12-0	
Magnesium	7439-95-4	Magnesium cation	22537-22-0	30
Strontium	7440-24-6	Strontium, ion (Sr+2)	22537-39-9	25
Silver	7440-22-4	Silver, ion (Ag1+)	14701-21-4	25
Potassium	7440-09-7	Potassium, ion (K1+)	24203-36-9	25
Calcium	7440-70-2	Calcium (2+)	17787-72-3	23
		Calcium cation	14127-61-8	
Sodium	7440-23-5	Sodium cation	17341-25-2	22
Barium	7440-39-3	Barium, ion (Ba2+)	22541-12-4	22
Iron	7439-89-6	Iron, ion (Fe2+)	15438-31-0	21
		Iron, ion (Fe3+)	20074-52-6	
Chromium	7440-47-3	Chromium, ion (Cr3+)	16065-83-1	18
		Chromate (CrO42−)	13907-45-4	
		Bichromate	13907-47-6	
Lithium	7439-93-2	Lithium, ion (Li1+)	17341-24-1	14
Beryllium	7440-41-7	Beryllium, ion (Be2+)	22537-20-8	12
Aluminum	7429-90-5	Aluminum, ion (Al3+)	22537-23-1	12
Lanthanum	7439-91-0	Lanthanum, ion (La3+)	16096-89-2	11
Gold	7440-57-5	Gold, ion (1+)	20681-14-5	11
		Gold, ion (3+)	16065-91-1	

(Continued)

TABLE 5.2 (*Continued*)
Chemical Abstract Service Registry Numbers (CASRNs) of Metals and Metal Ions That Appeared in Table 3 of Walker et al. (2003)

Metals	CASRNs	Ions	Ion CASRNs for Metals That Appeared More Than 10 Times in Table 3 of Walker et al. (2003)	Number of Times Appearing in Table 3 of Walker et al. (2003)
Tin	7440-31-5	Tin, ion (Sn2+)	22541-90-8	10
		Tin, ion (Sn4+)	22537-50-4	
Palladium	7440-05-3			8
Cesium	7440-46-2			8
Platinum	7440-06-4			7
Yttrium	7440-65-5			6
Rubidium	7440-17-7			5
Indium	7440-74-6			5
Arsenic	7440-38-2			5
Thallium	7440-28-0			4
Scandium	7440-20-2			4
Antimony	7440-36-0			3
Titanium	7440-32-6			2
Thorium	7440-29-1			2
Rhodium	7440-16-6			2
Niobium	7440-03-1			2
Vanadium	7440-62-2			1
Tungsten	7440-33-7			1
Tantalum	7440-25-7			1
Ruthenium	7440-18-8			1
Osmium	7440-04-2			1
Germanium	7440-56-4			1
Gallium	7440-55-3			1
Gadolinium	7440-54-2			1
Cerium	7440-45-1			1

Source: Appeared in J.D. Walker, M. Enache, and J.C. Dearden. "Quantitative Cationic Activity Relationships for Predicting Toxicity of Metals." *Environ. Toxicol. Chem.* 22 (2003):1916–1935, T-3.

previously reviewed by Walker et al. (2003), but the QSARs of Lewis et al. (1999) and Enache et al. (2003) were not discussed by Walker et al. Lewis et al. (1999) used the intraperitoneal injection dosing data from Venugopal and Luckey (1978) to develop the QSAR in Table 5.1 to predict the LD_{50} value of 30 cations to mice: $pT50 = 0.41$ $E° + 3.72$, $r^2 = 0.73$, $SE = 0.32$, $p = 0.0001$, $F = 75.0$ and $n = 30$, where SE is the standard error, r^2 is the coefficient of determination, F is the variance ratio (F test), p is the

TABLE 5.3

2004 to 2011 Studies to Predict the Bioconcentration, Biosorption, Binding, or Toxicity of Metal Ions

Author(s)	Year	Title	Citation	Metal ions	Summary[a]
Hickey	2005	Estimation of inorganic species aquatic toxicity	In *Techniques in Aquatic Toxicology: Volume 2.* CRC Press	Ag^{1+}, Al^{3+}, As^{3+}, B^{3+}, Ba^{2+}, Be^{2+}, Bi^{3+}, Ca^{2+}, Cd^{2+}, Ce^{3+}, Co^{2+}, Cr^{3+} Cr^{6+}, Cs^{1+}, Cu^{2+}, Fe^{2+}, Fe^{3+}, Hg^{2+}, K^{1+}, Li^{1+}, Mg^{2+}, Mn^{2+}, Mo, Na^{1+}, Ni^{2+}, Pb^{2+}, Pt^{2+}, Rb^{1+}, Sb^{3+}, Sn^{2+}, Sr^{2+}, Th^{6+}, Tl^{1+}, U^{6+}, V^{5+}, Yb^{3+}, Zn^{2+}	Acute aquatic toxicities of neutral inorganic species were estimated for three species using existing linear solvation energy relationship (LSER) equations developed for neutral organic compounds. Use of the whole inorganic species addresses ionic charge and elemental valence in an uncomplicated manner. The general LSER equation was Log(Property) = $mV_i/100 + s\pi^* + b\beta_m + a\alpha_m$, where V_i is the intrinsic (van der Waals) molecular volume, π^* is the solute ability to stabilize a neighboring charge or dipole by nonspecific dielectric interactions, and β_m and α_m are the solute ability to accept or donate a hydrogen in a hydrogen bond. The coefficients m, s, b, and a are constants for a particular set of conditions, determined by multiple linear regression of the LSER variable values for a series of chemicals with the measured value for a particular chemical property. This equation was used to estimate cation solubility and predict cation toxicity to *Vibrio fischeri, Daphnia magna,* and *Leuciscus idus melanotus.*
Can and Jianlong	2007	Correlating metal ionic characteristics with biosorption capacity using QSAR model	*Chemosphere* 69:1610–1616	Ag^+, Cd^{2+}, Co^{2+}, Cr^{3+}, Cs^+, Cu^{2+}, Ni^{2+}, Pb^{2+}, Sr^{2+}, Zn^{2+}	The relationship between metal ionic characteristics and the maximum biosorption capacity (q_{max}) was established using QSAR models based on the classification of metal ions (soft, hard, and borderline ions). Ten kinds of metal were selected and the waste biomass of *Saccharomyces cerevisiae* obtained from a local brewery was used as biosorbent. Eighteen parameters of physiochemical characteristics of metal ions were selected and correlated with q_{max}. The suggestion was made that classification of metal ions could improve the QSAR models and different characteristics were significant in correlating with q_{max}, such as polarizing power Z^2/r or the first hydrolysis constant $logK_{OH}$ or ionization potential IP.

(Continued)

TABLE 5.3 (Continued)
2004 to 2011 Studies to Predict the Bioconcentration, Biosorption, Binding, or Toxicity of Metal Ions

Author(s)	Year	Title	Citation	Metal Ions	Summary[a]
Chen and Wang	2007	Influence of metal ionic characteristics on their biosorption capacity by *Saccharomyces cerevisiae*	*Appl. Microbiol. Biotechnol.* 74:911–917	Ag^+, Cd^{2+}, Co^{2+}, Cr^{3+} Cs^+, Cu^{2+}, Ni^{2+}, Pb^{2+}, Sr^{2+}, Zn^{2+}	The influence of metal ionic characteristics on their biosorption capacity was analyzed using QSAR models. The waste biomass of *Saccharomyces cerevisiae* was used as biosorbent to adsorb 10 metal ions, and their maximum biosorption capacity (q_{max}) was predicted by the models. q_{max} values were correlated with 22 metal ionic characteristics. Among these, covalent index ($X_m^2 r$) was correlated well with q_{max} for all metal ions tested in the following equation: $q_{max} = 0.029 + 0.061 (X_m^2 r)$ ($R^2 = 0.70$). The biomass was shown to preferentially absorb soft ions, then borderline ions, and last hard ions. Classification of metal ions, for divalent ion or for soft-hard ion, could improve the linear relationship ($R^2 = 0.89$).
Kinraide and Yermiyahu	2007	A Scale of metal ion binding strengths correlating with ionic charge: Pauling electronegativity, toxicity, and other physiological effects	*J. Inorganic Biochem.* 101:1201–1213	Ag^{1+}, Al^{3+}, Ba^{2+}, Be^{2+}, Bk^{3+}, Ca^{2+}, Cd^{2+}, Ce^{3+}, Cf^{3+}, Cm^{3+}, Co^{2+}, Cr^{3+}, Cs^{1+}, Cu^{2+}, Dy^{3+}, Er^{3+}, Eu^{3+}, Fe^{2+}, Fe^{3+}, Ga^{3+}, Gd^{3+}, Hg^{2+}, Ho^{3+}, In^{3+}, K^{1+}, La^{3+}, Li^{1+}, Lu^{3+}, Mg^{2+}, Mn^{2+}, Mo, Na^{1+}, Nd^{3+}, Ni^{2+}, Pb^{2+}, Pm^{3+}, Pr^{3+}, Rb^{1+}, Sc^{3+}, Sm^{3+}, Sr^{2+}, Tb^{3+}, Th^{4+}, Tl^{1+}, Tm^{3+}, U^{4+}, Y^{3+}, Yb^{3+}, Zn^{2+}, Zr^{4+}	A scale of metal ion binding strength to inorganic and organic ligands was developed, which was closely related to ion charge and Pauling electronegativity. The toxic effects of metal ions on wheat root elongation correlated well with metal ion binding strength ($R^2 = 0.835$); the addition of ion softness to the equation increased the strength of the correlation ($R^2 = 0.950$). Correlations were also presented between metal ion binding strength and growth of sunflower callus tissue ($R^2 = 0.713$), 24-hour survival of *Caenorhabditis elegans* ($R^2 = 0.574$), and bioluminescence response of *Vibrio fischeri* ($R^2 = 0.508$).

Author	Year	Title	Reference	Ions	Description
Walker, Enache, and Dearden,	2007	Quantitative cationic activity relationships for predicting toxicity of metal ions from physicochemical properties and natural occurrence levels	*QSAR Combin. Sci.* 26:522–527	Ag^+, Al^{3+}, Ba^{2+}, Cd^{2+}, Co^{2+}, Cu^{2+}, Fe^{3+}, Pb^{2+}, Li^+, Mn^{2+}, Ni^{2+}, Zn^{2+}	Quantitative cationic activity relationships (QCARs) were developed to predict the toxicity of metal ions from physicochemical properties and natural occurrence levels. In vivo toxicity data for different concentrations of nitrate salts of 17 metal ions were developed based on germination of sunflower seeds (F.1. *Helianthus annuus* Sunspot) in distilled water. The EC50 data were reported as the concentration giving 50% inhibition of radicle growth one day after emergence. Stepwise regression of the toxicity data produced correlations with some physicochemical properties and natural occurrence levels. For physicochemical properties, good results were obtained with the density of the elements, enthalpy of formation of metal sulfides, and the stability constants of metal ions with sulfate (r^2adj.$=0.72$–0.81). For natural occurrence levels, good results were obtained with metal concentrations in soil, the median elemental composition of soils and the calculated mean of the elemental content in land plants (r^2adj.$=0.69$–0.83).
Van Kolck, Huijbregts, Veltman, and Hendriks	2008	Estimating bioconcentration factors, lethal concentrations and critical body residues of metals in the mollusks *Perna viridis* and *Mytilus edulis* using ion characteristics	*Environ. Toxicol. Chem.* 27:272–276	Ag^+, Cd^{2+}, Co^{2+}, Cr^{3+}, Cs^{1+}, Cu^{2+}, Hg^{2-}, Pb^{2+}, Zn^{2+}	Regression equations were developed between individual metal ion characteristics and toxicity data (LC_{50}s) for two species of mollusks. *Perna viridis* and *Mytilus edulis*. The four ion characteristics studied were: the covalent index, softness index, hydrolysis constant, and ionic index. A statistically significant relationship was observed between the covalent index and toxicity to each species of mollusks ($R^2=0.74$ – 0.79; $P=0.03$ – 0.04), suggesting that covalent index is a good predictor of toxicity. Regressions between the remaining metal ion characteristics and toxicity values did not produce significant relationships.

(Continued)

TABLE 5.3 (Continued)
2004 to 2011 Studies to Predict the Bioconcentration, Biosorption, Binding, or Toxicity of Metal Ions

Author(s)	Year	Title	Citation	Metal Ions	Summary[a]
Workentine, Harrison, Stenroos, Ceri, and Turner	2008	*Pseudomonas fluorescens'* view of the periodic table	*Environ. Microbiol.* 10:238–250	The toxicity and physicochemical characteristics of 44 metals were used, representing every group in the periodic table with the exception of halogens and noble gases.	The toxicity of 44 metals to the biofilms and planktonic cells of *Pseudomonas fluorescens* was measured and expressed as minimum inhibitory concentration, minimum bactericidal concentration, and minimum biofilm eradication concentration. Linear regression analyses were conducted to determine the relationships between the measured toxicity values and the following physicochemical parameters: standard reduction-oxidation potential, electronegativity, the solubility product of the corresponding metal–sulfide complex, the Pearson softness index, electron density, and the covalent index. Each of the physicochemical parameters was significantly ($P < 0.05$) correlated with one or more of the toxicity measurements. Heavy metal ions were found to show the strongest correlations between toxicity and physicochemical parameters.
Khangarot and Das	2009	Acute toxicity of metals and reference toxicants to a freshwater ostracod, *Cypris subglobosa* Sowerby 1840 and correlation to EC_{50} values of other test models	*J. Haz. Materials* 172:641–649	Ag^{1+}, Al^{3+}, As^{3+}, B^{3+}, Ba^{2+}, Be^{2+}, Bi^{3+}, Ca^{2+}, Cd^{2+}, Co^{2+}, Cr^{3+}, Cr^{6+}, Cu^{2+}, Fe^{3+}, Hg^{2+}, K^{1+}, La^{3+}, Li^{1+}, Mg^{2+}, Mn^{2+}, Mo^{2+}, Na^{1+}, Ni^{2+}, Os^{4+}, Pb^{2+}, Pd^{4+}, Pt^{4+}, Sb^{3+}, Se^{2+}, Sn^{2+}, Sr^{2+}, Te^{2+}, UO_2^{2+}, W^{6+}, Zn^{2+}, Zr^{2+}	The ostracod *Cypris subglobosa* Sowerby 1840 static bioassay test on the basis of 50% of immobilization (EC_{50}) was used to measure the toxicity of 36 metals and metalloids and 12 reference toxicants. EC_{50} values of this study revealed positive linear relationship with 7 other species. Coefficients of determination (r^2) for 17 physicochemical properties of metals or metal ions and EC_{50} values were examined by linear regression analysis. The electronegativity, ionization potential, melting point, solubility product of metal sulfides (pK_{sp}), softness parameter, and some other physicochemical characteristics were significantly correlated with EC_{50} values of metals to *C. subglobosa*. The reproducibility of the toxicity test was determined using 12 reference toxicants.

Author	Year	Title	Source	Data/Ions	Description
Kinraide	2009	Improved scales for metal ion softness and toxicity	*Environ. Toxicol. Chem.* 28:525–533	Data were compiled for 92 metal ions, but only 51 ions were used to develop the softness consensus scale.	A strong correlation was observed between a newly developed scale of metal ion softness and a scale of toxicity. The scale of metal ion softness was developed by normalizing and averaging the values from 8 previously developed scales of metal ion softness. The scale of toxicity was obtained by normalizing and averaging toxicity values from 10 previously published toxicity studies that used many different taxa. An equation was developed that predicted toxicity based on water softness and ion charge ($R^2 = 0.923$).
Lepădatu, Enache, and Walker	2009	Toward a more realistic QSAR approach to predicting metal toxicity	*QSAR Combin. Sci.* 28:520–525	Ag^{1+}, Al^{3+}, Ba^{2+}, Ca^{2+}, Cd^{2+}, Cu^{2+}, Co^{2+}, Fe^{3+}, K^{1+}, La^{3+}, Li^{1+}, Mg^{2+}, Mn^{2+}, Na^{1+}, Ni^{2+}, Pb^{2+}, Zn^{2+}	The paper presented molecular descriptors representing the electronegativity of occupied molecular orbital (OMO) and unoccupied molecular orbital (UMO) quantum molecular states that could be used to obtain information on the mechanism of electron transfer between metal compound and biological receptor. The metal descriptors that were used suggested that the (s,p) and d^N metal ions have different mechanisms of interaction with the receptor to explain why the correlation between activity–descriptor is poor or practically nonexistent when all metal ions are analyzed together, irrespective of their valence shell.
Roy, Giri, and Chattaraj	2009	Arsenic toxicity: An atom counting and electrophilicity-based protocol	*Molecular Diversity* 13:551–556	Training set to predict As^{3+} and As^{5+} toxicity included 10 of the following: Ca^{2+}, Co^{2+}, Cr^{3+}, Cu^{2+}, Fe^{3+}, K^+, Li^+, Mg^{2+}, Mn^{2+}, Na^+, Ni^{2+}, Zn^{2+}	Several equations were developed to predict the toxicity of arsenic and its derivatives using one global chemical reactivity parameter (either electrophilicity or the number of nonhydrogen atoms) and two local parameters (philicity and atomic charge on the arsenic atom). A training set of metal ions was used, along with a set of experimental toxicity values of these ions in the roundworm, *Caenorhabditis elegans*, to develop equations to predict the toxicity of arsenic ($R^2 = 0.865$ to 0.963).

(Continued)

TABLE 5.3 (Continued)
2004 to 2011 Studies to Predict the Bioconcentration, Biosorption, Binding, or Toxicity of Metal Ions

Author(s)	Year	Title	Citation	Metal Ions	Summary[a]
Sacan, Cecen, Erturk, and Semerci	2009	Modelling the relative toxicity of metals on respiration of nitrifiers using ion characteristics	*SAR QSAR Environ. Res.* 20:727–740	Ag^+, Cd^{2+}, Co^{2+}, Cu^{2+}, Cr^{3+}, Hg^{2+}, Ni^{2+}, Zn^{2+},	The effects of 8 transition metals were studied in a nitrifying system to investigate the relationship between the ionic characteristics of metals and their toxicity to nitrifiers. The cumulative oxygen consumption and the cumulative carbon dioxide production were monitored throughout each respirometric batch run to determine the toxicity of metals to nitrifiers. Several QCAR models were developed on the basis of these different toxicity endpoints using quantum chemical descriptors.
Zamil, Ahmad, Choi, Park, and Yoon	2009	Correlating metal ionic characteristics with biosorption capacity of *Staphylococcus saprophyticus* BMSZ711 using QICAR model	*Bioresource Technol.* 100:1895–1902	Cd^{2+}, Cr^{3+}, Co^{2+}, Cu^{2+}, Hg^{2+}, Pb^{2+}, Ni^{2+}, K^+, Zn^{2+}	QICAR was used for correlating metal ionic properties with maximum biosorption capacity (q_{max}). Heat inactivated biomass of *Staphylococcus saprophyticus* BMSZ711 was studied for biosorption of nine metal ions. Influence of contact time and initial pH was evaluated. q_{max} was determined by Langmuir isotherm. q_{max} values were modeled with 20 metal ionic characteristics. Classification of metal ions according to valence or soft/hard characteristics. Classification of metal ions according to valence or soft/hard improved QICARs modeling and more characteristics significantly correlated with q_{max} which revealed that covalent bonding played major role in biosorption of soft metal ions and ionic bonding for borderline and hard ions.
Mendes, Bastos, and Stevani	2010	Prediction of metal cation toxicity to the bioluminescent fungus *Gerronema viridilucens*	*Environ. Toxicol. Chem.* 29:2177–2181	Ag^{1+}, Ba^{2+}, Ca^{2+}, Cd^{2+}, Co^{2+}, Cs^{1+}, Cu^{2+}, Fe^{2+}, Hg^{2+}, K^{1+}, Li^{1+}, Mg^{2+}, Mn^{2+}, Na^{1+}, Ni^{2+}, Pb^{2+}, Sr^{2+}, Zn^{2+}	A correlation between the physicochemical properties of mono- [Li(I), K(I), Na(I)] and divalent [Cd(II), Cu(II), Mn(II), Ni(II), Co(II), Zn(II), Mg(II), Ca(II)] metal cations and their toxicity (evaluated by the free ion median effective concentration, EC_{50F}) to the naturally bioluminescent fungus *Gerronema viridilucens* has been studied using the QICAR approach. Among the 11 ionic parameters used in the current study, a univariate model based on the covalent index $(X_m^2 r)$ proved to be the most adequate for prediction of fungal metal toxicity evaluated by the logarithm of free ion median effective log EC_{50F}. Additional two- and three-variable models were also tested and proved

less suitable to fit the experimental data. These results indicate that covalent bonding is a good indicator of metal inherent toxicity to bioluminescent fungi. Furthermore, the toxicity of additional metal ions [Ag(I), Cs(I), Sr(II), Ba(II), Fe(II), Hg(II), and Pb(II)] to *G. viridilucens* was predicted, and Pb was found to be the most toxic metal to this bioluminescent fungus.

Authors	Year	Title	Source	Metal ions	Description
Su, Zhao, Yuan, Mu, Wang, and Yan	2010	Evaluation of combined toxicity of phenols and lead to *Photobacterium phosphoreum* and quantitative structure-activity relationships	*Bull. Environ. Contam. Toxicol.* 84:311–314	Lead	Each of nine phenols was combined individually with three concentrations of lead and these combinations were tested in *Photobacterium phosphoreum*. The result indicated that the combined toxicity of lead and phenol is not only dependent on the Pb concentrations but also on the positions of substituted groups of phenols. QSARs were built from the combined toxicity and physicochemical descriptors of phenols in the different Pb concentrations. The combined toxicity was related to water solubility and the third order molecular connectivity index ($3X$) in low Pb concentration, to solute excess molar refractivity (E) and ionization constant (pKa) in medium Pb concentration, and to dipolarity/polarizability (S) in high Pb concentration.
Zhou, Li, Peijnenburg, Ownby, Hendriks, Wang, and Li	2011	A QICAR approach for quantifying binding constants for metal-ligands complexes	*Ecotox. Environ. Safety* 74:1036–1042	Ag^{1+}, Ca^{2+}, Cd^{2+}, Co^{2+}, Cu^{2+}, K^{1+}, Mg^{2+}, Na^{1+}, Ni^{2+}, Pb^{2+}, Zn^{2+}	To find the necessary variety of metals and species to develop QICARs, this study evaluated 50 conditional binding constants (K) for biotic ligands from 27 studies. Species were included if the conditional binding constants were available for at least six different metals. For each species, the K values determined for similar exposure conditions and response types (i.e., immobilization, root growth inhibition, cocoon production, mortality) were combined to calculate mean ligand-specific conditional binding constants (K). log K QICARs were developed for the Pearson and Mawby softness parameter (σp), the covalent index ($\chi 2m\,r$), the absolute value of the first hydrolysis constant (log KOH), and the ionic index $Z2/r$) for five species. log K QICARs were developed for *Daphnia magna*, fathead minnows, rainbow trout, barley, and earthworms.

a The references are summarized with emphasis on QSARs, as they relate to bioavailability, bioconcentration, biosorption, binding, or toxicity of metal ions.

TABLE 5.4
QSARs from 2004–2011 References

Reference	Endpoint	QSAR
Hickey (2005)	Water solubility	$\log(Sw) = 0.05 - 5.85V_f/100 + 1.09\ \pi^* + 5.23\beta$
	Toxicity to the bioluminescent bacterium, *Vibrio fischeri* Beijerinck 1889	$\log(EC50) = 7.49 - 7.39V_f/100 - 1.38\ \pi^* + 3.70\beta - 1.66\alpha$
	Toxicity to the water flea, *Daphnia magna*	$\log(EC50) = 4.18 - 4.73V_f/100 - 1.67\ \pi^* + 1.48\beta - 0.93\alpha$
	Toxicity to the Golden Orfe, *Leuciscus idus melanotus*	$\log(LC50) = 2.90 - 5.71\ V_f/100 - 0.92\ \pi^* + 4.36\beta - 1.27\alpha$
Can and Jianlong (2007)	Cd^{2+}, Co^{2+}, Cr^{3+}, Cu^{2+}, Ni^{2+}, Zn^{2+} biosorption to the yeast *Saccharomyces cerevisae*	$qmax = 0.33 - 0.020(\log KOH)$
		$qmax = 0.38 - 0.93\left(X_m^2 r\right)$
		$qmax = 0.076 + 0.011(Z2/r)$
		$qmax = -0.018 + 0.008(IP)$
		$qmax = 0.065 + 0.022(Z/r2)$
		$qmax = 0.014 + 0.11(Z/AR2)$
		$qmax = 0.021 + 0.045(Z/r)$
		$qmax = -0.03 + 0.11(Z/AR)$
		$qmax = 0.28 - 0.050(Z2/r)$
	Cd^{2+}, Co^{2+}, Cr^{3+}, Cs^{+}, Cu^{2+}, Sr^{2+}, Zn^{2+} biosorption to *S. cerevisae*	$qmax = -0.008 + 0.077(OX)$
		$qmax = 0.28 - 0.013(\log KOH)$
		$qmax = 0.084 + 0.011(Z2/r)$
		$qmax = 1.90 + 0.008(AN/\Delta IP)$

$q_{max} = 0.050 + 0.006(IP)$

$q_{max} = 0.080 + 0.02(Z^2/r)$

$q_{max} = 0.069 + 0.078(Z/AR^2)$

$q_{max} = 0.055 + 0.037(Z/r)$

$q_{max} = 0.045 + 0.073(Z/AR)$

$q_{max} = 0.102 + 0.023(Z^{*2}/r)$

$q_{max} = 0.145 - 0.004(LogKOH) + 0.008(Z^2/r)$

$q_{max} = 0.086 + 0.011(Z^2r) - 0.0003(AN/\Delta IP)$

$q_{max} = -0.03 + 0.008(IP) + 0.006(AN/\Delta IP)$

$q_{max} = 0.03 - 0.018(LogKOH) + 0.005(AN/\Delta IP)$

Cd2+, Co2+, Cr3+,Cs+, Cu2+, Ni2+, Sr2+, Zn2+ biosorption to *S. cerevisae*

$q_{max} = -0.013 + 0.077(OX)$

$q_{max} = 0.27 - 0.013(logK_{OH})$

$q_{max} = 0.079 + 0.011(Z^2/r)$

$q_{max} = 0.047 + 0.006(IP)$

$q_{max} = 0.069 + 0.07(Z/AR^2)$

$q_{max} = 0.053 + 0.035(Z/r)$

$q_{max} = 0.044 + 0.069(Z/AR)$

$q_{max} = 0.10 + 0.02(Z^2/r)$

Ag$^+$, Cd^{2+}, Co^{2+}, Cr^{3+}, Cu^{2+}, Ni^{2+}, Pb^{2+}, Zn^{2+} biosorption to *S. cerevisae*

$q_{max} = 0.037 + 0.004(AN)$

$q_{max} = -0.16 + 0.44r$

$q_{max} = 0.017 + 0.064\left(X_m^2 r\right)$

$q_{max} = 0.087 + 0.028(AN/\Delta IP)$

$q_{max} = -0.71 + 0.68(AR)$

$q_{max} = 0.47 - 14.08(AR/AW)$

(Continued)

TABLE 5.4 (*Continued*)
QSARs from 2004–2011 References

Reference	Endpoint	QSAR
Chen and Wang (2007)	Ag^+, Cd^{2+}, Co^{2+}, $Cr^{3+}Cs^+$, Cu^{2+}, Ni^{2+}, Pb^{2+}, Sr^{2+}, Zn^{2+} biosorption to *S. cerevisae*	qmax = $0.029 + 0.061(X_m^2 r)$ qmax = $-0.127 + 0.023(Z^*)$ qmax = $0.47 - 14.9(AR/AW)$ qmax = $-0.026 + 0.005(AN)$ qmax = $0.029 + 0.061(X_m^2 r)$ qmax = $0.039 + 0.\,026(AN/\Delta IP)$ qmax = $0.002 + 0.\,002(AW)$ qmax = $0.457 - 15.56(AR/AW)$ qmax = $-0.024 + 0.028(Z^*)$ qmax = $-0.018 + 0.05(X_m^2 r) - 0.01(AN/\Delta IP)$
	Cd^{2+}, Co^{2+}, Cr^{3+}, Cu^{2+}, Ni^{2+}, Pb^{2+}, Sr^{2+}, Zn^{2+} biosorption to *S. cerevisae*	
Kinraide and Yermiyahu (2007)	Binding strength	$\log a_{PM,M} = 1.60 - 2.41 \exp [0.238(\text{Scale values})]$
Walker et al. (2007)	Toxicity to sunflower seeds (*Helianthus annuus* F.1. "Sunspot")	$-(\log EC_{50}) = 0.756 + 0.000140\ \rho$ $-(\log EC_{50}) = 0.534 + 0.000157\ \rho - 0.00166\ \Delta H_s$ $-(\log EC_{50}) = 0.636 + 0.000209\ \rho - 0.00367\ \Delta H_s - 0.242 \log K_1(\text{sulphate})$ $-(\log EC_{50}) = 1.17 - 0.254 \log M_{soil}$ $-(\log EC_{50}) = 0.916 - 0.337 \log M_{soil} + 0.000014\ \text{mg X/kg soil}$ $-(\log EC_{50}) = 1.04 - 0.302 \log M_{soil} + 0.000013\ \text{mg X/kg soil} - 0.000025\ \text{Land plants}$
Van Kolck et al. (2008)	Bioconcentration Factor (BCF) for the mussel, *Mytilus edulis*	$\text{Log BCF} = 1.4 - 0.5 X_m^2 r$ $\text{Log BCF} = 2.0 + 0.1 \log|K_{OH}|$ $\text{Log BCF} = 2.8 + 0.1\ \sigma_p$

Bioconcentration Factor (BCF) for the mussel, *Perna viridis*	$\text{Log BCF} = 3.6 + 0.1\, Z^2/r$ $\text{Log BCF} = 1.1 - 1.0\, X_m^2 r$ $\text{Log BCF} = 2.7 + 0.2\, \log|K_{OH}|$ $\text{Log BCF} = 5.2 - 6.5\sigma_p$ $\text{Log BCF} = 5.6 - 0.3 Z^2/r$
Toxicity to *Mytilus edulis*	$\text{Log LC50} = 2.8 - 0.7 X_m^2 r$ $\text{Log LC50} = 1.1 + 0.4\, \log|K_{OH}|$ $\text{Log LC50} = 1.0 + 0.24\, \sigma_p$ $\text{Log LC50} = -0.19 + 0.25 Z^2/r$
Toxicity to *Perna viridis*	$\text{Log LC50} = 3.6 - 0.8\, X_m^2 r$ $\text{Log LC50} = 0.99 + 0.05\, \log|K_{OH}|$ $\text{Log LC50} = -1.2 + 26\sigma_p$ $\text{Log LC50} = 0.6 + 0.13 Z^2/r$
Kinraide (2009) Generic Toxicity	$T_{\text{Con obs}} = a\sigma_{\text{Con comp}} + b\sigma_{\text{Con comp}} Z + cZ$
Lepădatu et al. (2009) Toxicity to sunflower seeds (*Helianthus annuus* F.1. "Sunspot") for s, p metals	$-\log\text{EC50} = 1.88 - 1.65\ \text{HELH}$ $-\log\text{EC50} = 2.74 - 0.246\ \text{HEO}$ $-\log\text{EC50} = 1.64 + 0.081\ \text{HEM}$ $-\log\text{EC50} = 2.74 - 0.049\ \text{O-EO}$ $-\log\text{EC50} = 1.64 + 0.016\ \text{O-EM}$ $-\log\text{EC50} = 2.44 - 0.033\ \text{U-ELH}$ $-\log\text{EC50} = 3.20 - 0.272\ \text{U-EO}$ $-\log\text{EC50} = 0.979 + 0.031\ \text{U-EM}$

(Continued)

TABLE 5.4 (Continued)
QSARs from 2004–2011 References

Reference	Endpoint	QSAR
	Toxicity to sunflower seeds (*Helianthus annuus* F.1. Sunspot) for dN-metals	$-\log EC50 = 1.38 + 1.55$ HELH
		$-\log EC50 = 1.81 - 0.021$ HEO
		$-\log EC50 = -0.39 + 0.152$ HEM
		$-\log EC50 = 1.75 + 0.009$ O-EO
		$-\log EC50 = 1.49 + 0.004$ O-EM
		$-\log EC50 = 1.00 + 0.033$ U-ELH
		$-\log EC50 = 0.975 + 0.124$ U-EO
		$-\log EC50 = 2.06 - 0.012$ U-EM
Roy et al. (2009)	Toxicity to the soil nematode, *Caenorhabditis elegans*	$\text{Calc.LC50} = -0.0181 \text{AN} + 0.5272$
		$\text{Calc.LC50} = -7.1 \times 10{-}04 \ \omega + 0.2987$
Sacan et al. (2009)	Toxicity to nitrifying bacteria	$pTO_2 = -1.18(\pm 0.14) \ Z + 3.47(\pm 0.29)$
		$pTO_2 = 0.07(\pm 0.02) \ E_{\text{LUMO(g)}} + 2.61(\pm 0.38)$
		$pTO_2 = 0.08(\pm 0.01) \ E_{\text{HOMO(g)}} + 3.74(\pm 0.47)$
		$pTO_2 = -0.10(\pm 0.02) \ IP + 2.87(\pm 0.32)$
		$pTO_2 = -0.14(\pm 0.04) \ Z^2/r + 1.92(\pm 0.25)$
		$pTO_2 = 0.06(\pm 0.01) \ E_{\text{PSS}} + 2.28(\pm 0.29)$
		$pTO_2 = 0.08(\pm 0.01) \ \mu_{(g)} + 3.39(\pm 0.32)$
		$pTO_2 = 0.22(\pm 0.06) \ \mu_{(aq)} + 2.96(\pm 0.52)$
		$pTCO_2 = -1.17(\pm 0.21) \ Z + 3.26(\pm 0.43)$
		$pTCO_2 = 0.06(\pm 0.02) \ E_{\text{LUMO(g)}} + 2.36(\pm 0.47)$
		$pTCO_2 = 0.08(\pm 0.02) \ E_{\text{HOMO(g)}} + 3.62(\pm 0.51)$
		$pTCO_2 = -0.09(\pm 0.02) \ IP + 2.63(\pm 0.43)$
		$pTCO_2 = -0.14(\pm 0.04) \ Z^2/r + 1.70(\pm 0.30)$

$pTCO_2 = 0.05(\pm0.01)\, E_{PSS} + 2.07(\pm0.35)$

$pTCO_2 = 0.08(\pm0.02)\, \mu_{(g)} + 3.18(\pm0.45)$

$pTCO_2 = 0.22(\pm0.07)\, \mu_{(aq)} + 2.72(\pm0.60)$

$pTO_2\text{-labile} = 0.27(\pm0.05)^a\, \mu_{aq} + 3.70(\pm0.47)$

$pTO_2\text{-labile} = 0.06(\pm0.01)\, E_{PSS} + 2.71(\pm0.27)$

$pTO_2\text{-labile} = -0.15(\pm0.05)\, Z^2/r + 2.29(\pm0.33)$

$pTO_2\text{-labile} = 0.08(\pm0.02)\, E_{HOMO(g)} + 3.93(\pm0.80)$

$pTO_2\text{-labile} = 0.07(\pm0.02)\, E_{LUMO(g)} + 3.06(\pm0.45)$

$pTO_2\text{-labile} = 0.09(\pm0.02)\, \mu_{(g)} + 3.75(\pm0.52)$

$pTCO_2\text{-labile} = 0.26(\pm0.07)\, \mu_{aq} + 3.41(\pm0.58)$

$pTCO_2\text{-labile} = 0.06(\pm0.01)\, E_{PSS} + 2.48(\pm0.33)$

$pTCO_2\text{-labile} = -0.15(\pm0.05)\, Z^2/r + 2.06(\pm0.37)$

$pTCO_2\text{-labile} = 0.08(\pm0.02)\, E_{HOMO(g)} + 3.74(\pm0.82)$

$pTCO_2\text{-labile} = 0.07(\pm0.02)\, E_{LUMO(g)} + 2.80(\pm0.52)$

$pTCO_2\text{-labile} = 0.08(\pm0.02)\, \mu_{(g)} + 3.50(\pm0.59)$

$pTO_2\text{-labile} = 0.23(\pm0.04)\, E_{PSS} + 5.82(\pm0.89)$

$pTCO_2\text{-labile} = 0.23(\pm0.07)\, E_{PSS} + 5.68(\pm1.45)$

$pTO_2 = 0.08(\pm0.01)^a\, E_{HOMO(g)} + 0.75(\pm0.24)\, \Delta E_0 + 3.16(\pm0.35)$

$pTO_2 = 0.037(\pm0.009)\, E_{PSS} + 0.002(\pm0.00)\, ESE + 1.06(\pm0.39)$

$pTO_2 = -0.07(\pm0.01)\, IP + 0.04(\pm0.01)\, ESE + 1.66(\pm0.46)$

$pTO_2 = 0.07(\pm0.01)\, E_{HOMO(g)} + 0.18(\pm0.06)\, E_{LUMO}\, C_{max} + 3.64(\pm0.29)$

$pTO_2 = 0.08(\pm0.01)\, E_{HOMO(g)} + 0.13(\pm0.04)\, E_{LUMO}\, C_{min} + 4.03(\pm0.32)$

$pTO_2 = 0.10(\pm0.03)\, logK_{OH} + 0.57(\pm0.09)\, X_m^2 r - 1.33(\pm0.34)$

$pTO_2 = 0.12(\pm0.05)\, E_{HOMO(aq)} + 0.34(\pm0.10)\, E_{LUMO}\, C_{max} + 3.21(\pm0.59)$

$pTCO_2 = 0.14(\pm0.05)\, E_{HOMO(aq)} + 0.31(\pm0.12)\, E_{LUMO}C_{max} + 3.14(\pm0.67)$

(Continued)

Fundamental QSARs for Metal Ions

TABLE 5.4 (Continued)
QSARs from 2004–2011 References

Reference	Endpoint	QSAR
Zamil et al. (2009)	Cd^{2+}, Cr^{3+}, Co^{2+}, Cu^{2+}, Hg^{2+}, Pb^{2+}, Ni^{2+}, K^+, Zn^{2+} biosorption to the bacterium, *Staphylococcus saprophyticus* BMSZ711	$qmax = 0.091 + 0.111(X_m^2 r)$ $qmax = 0.191 + 0.002(AW)$ $qmax = 0.159 + 0.006(AN)$ $qmax = -0.194 + 0.342(X_m)$ $qmax = 0.209 + 0.041(AN/\Delta IP)$ $qmax = 1.100 - 1.412(AR) + 1.534(IR)$ $qmax = -0.046 + 0.739(IR) - 0.309(\Delta E_0)$
Mendes et al. (2010)	Toxicity to the bioluminescent fungus *Gerronema viridilucens*	$Log\ EC_{50} = 3.57 - 0.07(AN)$ $Log\ EC_{50} = -1.91 + 2.65(AR)$ $Log\ EC_{50} = 0.75 + 0.84(\Delta E_0)$ $Log\ EC_{50} = 4.93 - 0.37(\Delta IP)$ $Log\ EC_{50} = -3.10 + 0.46(log|K_{OH}|)$ $Log\ EC_{50} = 0.21 + 2.15r$ $Log\ EC_{50} = -0.77 + 17.52(\sigma_p)$ $Log\ EC_{50} = 5.53 - 2.44(X_m)$ $Log\ EC_{50} = 3.05 - 0.40(AN/\Delta IP)$ $Log\ EC_{50} = 4.24 - 1.27(X_m\ r)$ $Log\ EC_{50} = 3.52 - 0.38(Z^{2/r})$ $Log\ EC_{50} = -1.80 + 0.18(log|K_{OH}|) + 11.39(\sigma_p)$ $Log\ EC_{50} = 4.47 - 2.15(X_m) + 0.06(log|K_{OH}|)$ $Log\ EC_{50} = 2.83 - 1.45(X_m) + 8.08(\sigma_p)$ $Log\ EC_{50} = 3.69 - 1.18(X_m^2 r) + 0.04(log\ |K_{OH}|)$

		$\text{Log EC}_{50} = 4.34 - 1.81\left(X_m^2 r\right) + 2.32(\sigma_p)$
		$\text{Log EC}_{50} = 4.34 - 1.81(X_m) - 0.11(\log |K_{OH}|) + 9.35(\sigma_p)$
		$\text{Log EC}_{50} = 3.47 - 1.10\left(X_m^2 r\right) - 0.01(\log |K_{OH}|) + 2.06(\sigma_p)$
		$\text{Log EC}_{50} = 2.77 - 1.00\left(X_m^2 r\right) + 2.82(\sigma_p) + 0.37(AR)$
Su et al. (2010)	Toxicity of phenols and Pb to the bioluminescent bacterium, *Vibrio fischeri*	$\log 1/EC50 = 0.808 + 2.207 \times 10\text{-}5WS + 0.8533X$
		$\log 1/EC50 = -12.039 + 12.315E + 0.502pKa$
		$\log 1/EC50 = 2.21 + 2.882S$
Zhou et al. (2011)	Binding constant for the water flea, *Daphnia magna*	$\text{Log } K = 11.69 - 43.9\sigma_p$
		$\text{Log } K = 2.36 + 1.71\, X_m^2 r$
		$\text{Log } K = 13.99 - 0.72$
		$\text{Log } K_{OH}$
		$\text{Log} K = 5.88 - 0.11\, Z^2/r$
	Binding constant for the fathead minnow, *Pimephales promelas*	$\text{Log } K = 10.57 - 37.62\sigma_p$
		$\text{Log } K = 2.18 + 1.54\, X_m^2 r$
		$\text{Log} K = 12.6 - 0.66$
		$\text{Log } K_{OH}$
		$\text{Log} K = 3.96 + 0.33\, Z^2/r$
	Binding constant for the rainbow trout, *Oncorhynchus mykiss*	$\text{Log } K = 11.08 - 42.73\, \sigma_p$
		$\text{Log } K = 3.9 + 0.69\, X_m^2 r$
		$\text{Log } K = 10.2 - 0.49$
		$\text{Log } K_{OH}$
		$\text{Log } K = 3.69 + 0.37\, Z^2/r$

(Continued)

TABLE 5.4 (*Continued*)
QSARs from 2004–2011 References

Reference	Endpoint	QSAR
	Binding constant for barley, *Hordeum vulgare*	$\text{Log } K = 0.51 + 1.91\ X_m^2 r$
		$\text{Log } K = 9.41 - 34.59\ \sigma_p$
		$\text{Log } K = 12.44 - 0.76$
		$\text{Log } K_{OH}$
		$\text{Log } K = 1.52 + 0.55\ Z^2/r$
	Binding constant for the earthworm, *Eisenia fetida*	$\text{Log } K = 0.82 + 1.68\ X_m^2 r$
		$\text{Log } K = 7.78 - 25.36\ \sigma_p$
		$\text{Log } K = 11.61 - 0.68$
		$\text{Log } K_{OH}$
		$\text{Log } K = 1.35 + 0.63\ Z^2/r$

TABLE 5.5
Studies Using Standard Electrode Potential to Predict Cation Toxicity

Reference	Test Systems	Cations
Mathews (1904)	Development of eggs of the fish *Fundulus heteroclitus* (determination of the concentration which prevents embryo formation)	Ag^{1+}, Al^{3+}, Au^{3+}, Ba^{2+}, Ca^{2+}, Cd^{2+}, Co^{2+}, Cu^{2+}, Fe^{2+}, Fe^{3+}, Hg^{2+}, K^{1+}, Li^{1+}, Mg^{2+}, Mn^{2+}, Na^{1+}, Ni^{2+}, Pb^{2+}, Sr^{2+}, Zn^{2+}.
McGuigan (1904)	Inhibition of the action of malt diastase on starch (determination of the concentration of each salt sufficient to prevent the formation of sugar in one hour at 40°C)	$AgNO_3$, Ag_2SO_4, $AuCl_3$, $HgCl_2$, $CuCl_2$, $PbNO_3$, $NiCl_2$, $CoCl_2$, $FeCl_3$, $CdCl_2$, $ZnCl_2$, $MnCl_2$, $AlCl_3$, $MgCl_2$, $LiCl$, $CaCl_2$, $SrCl_2$, $BaCl_2$, NaI, KI, $NaCl$, KCl, KOH, $NaOH$, $KHSO_4$.
Woodruff and Bunzel (1909)	Toxicity toward the *paramecium* (determination of the concentration necessary to kill within two seconds one half of the organisms tested at 20°C)	Ag^{1+}, Ca^{2+}, Cd^{2+}, Co^{2+}, Cu^{2+}, Fe^{3+}, Hg^{2+}, K^{1+}, Mg^{2+}, Mn^{2+}, Ni^{2+}, Pb^{2+}, Sr^{2+}, Zn^{2+}.
Jones (1939)	Determination of lethal concentration limits in the stickleback (*Gasterosteus aculeatus*).	Ag^{1+}, Al^{3+}, Au^{3+}, Ba^{2+}, Ca^{2+}, Cd^{2+}, Co^{2+}, Cr^{3+}, Cu^{2+}, Hg^{2+}, K^{1+}, Mg^{2+}, Mn^{2+}, Na^{1+}, Ni^{2+}, Pb^{2+}, Sr^{2+}, Zn^{2+}.
Jones (1940)	Determination of the concentration at which a survival time of 48 h is attained for the planarian *Polycelis nigra*.	Ag^{1+}, Al^{3+}, As^{3+}, Au^{3+}, Ca^{2+}, Cd^{2+}, Co^{2+}, Cr^{3+}, Cu^{2+}, Hg^{2+}, K^{1+}, Mg^{2+}, Mn^{2+}, Na^{1+}, Ni^{2+}, Pb^{2+}, Sr^{2+}, Zn^{2+}.
Turner et al. (1983)	Data from Williams et al. (1982) Intraperitoneal (ip) injections in mice to estimate 14-day LD_{50} values and dietary exposures in *Drosophila melanogaster* (4-day LC_{50} values)	Ba^{2+}, Be^{2+}, Cd^{2+}, Co^{2+}, Cu^{2+}, Hg^{2+}, Mg^{2+}, Mn^{2+}, Ni^{2+}, Pb^{2+}, Pd^{2+}, Pt^{2+}, Sr^{2+}, Zn^{2+}
Lewis et al. (1999)	Data from Venugopal and Luckey (1978) Acute toxicity to mice (ip data) (LD_{50} values)	Al^{3+}, Ag^{1+}, AsO_2^-, Ca^{2+}, Cd^{2+}, Co^{2+}, Cr^{3+}, Cu^{2+}, Fe^{3+}, Ga^{3+}, Hg^{2+}, In^{3+}, K^{1+}, La^{3+}, Mg^{2+}, Mn^{2+}, Na^{1+}, Nb^{3+}, Ni^{2+}, Li^{1+}, Pb^{2+}, SbO^+, Sc^{3+}, Sn^{2+}, Sr^{2+}, Ta^{3+}, Ti^{2+}, Tl^{3+}, Zn^{2+}, Y^{3+}.
Enache et al. (2003)	Determination of the concentration at which there is a 50% decrease in the rate of callus development in the sunflower (*Helianthus annuus* F.1. Sunspot)	Ca^{2+}, Cd^{2+}, Co^{2+}, Cu^{2+}, K^{1+}, Li^{1+}, Mg^{2+}, Mn^{2+}, Na^{1+}, Ni^{2+}, Zn^{2+}

probability of a chance correlation, and pT50 is the negative logarithm of toxicity expressed as a mouse LD_{50} value. The test system and the 30 cations used to develop the Lewis et al. (1999) QSAR are provided in Table 5.5.

Enache et al. (2003) determined the concentration at which there was a 50% decrease in the rate of callus development in the sunflower (*Helianthus annuus*

F.1. Sunspot). Enache et al. used these data to develop the QSAR in Table 5.1 that predicted the EC_{50} values of 11 cations: $-(\log EC_{50}) = 3.58 + 0.881\ E°$, r^2 adj.$= 0.86$, $s = 0.45$, $p = 0.000$, $F = 66.14$ and $n = 11$, where s is the standard deviation. The test system and the 11 cations used to develop the Enache et al. (1999) QSAR are provided in Table 5.5.

5.2.2 NEGATIVE LOGARITHM OF THE SOLUBILITY PRODUCT EQUILIBRIUM CONSTANT (pK_{sp})

The solubility product equilibrium constant (K_{sp}) is the constant for the equilibrium that exists between a solid ionic solute and its ions in a saturated aqueous solution. Eight studies used the negative logarithm of solubility product equilibrium constant of the corresponding metal sulfide (pK_{sp}) to predict metal ion toxicity (Table 5.6). The studies of Shaw (1954a, 1954b), Grushkin (1956), and Shaw and Grushkin (1957) were discussed in Walker et al. (2003). Of the 8 studies that used pK_{sp}, only Biesinger and Christensen (1972) had a figure from which a QSAR could be extrapolated. Biesinger and Christensen used a standard 3-week *Daphnia magna* reproductive test to determine the toxicity of 9 cations (Table 5.6). They used the concentration of ions causing a 16% reproductive impairment to develop the relationship between cation concentration and reproductive impairment. The following QSAR was extrapolated from Figure 1 of Biesinger and Christensen (1972): $pM = 0.13\ pK_{sp} + 2.5$, $r^2 = 0.36$, $p < 0.05$, $n = 9$ and $pM = -\log$ molarity at which a 16% decrease in reproduction occurs over a 3-week period. The test system and cations used to develop the Biesinger and Christensen (1972) QSAR are listed in Table 5.6.

Khangarot and Ray (1989) developed correlation coefficients to describe the relationship between the toxicity of 11 unspecified cations and 2-day *Daphnia magna* immobilization EC_{50} values. Their correlation coefficients were converted to coefficients of determination and the statistics for their linear regression analysis were $r^2 = 0.876$ and $p < 0.001$.

Workentine et al. (2008) exposed *Pseudomonas fluorescens* ATCC 13525 in a Calgary Biofilm Device to 44 metal ions. Toxicity was measured and expressed as the minimum inhibitory concentration to inhibit growth for at least 24 h (MIC_{24}), minimum bactericidal concentration to kill 100% (MBC_{100}) of the bacterial population, and minimum biofilm eradication concentration to kill 100% ($MBEC_{100}$) of the biofilm cell population. Linear regression analyses were conducted to determine the relationships between the measured toxicity values and the solubility product of the corresponding metal–sulfide complex (pK_{sp}): $r^2 = 0.55$, $p = 0.001$ (MIC_{24}), $r^2 = 0.45$, $p = 0.023$ (MBC_{100}) and $r^2 = 0.46$ $p = 0.096$ ($MBEC_{100}$).

While Khangarot and Das (2009) did not develop a QSAR, they did develop correlation coefficients to describe the relationship between the toxicity of 8 unspecified cations and 2-day ostracod, *Cypris subglobosa* Sowerby 1840 EC_{50} values based on immobilization. Their correlation coefficients were converted to coefficients of determination and the statistics for their linear regression analysis were $r^2 = 0.82$ and $p < 0.005$. Their test system is provided in Table 5.6.

TABLE 5.6
Studies Using the Negative Logarithm of Solubility Product Equilibrium Constant (pK_{sp}) to Predict Cation Toxicity

References	Test Systems	Cations
Shaw (1954a)	Inhibition of enzyme urease data from Schmidt (1928) and Dounce and Lan (1949)	Schmidt (1928): Ag^{1+}, Cd^{2+}, Co^{2+}, Cu^{2+}, Hg^{2+}, Mn^{2+}, Pb^{2+}, Zn^{2+}. Dounce and Lan (1949): Ag^{1+}, Cd^{2+}, Co^{2+}, Cu^{2+}, Hg^{2+}, Mn^{2+}
Shaw (1954b)	Literature data on effects of cations on the diastase enzyme, paramecium, planarian, and stickleback	Ag^{1+}, Cd^{2+}, Co^{2+}, Cu^{2+}, Hg^{2+}, Mn^{2+}, Ni^{2+}, Pb^{2+}, Zn^{2+}
Grushkin (1956)	Determination of LD_{50} values in guppies (*Lebistes reticulatus*) and toad (*Bufo valliceps*) tadpoles	Chlorides of Hg^{2+} and Ni^{2+}, nitrates of Ag^{1+}, Cd^{2+}, Co^{2+} and Pb^{2+} and sulfates of Cu^{2+}, Mn^{2+} and Zn^{2+}
Shaw and Grushkin (1957)	Determination of LD_{50} values in guppies (*Lebistes reticulatus*) and toad (*Bufo valliceps*) tadpoles	Ag^{1+}, Cd^{2+}, Co^{2+}, Cu^{2+}, Hg^{2+}, Mn^{2+}, Ni^{2+}, Pb^{2+}, Zn^{2+}
Biesinger and Christensen (1972)	Concentrations of ions effecting a 16% reproductive impairment of *Daphnia magna* after 3 weeks	Chlorides of Cd^{2+}, Co^{2+}, Cu^{2+}, Hg^{2+}, Mn^{2+}, Ni^{2+}, Sn^{2+}, Pb^{2+}, Zn^{2+}
Khangarot and Ray (1989)	48-hour tests with *Daphnia magna* to measure EC_{50} values based on immobilization	11 cations; not specified which ones
Workentine et al. (2008)	*Pseudomonas fluorescens* ATCC 13525 was grown in a Calgary Biofilm Device and exposed to different metal ions	44 metal ions (anions and cations)
Khangarot and Das (2009)	48-hr tests with ostracod *Cypris subglobosa* Sowerby 1840 to measure EC_{50} values based on immobilization	8 cations; not specified which ones

5.2.3 STANDARD REDUCTION-OXIDATION POTENTIAL

The standard reduction-oxidation potential represents the absolute difference in electrochemical potential between an ion and its first stable reduced state (ΔE_0), or in other terms, the ability of an ion to change its electronic state. Nine studies used standard reduction-oxidation potential to predict cation toxicity (Table 5.7). These studies were reviewed by Walker et al. (2003), except for Workentine et al. (2008) and Mendes et al. (2010). None of the QSARs were described or discussed by Walker et al. (2003).

Kaiser (1980) used data from Biesinger and Christensen (1972) to develop 2 QSARs that incorporated ΔE_0 (Table 5.8). Biesinger and Christensen used the concentration of ions causing a 16% reproductive impairment to develop the relationship

TABLE 5.7

Studies Using Standard Reduction-Oxidation Potential to Predict Cation Toxicity

Reference	Test Systems	Cations
Kaiser (1980)	Cation concentration causing a 16% decrease in reproduction of *Daphnia magna* over a 3-wk period. Data from Biesinger and Christensen (1972)	Kaiser's group I or group II; see Table 5.8
Kaiser (1985)	Intraperitoneal injections in mice to estimate 14-day LD_{50} values. Data from Williams et al. (1982)	See Table 5.8
McCloskey et al. (1996)	Relative decrease in bioluminescence of *Vibrio fischeri* (the total metal ion (unspeciated) 15-min EC_{50} values)	20 metal ions: Nitrate salts of Ag^+, Ba^{2+}, Ca^{2+}, Cd^{2+}, Co^{2+}, Cr^{3+}, Cs^+, Cu^{2+}, Fe^{3+}, Hg^{2+}, K^+, La^{3+}, Li^+, Mg^{2+}, Mn^{2+}, Na^+, Ni^{2+}, Pb^{2+}, Sr^{2+}, Zn^{2+}
Newman and McCloskey (1996)	Relative decrease in bioluminescence of *Vibrio fischeri* (the total metal ion (unspeciated) 15-min EC_{50} values)	9 metal ions: Chloride salts of Ca^{2+}, Cd^{2+}, Cu^{2+}, Hg^{2+}, Mg^{2+}, Mn^{2+}, Ni^{2+}, Pb^{2+}, Zn^{2+}
Tatara et al. (1997)	24h LC_{50} values (expressed as free ion) for the soil nematode, *Caenorhabditis elegans*	9 metal ions: Chloride salts of Ca^{2+}, Cd^{2+}, Cu^{2+}, Hg^{2+}, Mg^{2+}, Mn^{2+}, Ni^{2+}, Pb^{2+}, Zn^{2+}
Tatara et al. (1998)	24h LC_{50} values (expressed as free ion) for the soil nematode, *Caenorhabditis elegans*	17 metal ions: Nitrate salts of Ca^{2+}, Cd^{2+}, Co^{2+}, Cr^{3+}, Cs^+, Cu^{2+}, Fe^{3+}, K^+, La^{3+}, Li^+, Mg^{2+}, Mn^{2+}, Na^+, Ni^{2+}, Pb^{2+}, Sr^{2+}, Zn^{2+}.
Enache et al. (2000)	Data from Hara and Sonoda (1979). Cabbage plants (*Brassica oleracea* L. var. *capitata*) were grown in 4-L pots of culture solution plus 10 ppm of each cation for a maximum of 55 days. Concentrations to produce a 50% reduction of relative growth (Rrg_{50}) of outer leaves in cabbage plants were recorded	Cd^{2+}, Co^{2+}, Cr^{3+}, Cr^{6+}, Cu^{2+}, Fe^3, Hg^+, Hg^{2+}, Mn^{2+}, Ni^2, V^{3+}, Zn^{2+} or all above - Cu^{2+}, or Ca^{2+}, Co^{2+}, Cu^{2+}, K^{1+}, Mg^{2+}, Mn^{2+}, Na^{1+}, Ni^{2+}, Zn^{2+} or Ca^{2+}, Co^{2+}, Cu^{2+}, Mg^{2+}, Mn^{2+}, Ni^{2+}, Zn^{2+}
Workentine et al. (2008)	*Pseudomonas fluorescens* ATCC 13525 was grown in a Calgary Biofilm Device and exposed to different metal ions	44 metal ions (anions and cations)
Mendes et al. (2010)	Relative decrease in bioluminescence of fungus *Gerronema viridilucens*	Ag^{1+}, Ba^{2+}, Ca^{2+}, Cd^{2+}, Co^{2+}, Cs^{1+}, Cu^{2+}, Fe^{2+}, Hg^{2+}, K^{1+}, Li^{1+}, Mg^{2+}, Mn^{2+}, Na^{1+}, Ni^{2+}, Pb^{2+}, Sr^{2+}, Zn^{2+}

TABLE 5.8

Kaiser's (1980, 1985) QSARs for Predicting Cation Toxicity Using Atomic Number, Ionization Potential Differential, and Standard Reduction-Oxidation Potential

Reference	Test Substance(s)	QSARs	r^2	SE	n
Kaiser (1980)	Al^{3+}, Ba^{2+}, Ca^{2+}, K^{1+}, Mg^{2+}, Na^{1+}, Sr^{2+} (Kaiser's group I ions)	$pTm = 11.63(\pm0.09) + 4.87(\pm0.17)\log(AN/\Delta IP) - 4.29(\pm0.14)\Delta E_0$	0.962		7
	As^{5+}, Au^{3+}, Cd^{2+}, Co^{2+}, Cr^{3+}, Cu^{2+}, Fe^{3+}, Hg^{2+}, Mn^{2+}, Ni^{2+}, Pt^{4+}, Zn^{2+} + (Kaiser's group II ions)	$pTm = 4.80(\pm0.32) + 6.15(\pm0.45)\log(AN/\Delta IP) - 3.11(\pm0.32)\Delta E_0$	0.956		12
Kaiser (1985)	Kaiser's group II ions plus In^{3+}, Rh^{3+}, Pd^{2+}	$pT = -0.0004 + 2.47 \log(AN/\Delta IP) - 1.17\Delta E_0$	0.76	0.29	15
	Ba^{2+}, Be^{2+}, Mg^{2+}, Sr^{2+}	$pT = 13.94 + 5.57 \log(AN/\Delta IP) - 6.65\Delta E_0$	0.98	0.43	4
	Ba^{2+}, Be^{2+}, Mg^{2+}, Sr^{2+}, Y^{3+}	$pT = 6.38 + 2.39 \log(AN/\Delta IP) - 3.06\Delta E_0$	0.56	0.43	5

pTm is the negative logarithm of molar concentrations causing a 16% decrease in reproduction of *Daphnia magna* over a 3-week period. Standard deviations for the Kaiser (1980) QCARs are shown in parenthesis.

pT is the negative logarithm of metal ion LD50 value expressed in mmoles/kg.

AN is the atomic number.

ΔIP is the difference of the ionization potential in volts between its oxidation number (OX) and the next lower one (OX − 1).

ΔE_0 is the standard reduction-oxidation potential, also known as the absolute value of the electrochemical potential in volts between the ion with the oxidation number OX and its first stable, reduced form.

between cation concentrations and reproductive impairment. The 2 QSARs relied on standard reduction-oxidation potential (ΔE_0), atomic number (AN), and the difference of the ionization potential in volts between its oxidation number (OX) and the next lower one (OX – 1) (ΔIP). Both QSARs for 7 and 12 cations had low standard deviations and high coefficients of determination (Table 5.8).

Kaiser (1985) used data from Williams et al. (1982) to develop 3 QSARs that incorporated ΔE_0 (Table 5.8). Williams et al. (1982) used intraperitoneal injections in mice to develop the relationship between cation concentrations and mouse 14-day LD_{50} values. The 3 QSARs relied on ΔE_0, AN, and ΔIP. The most statistically significant QSAR was developed for Ba^{2+}, Be^{2+}, Mg^{2+}, and Sr^{2+}; when Y^{3+} was added to these 4 cations, r^2 decreased but SE remained the same (Table 5.8).

Newman's group published a series of papers from 1996 to 1998 describing quantitative ion-character activity relationships (QICARs) for predicting metal ion toxicity (McCloskey et al. 1996; Newman and McCloskey 1996; Tatara et al. 1997; Tatara et al. 1998; Newman et al. 1998). McCloskey et al. (1996) and Newman and McCloskey (1996) developed 8 and 10 QSARs, respectively, to predict decrease in bioluminescence of *Vibrio fischeri* (Table 5.1). This team (McCloskey et al. 1996; Newman and McCloskey 1996) developed 3 of these QSARs using ΔE_0 (Table 5.9). The 3 QSARs relied on ΔE_0, AN, and ΔIP. QSARs with ΔE_0 and AN/ΔIP had improved statistical significance compared to QSARs that only used ΔE_0; using the log AN/ΔIP did not improve statistical significance (Table 5.9). Tatara et al. 1997 and Tatara et al. 1998 then developed 9 and 10 QSARs, respectively, to predict 24-hour LC_{50} values for the soil nematode, *Caenorhabditis elegans* (Table 5.1). They developed 2 and 4 of these QSARs using ΔE_0 (Table 5.10). Their two common QSARs relied on ΔE_0, AN, and ΔIP. Adding AN/ΔIP to the QSAR with ΔE_0 increased statistical significance for 9 cations, but not for 17 cations (Table 5.10). Tatara et al. (1998) developed 2 additional QSARs that also used ΔE_0 to predict 24-hour LC_{50} values for the soil nematode, *Caenorhabditis elegans* (Table 5.10). One used the covalent index $\left(X_m^2 r\right)$ in addition to ΔE_0; the other used the absolute value of logarithm of the first hydrolysis constant ($|\log K_{OH}|$) in addition to ΔE_0. Adding $|\log K_{OH}|$ to ΔE_0 provided the most statistically significant QSAR (Table 5.10). Newman et al. (1998) provided a very comprehensive coverage of these studies and others.

Enache et al. (2000) developed 11 QSARs to predict a 50% reduction of relative growth of outer leaves in cabbage, *Brassica oleracea* L. var. *capitata* (Table 5.1). Of these 11 QSARs, 9 used ΔE_0 (Table 5.11). Six of these QSARs used 12 cations and combinations of ΔE_0, AN, atomic weight (AW), ΔIP, crystal ionic radius (r), Allred-Rochow electronegativity (X_{AR}), Pauling's electronegativity (X), and charge on the ion (Z). The simplest QSAR with ΔE_0 and both electronegativity properties was the most statistically significant (Table 5.11). Three of these QSARs used 11 cations and combinations of ΔE_0, AN, atomic weight (AW), ΔIP, Allred-Rochow electronegativity (X_{AR}), and Pauling's electronegativity (X) (Table 5.11). Eliminating Cu^{2+} produced the most robust QSAR, again with ΔE_0 and both electronegativity properties (Table 5.11). The outlier position of Cu^{2+} is due to its exceptionally high toxicity determined experimentally by Hara and Sonoda (1979). The test system of Enache et al. (2000) is listed in Table 5.7.

TABLE 5.9
McCloskey et al. (1996) and Newman and McCloskey (1996) QSARs for Predicting Cation Toxicity Using Standard Reduction-Oxidation Potential or a Combination of Standard Reduction-Oxidation Potential, Atomic Number, and Ionization Potential Differential or Pearson and Mawby Softness Parameter

Reference	Test Substance(s)	QSARs	r^2	MSE	AIC	n
McCloskey et al. (1996)	Ag^+, Ba^{2+}, Ca^{2+}, Cd^{2+}, Co^{2+}, Cr^{3+}, Cs^+, Cu^{2+}, Fe^{3+}, Hg^{2+}, K^+, La^{3+}, Li^+, Mg^{2+}, Mn^{2+}, Na^+, Ni^{2+}, Pb^{2+}, Sr^{2+}, Zn^{2+}	$\text{Log EC}_{50} = 0.14 + 1.62(\Delta E_0)$	0.57	1.696	82.06	20
		$\text{Log EC}_{50} = 0.95 - 0.19(AN/\Delta IP) + 1.73(\Delta E_0)$	0.65	1.571	78.50	20
		$\text{Log EC}_{50} = 1.30 - 2.03(\log AN/\Delta IP) + 1.64(\Delta E_0)$	0.65	1.569	78.44	20
		$\text{Log EC}_{50} = -3.25 + 39.95(\sigma_p)$	0.73	1.349	71.04	20
Newman and McCloskey (1996)	Ca^{2+}, Cd^{2+}, Cu^{2+}, Hg^{2+}, Mg^{2+}, Mn^{2+}, Ni^{2+}, Pb^{2+}, Zn^{2+}	$\text{Log EC}_{50} = -1.42 + 2.10(\Delta E_0)$	0.15	5.206	71.83	9
		$\text{Log EC}_{50} = 4.79 - 1.04(AN/\Delta IP) + 0.83(\Delta E_0)$	0.52	4.214	60.77	9
		$\text{Log EC}_{50} = 7.03 - 11.99(\log AN/\Delta IP) + 0.78(\Delta E_0)$	0.49	4.338	61.78	9
		$\text{Log EC}_{50} = 12.87 - 110.9(\sigma_p)$	0.60	3.554	56.84	9

EC_{50} is the 50% decrease in bioluminescence of *Vibrio fischeri* caused by the total metal ion (unspeciated) after 15-min exposure.

AN is the atomic number.

ΔIP is the difference of the ionization potential in volts between its oxidation number (OX) and the next lower one (OX −1).

ΔE_0 is the standard reduction-oxidation potential, also known as the absolute value of the electrochemical potential in volts between the ion with the oxidation number OX and its first stable, reduced form.

σ_p is the Pearson and Mawby softness parameter.

MSE is mean square error and AIC is Akaike's Information Criterion. QSARs with the smallest AIC are likely to have the most information regardless of the number of independent variables.

TABLE 5.10

Tatara et al. (1997, 1998) QSARs for predicting cation toxicity using standard reduction-oxidation potential alone or in combinations with standard reduction-oxidation potential, atomic number, ionization potential differential, covalent index, logarithm of the first hydrolysis constant, or Pearson and Mawby softness parameter alone or in combination with logarithm of the first hydrolysis constant

Reference	Test substance(s)	QSARs	r^2	MSE	AIC	n
Tatara et al. (1997)	Ca^{2+}, Cd^{2+}, Cu^{2+}, Hg^{2+}, Mg^{2+}, Mn^{2+}, Ni^{2+}, Pb^{2+}, Zn^{2+}	$\text{Log } LC_{50} = -1.27 + 0.98(\Delta E_0)$	0.05	20.727		9
		$\text{Log } LC_{50} = 3.93 - 0.87(AN/\Delta IP) + 0.08(\Delta E_0)$	0.43	14.458		9
		$\text{Log } LC_{50} = -10.47 + 83.49(\sigma_p)$	0.50	0.743		9
Tatara et al. (1998)	Ca^{2+}, Cd^{2+}, Co^{2+}, Cr^{3+}, Cs^+, Cu^{2+}, Fe^{3+}, K^+, La^{3+}, Li^+, Mg^{2+}, Mn^{2+}, Na^+, Ni^{2+}, Pb^{2+}, Sr^{2+}, Zn^{2+}.	$\text{Log } LC_{50} = 0.29 + 0.69(\Delta E_0)$	0.49	0.759	43.58	17
		$\text{Log } LC_{50} = 0.53 - 0.06(AN/\Delta IP) + 0.72(\Delta E_0)$	0.53	0.752	44.40	17
		$\text{Log } LC_{50} = 1.61 + 0.35(\Delta E_0) - 0.41(X_m^2 r)$	0.60	0.647	41.86	17
		$\text{Log } LC_{50} = -1.21 + 0.22(\Delta E_0) + 0.22(\text{llog KOHI})$	0.69	0.500	37.44	17
		$\text{Log } LC_{50} = -1.23 + 16.72(\sigma_p)$	0.50	10.890	40.18	17
		$\text{Log } LC_{50} = -1.63 + 4.58(\sigma_p) + 0.22(\text{llog KOHI})$	0.68	0.512	37.88	17

LC_{50} is the 50% decrease in the number of soil nematodes, *Caenorhabditis elegans* caused by the free ions after 24 hours exposure.

AN is the Atomic Number.

ΔIP is the difference of the ionization potential in volts between its oxidation number (OX) and the next lower one (OX − 1)

ΔE_0 is the standard reduction-oxidation potential, also known as the absolute value of the electrochemical potential in volts between the ion with the oxidation number OX and its first stable, reduced form.

$X_m^2 r$ is the covalent index.

llog KOHI is the absolute value of logarithm of the first hydrolysis constant.

σ_p is the Pearson and Mawby softness parameter.

MSE is mean square error and AIC is Akaike's Information Criterion, QCARs with the smallest AIC are likely to have the most information regardless of the number of independent variables.

TABLE 5.11

Enache et al. (2000) QCARs and Enache et al. (2003) QCAR for predicting cation toxicity using atomic number, atomic weight, ionization potential differential and standard reduction-oxidation potential

Test substance(s)	Enache et al. (2000) QCARs	r^2 adj.	s	p	F	n
Cd^{2+}, Co^{2+}, Cr^{3+}, Cr^{6+}, Cu^{2+}, Fe^3, Hg^+, Hg_2^{2+}, Mn^{2+}, Ni^2, V^{3+}, Zn^{2+}	$-(\log Rrg_{50}) = 8.57 - 7.26\ X_{AR} - 2.51\ \Delta E_0 + 4.45\ X$	0.66	0.78	0.009	7.98	12
	$-(\log Rrg_{50}) = -1.36 - 1.53\ \Delta E_0 + 1.20\ X^2 r + 0.239\ \Delta IP$	0.55	0.90	0.024	5.51	12
	$-(\log Rrg_{50}) = -0.74 + 0.0195\ AW - 1.94\ \Delta E_0 + 0.295\ \Delta IP$	0.55	0.90	0.025	5.42	12
	$-(\log Rrg_{50}) = -1.07 + 0.0519\ AN - 1.91\ \Delta E_0 + 0.299\Delta IP$	0.54	0.90	0.026	5.36	12
	$-(\log Rrg_{50}) = -0.39 + 0.713\ Z - 2.47\ \Delta E_0 + 1.43\ X^2 r$	0.54	0.90	0.025	5.39	12
	$-(\log Rrg_{50}) = 0.58 + 0.250\ Zr - 2.41\ \Delta E_0 + 1.35\ X^2 r$	0.51	0.93	0.033	4.84	12
Cd^{2+}, Co^{2+}, Cr^{3+}, Cr^{6+}, Fe^3, Hg^+, $Hg2+$, Mn^{2+}, Ni^2, V^{3+}, Zn^{2+}	$-(\log Rrg50) = 12.2 - 8.54\ XAR - 1.59\ \Delta E_0 - 3.04\ X$	0.81	0.49	0.002	15.06	11
	$-(\log Rrg50) = -1.02 + 0.0187\ AW - 1.10\ \Delta E0 + 0.264\ \Delta IP$	0.73	0.59	0.006	10.04	11
	$-(\log Rrg50) = -1.33 + 0.0498\ AN - 1.07\ \Delta E0 + 0.268\ \Delta IP$	0.72	0.60	0.007	9.65	11
	Enache et al. (2003) QCAR					
Ca^{2+}, Cd^{2+}, Co^{2+}, Cu^{2+}, K^{1+}, Li^{1+}, Mg^{2+}, Mn^{2+}, Na^{1+}, Ni^{2+}, Zn^{2+}	$-(\log EC_{50}) = -1.74 + 2.91\ X$	0.86	0.45	0.000	66.14	11

Rrg_{50} is a 50% reduction of relative growth of outer leaves in cabbage, *Brassica oleracea* L. var. *capitata*.

AN is the Atomic Number.

AW is the Atomic Weight.

ΔIP is the difference of the ionization potential in volts between its oxidation number (OX) and the next lower one (OX−1).

ΔE_0 is the standard reduction-oxidation potential, also known as the absolute value of the electrochemical potential in volts between the ion with the oxidation number OX and its first stable, reduced form.

X_{AR} is the Allred-Rochow electronegativity.

X is the Pauling's electronegativity.

r is the crystal ionic radius.

Z is the ion charge.

TABLE 5.12
Mendes et al. (2010) QCARs for Predicting Cation Toxicity Using Standard Reduction-Oxidation Potential, Electronegativity, Pearson and Mawby (1967) softness parameter and in combination with the Logarithm of the First Hydrolysis Constant, Covalent Index, and Atomic Radius

QCAR	adj. r^2	p	AIC
$\text{Log EC}_{50} = 0.75 + 0.84(\Delta E_0)$	0.893	0.0001	−58.38
$\text{Log EC}_{50} = 5.53 - 2.44(X_m)$	0.911	0.0001	53.08
$\text{Log EC}_{50} = 4.47 - 2.15(X_m) + 0.06(\log|K_{OH}|)$	0.807	0.0006	−47.93
$\text{Log EC}_{50} = 2.83 - 1.45(X_m) + 8.08(\sigma_p)$	0.85	0.0002	−50.69
$\text{Log EC}_{50} = 4.34 - 1.81(X_m) - 0.11(\log|K_{OH}|) + 9.35(\sigma_p)$	0.833	0.0012	−43.66
$\text{Log EC}_{50} = -0.77 + 17.52(\sigma_p)$	0.799	0.0001	−51.42
$\text{Log EC}_{50} = -1.80 + 0.18(\log|K_{OH}|) + 11.39(\sigma_p)$	0.799	0.0007	−47.48
$\text{Log EC}_{50} = 4.34 - 1.81(X_m^2 r) + 2.32(\sigma_p)$	0.903	0.0001	−55.47
$\text{Log EC}_{50} = 3.47 - 1.10(X_m^2 r) - 0.01(\log|K_{OH}|) + 2.06(\sigma_p)$	0.889	0.003	−48.16
$\text{Log EC}_{50} = 2.77 - 1.00(X_m^2 r) + 2.82(\sigma_p) + 0.37(AR)$	0.896	0.0002	−48.89

Source: Data from L.F. Mendes, E.L. Bastos, and C.V. Stevani. "Prediction of Metal Cation Toxicity to the Bioluminescent Fungus *Gerronema viridilucens*." *Environ. Toxicol. Chem.* 29 (2010):2177–2181.

ΔE_0 is the standard reduction-oxidation potential, also known as the absolute value of the electrochemical potential in volts between the ion with the oxidation number OX and its first stable, reduced form.

X_m is the electronegativity.

σ_p is the Pearson and Mawby softness parameter.

$\log|K_{OH}|$ is the logarithm of the first hydrolysis constant.

$X_m^2 r$ is the covalent index.

AR is the atomic radius.

AIC is Akaike's Information Criterion; QCARs with the smallest AIC are likely to have the most information regardless of the number of independent variables.

Workentine et al. (2008) used the test system listed in Table 5.7 and discussed in Section 5.2.2. Linear regression analyses were conducted to determine the relationships between the measured toxicity values and ΔE_0: $r^2 = 0.38$, p < 0.001 (MIC_{24}), $r^2 = 0.10$, p = 0.167 (MBC_{100}) and $r^2 = 0.03$ p = 0.537 (MBEC_{100}).

Mendes et al. (2010) developed 19 QSARs to predict EC_{50} values to the fungus *Gerronema viridilucens* for 18 cations (Table 5.4). Of these 19 QSARs, 1 used ΔE_0 (Table 5.12). The test system and cations used to develop the Mendes et al. (2010) QSAR are listed in Table 5.7.

In addition to the 27 QSARs that used ΔE_0 to predict cation toxicities, Zamil et al. (2009) used the following QSAR to predict biosorption of Cd^{2+}, Cr^{3+}, Co^{2+}, Cu^{2+}, Hg^{2+}, Pb^{2+}, Ni^{2+}, K^+, Zn^{2+} to *Staphylococcus saprophyticus* BMSZ711: $q_{max} = -0.046 + 0.739 (IR) - 0.309 (\Delta E_0)$. This QSAR was statistically significant with an adjusted $r^2 = 0.70$, SE = 0.111, F = 10.48, p = 0.011, Akaike's Information Criterion (AIC) = −4.12 and mean absolute percent error between observed and predicted values in the prediction (MAPE) = 21.63. QSARs with the smallest AIC are likely to have the most information regardless of the number of independent variables.

Of the 27 QSARs that used ΔE_0 to predict cation toxicities, 13 used combinations of ΔE_0, AN, and ΔIP (labeled as 1 in Table 5.13), 5 only used ΔE_0 (labeled as 2 in Table 5.13), and 2 used combinations of ΔE_0, AW, and ΔIP (labeled as 3 in Table 5.13). Of the 13 QSARs that used combinations of ΔE_0, AN, and ΔIP, Kaiser's (1980) QSAR for 12 cations had the highest r^2; adding In^{3+}, Rh^{3+}, and Pd^{2+} to those 12 cations decreased r^2 (Table 5.13). Of the 5 QSARs that only used ΔE_0, the Mendes et al. (2010) QSAR not only had the highest coefficient of determination, but the second-highest number of cations (18), only missing Cr^{3+} and La^{3+} from the McCloskey et al. (1996) QSAR for 20 cations (Table 5.13). Of the 2 that used combinations of ΔE_0, AW, and ΔIP, reducing the number of cations from 12 to 11 by eliminating Cu^{2+} produced a more robust QSAR as explained previously (Table 5.13).

5.2.4 ELECTRONEGATIVITY

Electronegativity is a chemical property that describes the power of an atom in a molecule to attract electrons to itself (Pauling 1932). Eleven studies used electronegativity (X, X_{AR}, or X_m) to predict cation toxicity (Table 5.14). The studies of Seifriz (1949), Danielli and Davies (1951), and Somers (1959) were discussed in Walker et al. (2003). The Biesinger and Christensen (1972) test system is discussed in Section 5.2.2. The following QSAR was extrapolated from Figure 1 of Biesinger and Christensen (1972): $pM = -3 + 5\ X_{AR}$, $r^2 = 0.61$, $p < 0.001$, $n = 20$. The test system and cations used to develop the Biesinger and Christensen (1972) QSAR are listed in Table 5.14.

Turner et al. (1985) and Khangarot and Ray (1989) did not report QSARs for electronegativity. However, Khangarot and Ray did use linear regression analysis to describe the relationship between 23 cation toxicities and standard 2-day *Daphnia magna* EC_{50} values based on immobilization: $r^2 = 0.542$ and $p < 0.001$. The Turner et al. (1985) and Khangarot and Ray (1989) test systems and cations are listed in Table 5.14.

Enache et al. (2000) developed 2 electronegativity QSARs. The test system for both QSARs is listed in Table 5.14. Both QSARs relied on Allred-Rochow electronegativity (X_{AR}), standard reduction-oxidation potential (ΔE_0), and Pauling's electronegativity (X) (Table 5.11). The first QSAR is for 12 cations listed in Table 5.11 and the second is for 11 cations listed in Table 5.11. The statistical analysis for the 12 cations produced a reliable QSAR, but improved results were obtained when the copper ion was omitted from the analysis as explained in Section 5.2.3.

Enache et al. (2003) developed an electronegativity-based QSAR to predict the toxicity of 11 cations (Table 5.11). The test system for the QSAR is listed in Table 5.14.

Workentine et al. (2008) used the test system listed in Table 5.14 and discussed in Section 5.2.2. Linear regression analyses were conducted to determine the relationships between the measured toxicity values and X_m: $r^2 = 0.28$, $p = 0.006$ (MIC_{24}), $r^2 = 0.04$, $p = 0.397$ (MBC_{100}) and $r^2 = 0.01$ $p = 0.725$ ($MBEC_{100}$).

TABLE 5.13
Standard Reduction-Oxidation Potential (ΔE0) QCARs

#	Reference	Test Substance(s)	QCARs	r^2	n
1	Kaiser (1980)	Al^{3+}, Ba^{2+}, Ca^{2+}, K^{1+}, Mg^{2+}, Na^{1+}, Sr^{2+}	$pTm = 11.63 + 4.87\log(AN/\Delta IP) - 4.29\Delta E_0$	0.962	7
1		As^{5+}, Au^{3+}, Cd^{2+}, Co^{2+}, Cr^{3+}, Cu^{2+}, Fe^{3+}, Hg^{2+}, Mn^{2+}, Ni^{2+}, Pt^{4+}, Zn^{2+}	$pTm = 4.80 + 6.15\log(AN/\Delta IP) - 3.11\Delta E_0$	0.956	12
1	Kaiser (1985)	As^{5+}, Au^{3+}, Cd^{2+}, Co^{2+}, Cr^{3+}, Cu^{2+}, Fe^{3+}, Hg^{2+}, Mn^{2+}, Ni^{2+}, Pt^{4+}, Zn^{2+}, In^{3+}, Rh^{3+}, Pd^{2+}	$pT = -0.0004 + 2.47\log(AN/\Delta IP) - 1.17\Delta E_0$	0.76	15
1		Ba^{2+}, Be^{2+}, Mg^{2+}, Sr^{2+}	$pT = 13.94 + 5.57\log(AN/\Delta IP) - 6.65\Delta E_0$	0.98	4
1		Ba^{2+}, Be^{2+}, Mg^{2+}, Sr^{2+}, Y^{3+}	$pT = 6.38 + 2.39\log(AN/\Delta IP) - 3.06\Delta E_0$	0.56	5
2	McCloskey et al. (1996)	Ag^+, Ba^{2+}, Ca^{2+}, Cd^{2+}, Co^{2+}, Cr^{3+}, Cs^+, Cu^{2+}, Fe^{3+}, Hg^{2+}, K^+, La^{3+}, Li^+, Mg^{2+}, Mn^{2+}, Na^+, Ni^{2+}, Pb^{2+}, Sr^{2+}, Zn^{2+}	$\log EC_{50} = 0.14 + 1.62(\Delta E_0)$	0.57	20
1			$\log EC_{50} = 0.95 - 0.19(AN/\Delta IP) + 1.73(\Delta E_0)$	0.65	20
1			$\log EC_{50} = 1.30 - 2.03(\log AN/\Delta IP) + 1.64(\Delta E_0)$	0.65	20
2	Newman and McCloskey (1996)	Ca^{2+}, Cd^{2+}, Cu^{2+}, Hg^{2+}, Mg^{2+}, Mn^{2+}, Ni^{2+}, Pb^{2+}, Zn^{2+}	$\log EC_{50} = -1.42 + 2.10(\Delta E_0)$	0.15	9
1			$\log EC_{50} = 4.79 - 1.04(AN/\Delta IP) + 0.83(\Delta E_0)$	0.52	9
1			$\log EC_{50} = 7.03 - 11.99(\log AN/\Delta IP) + 0.78(\Delta E_0)$	0.49	9
2	Tatara et al. (1997)	Ca^{2+}, Cd^{2+}, Cu^{2+}, Hg^{2+}, Mg^{2+}, Mn^{2+}, Ni^{2+}, Pb^{2+}, Zn^{2+}	$\log EC_{50} = -1.27 + 0.98(\Delta E_0)$	0.05	9
1			$\log EC_{50} = 3.93 - 0.87(AN/\Delta IP) + 0.08(\Delta E_0)$	0.43	9
2	Tatara et al. (1998)	Ca^{2+}, Cd^{2+}, Co^{2+}, Cr^{3+}, Cs^+, Cu^{2+}, Fe^{3+}, K^+, La^{3+}, Li^+, Mg^{2+}, Mn^{2+}, Na^+, Ni^{2+}, Pb^{2+}, Sr^{2+}, Zn^{2+}.	$\log EC_{50} = 0.29 + 0.69(\Delta E_0)$	0.49	17
1			$\log EC_{50} = 0.53 - 0.06(AN/\Delta IP) + 0.72(\Delta E_0)$	0.53	17
			$\log EC50 = 1.61 + 0.35(\Delta E_0) - 0.41(X^2_m r)$	0.60	17
			$\log EC50 = -1.21 + 0.22(\Delta E_0) + 0.22(|\log K_{OH}|)$	0.69	17

Enache et al. (2000)	Cd^{2+}, Co^{2+}, Cr^{3+}, Cr^{6+}, Cu^{2+}, Fe^3, Hg^+, Hg^{2+}, Mn^{2+}, Ni^2, V^{3+}, Zn^{2+}	$-(\log Rrg_{50}) = 8.57 - 7.26\ X_{AR} - 2.51\ \Delta E_0 + 4.45\ X$	0.66	12
		$-(\log Rrg_{50}) = -1.36 - 1.53\ \Delta E_0 + 1.20\ X^2r + 0.239\ \Delta IP$	0.55	12
3		$-(\log Rrg_{50}) = -0.74 + 0.0195\ AW - 1.94\ \Delta E_0 + 0.295\ \Delta IP$	0.55	12
1		$-(\log Rrg_{50}) = -1.07 + 0.0519\ AN - 1.91\ \Delta E_0 + 0.299\ \Delta IP$	0.54	12
		$-(\log Rrg_{50}) = -0.39 + 0.713\ Z - 2.47\ \Delta E_0 + 1.43\ X^2r$	0.54	12
		$-(\log Rrg_{50}) = 0.58 + 0.250\ Z/r - 2.41\ \Delta E_0 + 1.35\ X^2r$	0.51	12
	Cd^{2+}, Co^{2+}, Cr^{3+}, Cr^{6+}, Fe^3, Hg^+, Hg^{2+}, Mn^{2+}, Ni^2, V^{3+}, Zn^{2+}	$-(\log Rrg_{50}) = 12.2 - 8.54\ X_{AR} - 1.59\ \Delta E_0 - 3.04\ X$	0.81	11
3		$-(\log Rrg_{50}) = -1.02 + 0.0187\ AW - 1.10\ \Delta E_0 + 0.264\ \Delta IP$	0.73	11
1		$-(\log Rrg_{50}) = -1.33 + 0.0498\ AN - 1.07\ \Delta E_0 + 0.268\ \Delta IP$	0.72	11
Mendes et al. (2010)	Ag^{1+}, Ba^{2+}, Ca^{2+}, Cd^{2+}, Co^{2+}, Cs^{1+}, Cu^{2+}, Fe^{2+}, Hg^{2+}, K^{1+}, Li^{1+}, Mg^{2+}, Mn^{2+}, Na^{1+}, Ni^{2+}, Pb^{2+}, Sr^{2+}, Zn^{2+}	$Log\ EC_{50} = 0.75 + 0.84(\Delta E_0)$	0.893	18
2				

TABLE 5.14

Studies Using Electronegativity to Predict Cation Toxicity

Reference	Test Systems	Cations
Seifriz (1949)	Toxicity to slime molds (*Myxomycetes*) determined by a number of pathological changes in the protoplasm	Ag^{1+}, Au^{3+}, Ba^{2+}, Ca^{2+}, Cs^{1+}, K^{1+}, La^{3+}, Li^{1+}, Mg^{2+}, Na^{1+}, Pb^{2+}, Rb^{1+}, Sr^{2+}, Th^{4+}
Danielli and Davies (1951)	Toxicity to planaria *Polycelis nigra*	Ag^{1+}, Al^{3+}, As^{3+}, Ca^{2+}, Cd^{2+}, Co^{2+}, Cr^{3+}, Cu^{2+}, Fe^{3+}, Hg^{2+}, K^{1+}, Mg^{2+}, Mn^{2+}, Na^{1+}, Ni^{2+}, Pb^{2+}, Sr^{2+}, Zn^{2+}
Somers (1959)	Studies with conidia of *Alternaria tenuis* to determine ED_{50} using the standard test-tube dilution spore germination technique	Ag^{1+}, Ba^{2+}, Be^{2+}, Bi^{3+}, Ce^{3+}, Cr^{3+}, Cu^{2+}, Hg^{2+}, K^{1+}, Li^{1+}, Mg^{2+}, Mn^{2+}, Na^{1+}, Ni^{2+}, Os^{4+}, Pb^{2+}, Pd^{4+}, Ru^{3+}, Sr^{2+}, Tl^{+1}, Zn^{2+}, Zr^{2+}, Y^{3+}
Biesinger and Christensen (1972)	Concentrations of ions effecting a 16% reproductive impairment of *Daphnia magna* after 3 weeks	Al^{3+}, Au^{3+}, Ba^{2+}, Ca^{2+}, Cd^{2+}, Co^{2+}, Cr^{3+}, Cu^{2+}, Fe^{3+}, Hg^{2+}, K^{1+}, Mg^{2+}, Mn^{2+}, Na^{1+}, Ni^{2+}, Pb^{2+}, Pt^{4+}, Sn^{2+}, Sr^{2+}, Zn^{2+}
Turner et al. (1985)	14-day LD_{50} (IP) in mice	Ag^{+1}, Au^{3+}, Ba^{2+}, Be^{2+}, Cd^{2+}, Co^{2+}, Cr^{3+}, Cu^{2+}, Fe^{3+}, Gd^{3+}, Hg^{2+}, In^{3+}, Mg^{2+}, Mn^{2+}, Ni^{2+}, Pb^{2+}, Pd^{2+}, Pt^{2+}, Rh^{3+}, Sn^{4+}, Sr^{2+}, Tl^{+1}, Y^{3+}, Zn^{2+}
Khangarot and Ray (1989)	48-hour tests with *Daphnia magna* to measure EC_{50} values based on immobilization	Ag^{1+}, Al^{3+}, As^{3+}, Ba^{2+}, Be^{2+}, Ca^{2+}, Cd^{2+}, Co^{2+}, Cr^{6+}, Cu^{2+}, Fe^{2+}, Hg^{2+}, K^{1+}, Na^{1+}, Ni^{2+}, Mg^{2+}, Mn^{2+}, Pb^{2+}, Sb^{3+}, Sn^{2+}, Sr^{2+}, Zn^{2+}, W^{6+}
Enache et al. (2000)	Data from Hara and Sonoda (1979). Cabbage plants (*Brassica oleracea* L. var. *capitata*) were grown in 4 L pots of culture solution plus 10 ppm of each cation for a maximum of 55 days. Concentrations to produce a 50% reduction of relative growth (Rrg_{50}) of outer leaves in cabbage plants were recorded	Cd^{2}, Co^{2+}, Cr^{3+}, Cr^{6+}, Cu^{2+}, Fe^{3}, Hg^{+}, Hg^{2+}, Mn^{2+}, Ni^{2}, V^{3+}, Zn^{2+}
Enache et al. (2003)	Determination of the concentration at which there is a 50% decrease in the rate of callus development in the sunflower (*Helianthus annuus* F.1. Sunspot)	Ca^{2+}, Cd^{2+}, Co^{2+}, Cu^{2+}, K^{1+}, Li^{1+}, Mg^{2+}, Mn^{2+}, Na^{1+}, Ni^{2+}, Zn^{2+}
Workentine et al. (2008)	*Pseudomonas fluorescens* ATCC 13525 was grown in a Calgary Biofilm Device and exposed to different metal ions	44 metal ions (anions and cations)

TABLE 5.14 (*Continued*)
Studies Using Electronegativity to Predict Cation Toxicity

Reference	Test Systems	Cations
Khangarot and Das (2009)	48-hour tests with ostracod *Cypris subglobosa* Sowerby 1840 to measure EC_{50} values based on immobilization	Ag^{1+}, Al^{3+}, As^{3+}, B^{3+}, Ba^{2+}, Be^{2+}, Bi^{3+}, Ca^{2+}, Cd^{2+}, Co^{2+}, Cr^{3+}, Cr^{6+}, Cu^{2+}, Fe^{3+}, Hg^{2+}, K^{1+}, La^{3+}, Li^{1+}, Mg^{2+}, Mn^{2+}, Mo^{2+}, Na^{1+}, Ni^{2+}, Os^{4+}, Pb^{2+}, Pd^{4+}, Pt^{4+}, Sb^{3+}, Se^{2+}, Sn^{2+}, Sr^{2+}, Te^{2+}, UO_2^{2+}, W^{6+}, Zn^{2+}, Zr^{2+}
Mendes et al. (2010)	Relative decrease in bioluminescence of fungus *Gerronema viridilucens*	Ag^{1+}, Ba^{2+}, Ca^{2+}, Cd^{2+}, Co^{2+}, Cs^{1+}, Cu^{2+}, Fe^{2+}, Hg^{2+}, K^{1+}, Li^{1+}, Mg^{2+}, Mn^{2+}, Na^{1+}, Ni^{2+}, Pb^{2+}, Sr^{2+}, Zn^{2+}

Khangarot and Das (2009) developed correlation coefficients to describe the relationship between the toxicity of 36 cations and the ostracod 2-day *Cypris subglobosa* EC_{50} values based on immobilization. Their correlation coefficients were converted to coefficients of determination and the statistics for their linear regression analysis were $r^2 = 0.501$ and $p < 0.005$. Their test system and the 36 cations are provided in Table 5.14.

Mendes et al. (2010) developed 4 electronegativity QSARs to predict EC_{50} values to the fungus *Gerronema viridilucens* for 18 cations (Table 5.12). Of the 4, the QSAR that only used electronegativity was the most robust (Table 5.12). The test system and cations used to develop the Mendes et al. (2010) QSAR are listed in Table 5.14.

5.2.5 PEARSON AND MAWBY SOFTNESS PARAMETER

Twenty studies used the Pearson and Mawby (1967) softness parameter (σ_p) to predict cation toxicity (Table 5.15). Only 5 of the 17 studies did not provide QSARs: Babich et al. (1986), Khangarot and Ray (1989), Segner et al. (1994), Magwood and George (1996), and Khangarot and Das (2009). However, QSARs were developed for Babich et al. (1986) and Magwood and George (1996) by extrapolating data from figures in those publications.

Jones and Vaughn (1978) used intraperitoneal LD_{50} data in mice and oral LD_{50} data in rats (Hg^{2+}) and mice (Ag^{2+}, Au^{1+}) to develop 4 QSARs for the Pearson and Mawby softness parameter (σ_p) that predicted mouse LD_{50} values for groups of 4, 7, 17, and 28 divalent cations (Table 5.16). The coefficients of determination decreased and the standard error increased as the number divalent cations increased from 4 to 28 (Table 5.16). The test system used to develop the Jones and Vaughn (1978) QSARs is listed in Table 5.15.

Williams and Turner (1981) used a compilation of intraperitoneal-dosed mouse LD_{50} data available in the literature to develop a QSAR for the Pearson and Mawby softness parameter (σ_p) to predict the LD_{50} value of 25 mono-, di-, and trivalent cations to mice (Table 5.1). The 25 cations used to develop the Williams and Turner (1981) QSAR are listed in Table 5.15.

TABLE 5.15
Studies Using the Pearson and Mawby (1967) Softness Parameter (σ_p) to Predict Metal Ion Toxicity

Reference	Test Systems	Metal Ions
Jones and Vaughn (1978)	Intraperitoneal (ip) LD_{50} data in mice and oral LD_{50} data in rats (Hg^{2+}) and mice (Ag^{1+}, Au^{1+})	The salts used were the chlorides (Ba, Be, Ca, Cr, Cs, Fe, H, In, K, La, Li, Mg, Na, Rb, Sc, Sn, Sr, Y), nitrates (Ag) and sulphates (Al, Cd, Co, Cu, Fe, Hg, Ni, Tl, Zn), $[Na_3[Au(S_2O_3)_2]]$ for the gold ion and $[Pb(OAc)_2]$ for lead ion
Williams and Turner (1981)	Compilation of intraperitoneal (ip) mouse LD_{50} data available in the literature	Mono, di, and trivalent metal ions: Ag^{1+}, Cs^{1+}, K^{1+}, Li^{1+}, Na^{1+}, Rb^{1+}, Ba^{2+}, Be^{2+}, Ca^{2+}, Cd^{2+}, Cu^{2+}, Co^{2+}, Fe^{2+}, Mg^{2+}, Ni^{2+}, Sn^{2+}, Sr^{2+}, Zn^{2+}, Al^{3+}, Cr^{3+}, Fe^{3+}, La^{3+}, In^{3+}, Sc^{3+}, Y^{3+}
Williams et al. (1982)	14-day ip LD_{50} data in mice and 4-day LC_{50} data from dietary exposures to *Drosophila melanogaster*	The salts used were the chlorides of 11 ions (Ba, Be, Cd, Co, Cu, Hg, Mg, Mn, Ni, Sr and Zn) for *Drosophila melanogaster* and 14 ions (Ba, Be, Cd, Co, Cu, Hg, Mg, Mn, Ni, Sr, Zn, Pd, Pt and Pb) for mice
Jacobson et al. (1983)	Dietary exposures of *Drosophila melanogaster* (4-day LC_{50})	Chloride salts of Ba^{2+}, Be^{2+}, Cd^{2+}, Co^{2+}, Cu^{2+}, Hg^{2+}, Mg^{2+}, Mn^{2+}, Ni^{2+}, Sr^{2+}, Zn^{2+}
Turner et al. (1983)	Data from Williams et al. (1982) ip injections in mice to estimate 14-day LD_{50} values for 8, 11 and 14 divalent cations	8 divalent cations Cd^{2+}, Co^{2+}, Cu^{2+}, Hg^{2+}, Mn^{2+}, Ni^{2+}, Pb^{2+}, Zn^{2+} 11 divalent cations Ba^{2+}, Cd^{2+}, Co^{2+}, Cu^{2+}, Hg^{2+}, Mg^{2+}, Mn^{2+}, Ni^{2+}, Pb^{2+}, Sr^{2+}, Zn^{2+} 14 divalent cations Ba^{2+}, Be^{2+}, Cd^{2+}, Co^{2+}, Cu^{2+}, Hg^{2+}, Mg^{2+}, Mn^{2+}, Ni^{2+}, Pb^{2+}, Pd^{2+}, Pt^{2+}, Sr^{2+}, Zn^{2+}
Tan et al. (1984)	Chinese hamster ovary (CHO) cells to measure 50% reduction in cloning efficiency (CE_{50})	11 divalent cations; except for berillium ($BeSO_4$) and palladium (K_2PdCl_4) ions, Cd^{2+}, Co^{2+}, Cu^{2+}, Hg^{2+}, Mg^{2+}, Mn^{2+}, Ni^{2+}, Sr^{2+}, Zn^{2+} were chloride salts
Turner et al. (1985)	Data from Williams et al. (1982) ip injections in mice to estimate 14-day LD_{50} values	8 divalent cations Cd^{2+}, Co^{2+}, Cu^{2+}, Hg^{2+}, Mn^{2+}, Ni^{2+}, Pb^{2+}, Zn^{2+}
Babich et al. (1986)	Bluegill (BF-2) cell cultures to measure reduction of uptake of neutral red dye by 50% (NR_{50})	9 metal ions: Chloride salts of Cd^{2+}, Co^{2+}, Cu^{2+}, Hg^{2+}, Mn^{2+}, Ni^{2+}, Sn^{2+}, Zn^{2+} and nitrate salt of Pb^{2+}
Khangarot and Ray (1989)	48-hr tests with *Daphnia magna* to measure EC_{50} values based on immobilization	17 metal ions; not specified which ones
Segner et al. (1994)	Rainbow trout cell cultures (R-1) to measure reduction of uptake of neutral red dye by 50% (NR_{50})	6 divalent metal ions: Chloride salts of Cd^{2+}, Hg^{2+}, Ni^{2+}, Zn^{2+}, nitrate salt of Pb^{2+} sulfate salt of Cu^{2+}
Magwood and George (1996)	Turbot (TF) cell cultures to measure reduction of uptake of neutral red dye by 50% (NR_{50})	7 divalent metal ions: Chloride salts of Cd^{2+}, Co^{2+}, Cu^{2+}, Mn^{2+}, Ni^{2+}, Pb^{2+}, Zn^{2+}

TABLE 5.15 (*Continued*)

Studies Using the Pearson and Mawby (1967) Softness Parameter (σ_p) to Predict Metal Ion Toxicity

Reference	Test Systems	Metal Ions
McCloskey et al. (1996)	Relative decrease in bioluminescence of *Vibrio fischeri* (unspeciated total metal ion 15-min EC_{50} values)	20 metal ions: Nitrate salts of Ag^+, Ba^{2+}, Ca^{2+}, Cd^{2+}, Co^{2+}, Cr^{3+}, Cs^+, Cu^{2+}, Fe^{3+}, Hg^{2+}, K^+, La^{3+}, Li^+, Mg^{2+}, Mn^{2+}, Na^+, Ni^{2+}, Pb^{2+}, Sr^{2+}, Zn^{2+}
Newman and McCloskey (1996)	Relative decrease in bioluminescence of *Vibrio fischeri* (unspeciated total metal ion 15-min EC_{50} values)	9 metal ions: Chloride salts of Ca^{2+}, Cd^{2+}, Cu^{2+}, Hg^{2+}, Mg^{2+}, Mn^{2+}, Ni^{2+}, Pb^{2+}, Zn^{2+}
Tatara et al. (1997)	24h LC_{50} values (expressed as free ion) for the soil nematode, *Caenorhabditis elegans*	9 metal ions: Chloride salts of Ca^{2+}, Cd^{2+}, Cu^{2+}, Hg^{2+}, Mg^{2+}, Mn^{2+}, Ni^{2+}, Pb^{2+}, Zn^{2+}
Tatara et al. (1998)	24h LC_{50} values (expressed as free ion) for the soil nematode, *Caenorhabditis elegans*	17 metal ions: Nitrate salts of Ca^{2+}, Cd^{2+}, Co^{2+}, Cr^{3+}, Cs^+, Cu^{2+}, Fe^{3+}, K^+, La^{3+}, Li^+, Mg^{2+}, Mn^{2+}, Na^+, Ni^{2+}, Pb^{2+}, Sr^{2+}, Zn^{2+}.
Enache et al. (1999)	EC_{50} values in sunflower (*Helianthus annuus* Sunspot) (determination of the concentration giving 50% inhibition of radicle growth one day after emergence)	10 divalent metal ions: Nitrate salts of Ba^{2+}, Ca^{2+}, Cd^{2+}, Co^{2+}, Cu^{2+}, Mg^{2+}, Mn^{2+}, Ni^{2+}, Pb^{2+}, Zn^{2+}
van Kolck et al. (2008)	96 hr LC_{50} values for the mussels, *Mytilis edulis* and *Perna viridis*	Ag^+, Cd^{2+}, Cu^{2+}, Hg^{2+}, Zn^{2+} for *Mytilis edulis* Cd^{2+}, Cr^{3+}, Cu^{2+}, Hg^{2+}, Pb^{2+}, Zn^{2+} for *Perna viridis*
Workentine et al. (2008)	*Pseudomonas fluorescens* ATCC 13525 was grown in a Calgary Biofilm Device and exposed to different metal ions	44 metal ions (anions and cations)
Khangarot and Das (2009)	48-hour tests with ostracod *Cypris subglobosa* Sowerby 1840 to measure EC_{50} values based on immobilization	19 metal ions; not specified which ones
Mendes et al. (2010)	Relative decrease in bioluminescence of fungus *Gerronema viridilucens*	Ag^{1+}, Ba^{2+}, Ca^{2+}, Cd^{2+}, Co^{2+}, Cs^{1+}, Cu^{2+}, Fe^{2+}, Hg^{2+}, K^{1+}, Li^{1+}, Mg^{2+}, Mn^{2+}, Na^{1+}, Ni^{2+}, Pb^{2+}, Sr^{2+}, Zn^{2+}

Williams et al. (1982) used intraperitoneal-dosed mouse LD_{50} data and dietary exposures to the fruit fly, *Drosophila melanogaster*, to develop QSARs for the Pearson and Mawby softness parameter (σ_p) to predict the 14-day LD_{50} value for mice and the 4-day LC_{50} value for *D. melanogaster* (Table 5.1). Williams et al. calculated the standard deviation for both QSARs, not the standard error of the estimate (SE) used by Jones and Vaughn (1978). Neither QSAR had very high r^2 values

TABLE 5.16

Jones and Vaughn (1978) Pearson and Mawby Softness Parameter (σ_p) QCARs

Test Substance(s)	QCARs	r^2	SE	p	n
Ag^+, Au^+, Cd^{2+}, Hg^{2+}	$\log LD_{50} = -1.95 + 18.76\sigma_p$	0.974	0.059	0.00628	4
Co^{2+}, Cu^{2+}, Fe^{2+}, Ni^{2+}, Pb^{2+}, Sn^{2+}, Zn^{2+}	$\log LD_{50} = -.999 + 18.79\sigma_p$	0.464	0.304		7
Al^{3+}, Ba^{2+}, Be^{2+}, Ca^{2+}, Cr^{3+}, Cs^{1+}, Fe^{3+}, H^{1+}, K^{1+}, La^{3+}, Li^{1+}, Mg^{2+}, Na^{1+}, Rb^{1+}, Sc^{3+}, Sr^{2+}, Y^{3+}	$\log LD_{50} = -1.706 + 11.35\sigma_p$	0.441	0.548	0.25	17
All cations	$\log LD_{50} = -1.64 + 9.94\sigma_p$	0.436	0.61	0.00004	28

Source: Data from M.M. Jones and W.K. Vaughn, "HSAB Theory and Acute Metal Ion Toxicity and Detoxification Processes. *J. Inorg. Nucl. Chem.* 40 (1978):2081–2088.

(Table 5.17). The test system and cations used to develop the Williams et al. (1982) QSARs are listed in Table 5.15.

Jacobsen et al. (1983) used dietary exposures in *Drosophila melanogaster* to develop a QSAR for the Pearson and Mawby softness parameter (σ_p) to predict the 4-day LC_{50} value for *D. melanogaster* (Table 5.1). The test system and cations used to develop the Jacobsen et al. (1983) QSAR are listed in Table 5.15.

Turner et al. (1983) used the intraperitoneal dosing data from Williams et al. (1982) to develop 3 QSARs for the Pearson and Mawby softness parameter (σ_p) that predicted mouse LD_{50} values for groups of 8, 11, and 14 divalent cations (Table 5.1). The coefficients of determination, standard deviations, and probabilities of chance correlations for groups of 8, 11, and 14 divalent cations decreased as the number of divalent cations increased (Table 5.17). The Turner et al. (1983) QSAR for 14 divalent cations is identical to the Williams et al. (1982) mouse QSAR for 14 divalent cations (Table 5.1). The test system used to develop the Turner et al. (1983) QSARs is listed in Table 5.15.

Tan et al. (1984) used Chinese hamster ovary (CHO) cell data to develop a QSAR with the Pearson and Mawby softness parameter (σ_p) to predict 50% reduction in cloning efficiency (CE_{50}) for 11 divalent cations (Table 5.1). The test system and cations used to develop the Tan et al. (1984) QSAR are listed in Table 5.15.

Turner et al. (1985) used the Williams et al. (1982) data to report the same QSAR that Turner et al. (1983) developed for 8 divalent cations (Table 5.1). The test system and cations used to develop the Turner et al. (1985) QSAR are listed in Table 5.15.

Babich et al. (1986) described the relationship between the toxicity of 9 cations and a 50% reduction in uptake of neutral red dye by fish cell cultures. A QSAR was extrapolated from Figure 6 of Babich et al. (1986) (Table 5.1). Their fish cultures were BF-2 cells, an established fibroblastic cell line derived from the caudal fin of the bluegill (*Lepomis macrochirus*). The 9 cations used to develop the Babich et al. (1986) QSAR are listed in Table 5.15.

Khangarot and Ray (1989) did not report any QSARs for the Pearson and Mawby softness parameter (σ_p). However, they did use linear regression analysis to develop

TABLE 5.17
Williams et al. (1982) and Turner et al. (1983) Pearson and Mawby Softness Parameter (σ_p) QSARs

Source	QSARs	Cations	r^2	sd	p	n
Williams et al (1982)	Mouse $\log LD_{50} = 9.56\sigma_p - 1.71$	Ba^{2+}, Be^{2+}, Cd^{2+}, Co^{2+}, Cu^{2+}, Hg^{+}, Mg^{2+}, Mn^{2+}, Ni^{2+}, Pb^{2+}, Pd^{2+}, Pt^{2+}, Sr^{2+}, Zn^{2+}	0.360	0.574	0.05	14
	Drosophila $\log LC_{50} = 9.77\sigma_p + 0.0993$	Ba^{2+}, Be^{2+}, Cd^{2+}, Cu^{2+}, Co^{2+}, Hg^{2+}, Mg^{2+}, Mn^{2+}, Ni^{2+}, Sr^{2+}, Zn^{2+}	0.503	0.404	0.02	11
Turner et al. (1983)	$\log LD_{50} = 23.1\sigma_p - 3.35$	Mn^{2+}, Co^{2+}, Ni^{2+}, Cu^{2+}, Zn^{2+}, Cd^{2+}, Hg^{2+}, Pb^{2+}	0.879	0.228	0.001	8
	$\log LD_{50} = 15.0\sigma_p - 2.56$	Mn^{2+}, Co^{2+}, Ni^{2+}, Cu^{2+}, Zn^{2+}, Cd^{2+}, Hg^{2+}, Pb^{2+}, Mg^{2+}, Sr^{2+}, Ba^{2+}	0.588	0.518	0.01	11
	$\log LD_{50} = 9.56\sigma_p - 1.71$	Ba^{2+}, Be^{2+}, Cd^{2+}, Co^{2+}, Cu^{2+}, Hg^{2+}, Mg^{2+}, Mn^{2+}, Ni^{2+}, Pb^{2+}, Pd^{2+}, Pt^{2+}, Sr^{2+}, Zn^{2+}	0.360	0.574	0.05	14

Source: Data from M.W. Williams, D. Hoeschele, J.E. Turner, K.B. Jacobson, N.T. Christie, C.L. Paton, L.H. Smith, H.R. Witschi, and E.H. Lee, "Chemical Softness and Acute Metal Toxicity in Mice and *Drosophila*." *Toxicol. Appl. Pharmacol.* 63 (1982):461–469; and J.E. Turner, E.H. Lee, K.B. Jacobson, N.T. Christie, M.W. Williams, and J.D. Hoeschele, "Investigation of Correlations between Chemical Parameters of Metal Ions and Acute Toxicity in Mice and *Drosophila*." *Sci. Total Environ.* 28 (1983):343–354.

correlation coefficients to describe the relationship between the toxicity of 17 unspecified cations and 2-day *Daphnia magna* EC_{50} values based on immobilization. Their correlation coefficients were converted to coefficients of determination and the statistics for their linear regression analysis were $r^2 = 0.682$ and $p < 0.001$. Their test system is provided in Table 5.15.

Segner et al. (1994) did not report any QSARs for the Pearson and Mawby softness parameter (σ_p). However, they did provide a correlation coefficient to describe the relationship between the toxicity of 6 divalent cations and reduction in uptake of neutral red dye (NR_{50}) for fish culture cells. Their correlation coefficients were converted to a coefficient of determination, $r^2 = 0.86$. Their fish cultures were a fibroblast-like cell line (R1 cells) from rainbow trout (*Oncorhynchus mykiss*) liver tissue. Their 6 cations are listed in Table 5.15.

Magwood and George (1996) described the relationship between the toxicity of 7 cations and a 50% reduction in uptake of neutral red dye by fish cell cultures. A QSAR was extrapolated from Figure 2B of Magwood and George (1996) (Table 5.1). Their fish cultures were TF (turbot, a flatfish) cells. The 7 cations used to develop the Magwood and George (1996) QSAR are listed in Table 5.15.

McCloskey et al. (1996) developed a QSAR for the Pearson and Mawby softness parameter (σ_p) to predict a decrease in bioluminescence of *Vibrio fischeri* for 20 cations (Table 5.9). The test system and cations used to develop the McCloskey et al. (1996) QSAR are listed in Table 5.15.

Newman and McCloskey (1996) also developed a QSAR for the Pearson and Mawby softness parameter (σ_p) to predict a decrease in bioluminescence of *Vibrio fischeri* for 9 cations (Table 5.9). The test system and cations used to develop the Newman and McCloskey (1996) QSAR are listed in Table 5.15.

Tatara et al. (1997) developed a QSAR for the Pearson and Mawby softness parameter (σ_p) to predict 24-hour LC_{50} values for the soil nematode *Caenorhabditis elegans* for 9 cations (Table 5.10). The test system and cations used to develop the Tatara et al. (1997) QSAR are listed in Table 5.15.

Tatara et al. (1998) developed 2 QSARs for the Pearson and Mawby softness parameter (σ_p) to predict 24-hour LC_{50} values for the soil nematode *Caenorhabditis elegans* for 17 cations (Table 5.10). Incorporating the logarithm of the first hydrolysis constant into the QSAR with the Pearson and Mawby softness parameter increased r^2 and decreased AIC (Table 5.10). The test system and cations used to develop the Tatara et al. (1998) QSAR are listed in Table 5.15. The studies of McCloskey et al. (1996), Newman and McCloskey (1996), Tatara et al. (1997), and Tatara et al. (1998) were all conducted in Newman's laboratory and discussed with other relevant information in Newman et al. (1998).

Enache et al. (1999) developed a QSAR for the Pearson and Mawby softness parameter (σ_p) to predict EC_{50} values in sunflower for 10 cations (Table 5.1). The test system and cations used to develop the Enache et al. (1999) QSAR are listed in Table 5.15.

Van Kolck et al. (2008) developed 4 QSARs to predict 96-hour LC_{50} values of 5 cations to the mussel, *Mytilis edulis* and 4 QSARs to predict the 96-hour LC_{50} values of 6 cations to the mussel *Perna viridis* (Table 5.17). The Pearson and Mawby softness parameter (σ_p) was used to develop 2 of these 8 QSARs, but neither QSAR produced very high r^2 values (Table 5.18).

TABLE 5.18

QCARs for Predicting Cation Toxicity and Bioconcentration Factors for the Mussels, *Mytilis edulis* and *Perna viridis* Using the Covalent Radius, Logarithm of the First Hydrolysis Constant, Pearson and Mawby Softness Parameter, and the Ionic Index

Species	Test Substance(s)	QCARs	r^2	SE	p	n
Mytilis edulis	Ag^+, Cd^{2+}, Cu^{2+}, Hg^{2+}, Zn^{2+}	$LogLC_{50} = 2.8 - 0.7 X_m^2 r$	0.79	0.28	0.04	5
		$LogLC_{50} = 1.1 + 0.4 \log|K_{OH}|$	0.05	0.81	0.71	5
		$LogLC_{50} = 1.0 + 0.24 \sigma_p$	0.31	0.69	0.33	5
		$LogLC_{50} = -0.19 + 0.25 Z^2/r$	0.42	0.63	0.24	5
Perna viridis	Cd^{2+}, Cr^{3+}, Cu^{2+}, Hg^{2+}, Pb^{2+}, Zn^{2+}	$LogLC_{50} = 3.6 - 0.8\, X_m^2 r$	0.74	0.47	0.03	6
		$LogLC_{50} = 0.99 + 0.05 \log|K_{OH}|$	0.03	0.91	0.74	6
			0.55	0.62	0.09	6
		$LogLC_{50} = -1.2 + 26\sigma_p$	0.41	0.71	0.17	6
		$LogLC_{50} = 0.6 + 0.13 Z^2/r$				
Mytilis edulis[a]	Ag^+, Cd^{2+}, Co^{2+}, Cr^{3+}, Cs^+, Hg^{2+}, Pb^{2+}, Zn^{2+}	$LogBCF = 1.4 - 0.5 X_m^2 r$	0.56	0.56	0.003	8
		$LogBCF = 2.0 + 0.1 \log|K_{OH}|$	0.44	0.64	0.07	8
		$LogBCF = 2.8 + 0.1\, \sigma_p$	0.02	0.85	0.76	8
		$LogBCF = 3.6 + 0.1 Z^2/r$	0.58	0.55	0.03	8
Perna viridis	Ag^+, Cd^{2+}, Cs^+, Cu^{2+}, Hg^{2+}, Pb^{2+}, Zn^{2+}	$LogBCF = 1.1 - 1.0\, X_m^2 r$	0.74	0.76	0.01	7
		$LogBCF = 2.7 + 0.2 \log|K_{OH}|$	0.29	1.3	0.21	7
		$Log\, BCF = 5.2 - 6.5\sigma_p$	0.77	0.73	0.01	7
		$Log\, BCF = 5.6 - 0.3 Z^2/r$	0.76	0.75	0.01	7

Source: Data from M. van Kolck, M.A.J. Huijbregts, K. Veltman, and A.J. Hendriks, "Estimating Bioconcentration Factors, Lethal Concentrations and Critical Body Residues of Metals in the Mollusks *Perna viridis* and *Mytilus edulis* Using Ion Characteristics." *Environ. Toxicol. Chem.* 27 (2008):272–276. Supplementary information available at http://onlinelibrary.wiley.com/store/10.1897/07-224R.1/asset/supinfo/10.1897_07-224.S1.pdf?v=1&s=fda547dc5689332031 992621f5e3833819a8e96f.

[a] Van Kolck et al. (2008) reported that the BCF QCARs for *Mytilis edulis* were developed for 7 cations. However, the supplementary information cited in their publication listed 8 cations for which *Mytilis edulis* BCFs were developed. The 8 cations above are from the supplementary information, http://onlinelibrary. wiley.com/store/10.1897/07-224R.1/asset/supinfo/10.1897_07-224.S1.pdf?v=1&s=fda547 dc5689332031992621f5e3833819a8e96f.

$X_m^2 r$ is the covalent index.

$\log|K_{OH}|$ is the logarithm of the first hydrolysis constant.

σ_p is the Pearson and Mawby softness parameter.

Z^2/r is the ionic index.

Workentine et al. (2008) used the test system discussed in Section 5.2.2 and listed in Table 5.15. Linear regression analyses were conducted to determine the relationships between the measured toxicity values and σ_p: $r^2 = 0.23$, $p = 0.032$ (MIC_{24}), $r^2 = 0.30$, $p = 0.043$ (MBC_{100}) and $r^2 = 0.32$, $p = 0.113$ ($MBEC_{100}$).

Khangarot and Das (2009) did not report any QSARs for the Pearson and Mawby softness parameter (σ_p). However, they did use linear regression analysis to develop correlation coefficients to describe the relationship between the toxicity of 19 unspecified cations and 2-day *Cypris subglobosa* Sowerby 1840 EC_{50} values based on immobilization. Their correlation coefficients were converted to coefficients of determination and the statistics for their linear regression analysis were $r^2 = 0.64$ and $p < 0.005$. Their test system is provided in Table 5.15.

Mendes et al. (2010) developed 7 QSARs using the Pearson and Mawby softness parameter (σ_p) alone or combined with electronegativity, logarithm of the first hydrolysis constant, covalent index, or atomic radius to predict EC_{50} values to the fungus *Gerronema viridilucens* for 18 cations (Table 5.12). Of the 7 QSARs, the combination of the Pearson and Mawby softness parameter, electronegativity, and the logarithm of the first hydrolysis constant had a high coefficient of determination and the lowest AIC (Table 5.12). The test system and cations used to develop the Mendes et al. (2010) QSAR are listed in Table 5.15.

Table 5.19 lists the 21 QSARs that only used the Pearson and Mawby softness parameter (σ_p) to predict cation toxicity. Two of these QSARs are duplicates. The Turner et al. (1983) QSAR with $r^2 = 0.360$ is a duplicate of the Williams et al. (1982) QSAR with the same r^2 value. The Turner et al. (1985) QSAR with $r^2 = 0.879$ is a duplicate of the Turner et al. (1983) QSAR with the same r^2 value. The Jones and Vaughn (1978) QSAR for Ag^+, Au^+, Cd^{2+} and Hg^{2+} had the highest r^2, perhaps because of their proximity in the periodic table. The Turner et al. (1983) QSAR for Mn^{2+}, Co^{2+}, Ni^{2+}, Cu^{2+}, Zn^{2+}, Cd^{2+}, Hg^{2+} and Pb^{2+} had the second highest r^2, perhaps because Mn^{2+}, Co^{2+}, Ni^{2+}, Cu^{2+} and Zn^{2+} were in row 4 of the periodic table and Cd^{2+}, Hg^{2+} and Pb^{2+} were in close proximity. For the Jones and Vaughn (1978) and Turner et al. (1983) references with 4 and 3 QSARs, respectively, the r^2 decreased as the number of divalent cations increased (Table 5.18). The Babich et al. (1986) and Magwood and George (1996) QSARs both had high r^2 and almost identical cations. The Babich et al. (1986) QSAR had almost identical cations to those used for the Turner et al. (1983) QSAR with $r^2 = 0.879$ (Table 5.18). The Enache et al. (1999) QSAR also had a high r^2. However, the Mendes et al. (2010) QSAR with a high r^2 was the best for the highest number (18) of cations (Table 5.19).

5.3 LESS COMMON PHYSICOCHEMICAL PROPERTIES USED TO PREDICT CATION TOXICITY

Table 5.20 lists the QSARs with less common physicochemical properties used to predict cation toxicity. Only a few of the test systems for the references listed in Table 5.20 have not been previously described. These test systems are described here.

Sauvant et al. (1997) used the ionization potential (IP) of 16 cations and five bioassays (RNA synthesis rate assay, MTT reduction assay, neutral red incorporation

TABLE 5.19

QCARs That Used Only the Pearson and Mawby (1967) Softness Parameter (σ_p) to Predict Cation Toxicity

Reference	QCAR	r^2	Cations
Jones and Vaughn (1978)	$LD_{50} = -1.706 + 11.35\sigma_p$	0.974	Ag^+, Au^+, Cd^{2+}, Hg^{2+}
	$LD_{50} = -2.999 + 18.79\sigma_p$	0.464	Co^{2+}, Cu^{2+}, Fe^{2+}, Ni^{2+}, Pb^{2+}, Sn^{2+}, Zn^{2+}
	$LogLD_{50} = -1.95 + 18.76\sigma_p$	0.441	Al^{3+}, Ba^{2+}, Be^{2+}, Ca^{2+}, Cr^{3+}, Cs^{1+}, Fe^{3+}, H^{1+}, K^{1+}, La^{3+}, Li^{1+}, Mg^{2+}, Na^{1+}, Rb^{1+}, Sc^{3+}, Sr^{2+}, Y^{3+}
	$LogLD_{50} = -1.64 + 9.94\sigma_p$	0.436	Ag^+, Al^{3+}, Au^+, Ba^{2+}, Be^{2+}, Ca^{2+}, Cd^{2+}, Co^{2+}, Cr^{3+}, Cs^{1+}, Cu^{2+}, Fe^{3+}, Fe^{2+}, H^{1+}, Hg^{2+}, K^{1+}, La^{3+}, Li^{1+}, Mg^{2+}, Na^{1+}, Ni^{2+}, Pb^{2+}, Rb^{1+}, Sc^{3+}, Sn^{2+}, Sr^{2+}, Zn^{2+}, Y^{3+}
Williams and Turner (1981)	$logLD_{50} = -2.30 + 14.1\sigma_p$	0.656	Ag^{1+}, Cs^{1+}, K^{1+}, Li^{1+}, Na^{1+}, Rb^{1+}, Ba^{2+}, Be^{2+}, Ca^{2+}, Cd^{2+}, Cu^{2+}, Co^{2+}, Fe^{2+}, Mg^{2+}, Ni^{2+}, Sn^{2+}, Sr^{2+}, Zn^{2+}, Al^{3+}, Cr^{3+}, Fe^{3+}, La^{3+}, In^{3+}, Sc^{3+}, Y^{3+}
Williams et al. (1982)	$logLC_{50} = 0.0993 + 9.77\sigma_p$	0.503	Ba^{2+}, Be^{2+}, Cd^{2+}, Cu^{2+}, Co^{2+}, Hg^{2+}, Mg^{2+}, Mn^{2+}, Ni^{2+}, Sr^{2+}, Zn^{2+}
	$logLD_{50} = -1.71 + 9.56\sigma_p$	0.360	Ba^{2+}, Be^{2+}, Cd^{2+}, Co^{2+}, Cu^{2+}, Hg^{2+}, Mg^{2+}, Mn^{2+}, Ni^{2+}, Pb^{2+}, Pd^{2+}, Pt^{2+}, Sr^{2+}, Zn^{2+}
Jacobson et al. (1983)	$logLC_{50} = 0.109 + 9.74\sigma_p$	0.494	Ba^{2+}, Be^{2+}, Cd^{2+}, Co^{2+}, Cu^{2+}, Hg^{2+}, Mg^{2+}, Mn^{2+}, Ni^{2+}, Sr^{2+}, Zn^{2+}
Turner et al. (1983)	$logLD_{50} = -3.35 + 23.1\sigma_p$	0.879	Cd^{2+}, Co^{2+}, Cu^{2+}, Hg^{2+}, Mn^{2+}, Ni^{2+}, Pb^{2+}, Zn^{2+}
	$logLD_{50} = -2.56 + 15.0\sigma_p$	0.588	Ba^{2+}, Cd^{2+}, Co^{2+}, Cu^{2+}, Hg^{2+}, Mg^{2+}, Mn^{2+}, Ni^{2+}, Pb^{2+}, Sr^{2+}, Zn^{2+}
	$logLD_{50} = -1.71 + 9.56\sigma_p$	0.360	Ba^{2+}, Be^{2+}, Cd^{2+}, Co^{2+}, Cu^{2+}, Hg^{2+}, Mg^{2+}, Mn^{2+}, Ni^{2+}, Pb^{2+}, Pd^{2+}, Pt^{2+}, Sr^{2+}, Zn^{2+}
Tan et al. (1984)	$logCE_{50} = -0.28 + 24.1\sigma$	0.449	Be^{2+}, Cd^{2+}, Co^{2+}, Cu^{2+}, Hg^{2+}, Mg^{2+}, Mn^{2+}, Ni^{2+}, Pd^{2+}, Sr^{2+}, Zn^{2+}
Turner et al. (1985)	$logLD_{50} = -3.35 + 23.1\sigma_p$	0.879	Cd^{2+}, Co^{2+}, Cu^{2+}, Hg^{2+}, Mn^{2+}, Ni^{2+}, Pb^{2+}, Zn^{2+}
Babich et al. (1986)	$NR_{50} = 0.01 + 31.9\sigma_p$	0.826	Cd^{2+}, Co^{2+}, Cu^{2+}, Hg^{2+}, Mn^{2+}, Ni^{2+}, Pb^{2+}, Sn^{2+}, Zn^{2+}
Magwood and George (1996)	$NR_{50} = -0.45 + 18.9\sigma_p$	0.865	Cd^{2+}, Co^{2+}, Cu^{2+}, Mn^{2+}, Ni^{2+}, Pb^{2+}, Zn^{2+}
McCloskey et al. (1996)	$Log\ EC50 = -3.25 + 39.95(\sigma_p)$	0.73	Ag^+, Ba^{2+}, Ca^{2+}, Cd^{2+}, Co^{2+}, Cr^{3+}, Cs^+, Cu^{2+}, Fe^{3+}, Hg^{2+}, K^+, La^{3+}, Li^+, Mg^{2+}, Mn^{2+}, Na^+, Ni^{2+}, Pb^{2+}, Sr^{2+}, Zn^{2+}
Newman and McCloskey (1996)	$Log\ EC50 = 12.87 - 110.9(\sigma_p)$	0.60	Ca^{2+}, Cd^{2+}, Cu^{2+}, Hg^{2+}, Mg^{2+}, Mn^{2+}, Ni^{2+}, Pb^{2+}, Zn^{2+}

(Continued)

TABLE 5.19 (*Continued*)

QCARs That Used Only the Pearson and Mawby (1967) Softness Parameter (σ_p) to Predict Cation Toxicity

Reference	QCAR	r^2	Cations
Tatara et al. (1997)	Log EC50 = −10.47 + 83.49(σ_p)	0.50	Ca^{2+}, Cd^{2+}, Cu^{2+}, Hg^{2+}, Mg^{2+}, Mn^{2+}, Ni^{2+}, Pb^{2+}, Zn^{2+}
Tatara et al. (1998)	Log EC50 = −1.23 + 16.72(σ_p)	0.50	Ca^{2+}, Cd^{2+}, Co^{2+}, Cr^{3+}, Cs^+, Cu^{2+}, Fe^{3+}, K^+, La^{3+}, Li^+, Mg^{2+}, Mn^{2+}, Na^+, Ni^{2+}, Pb^{2+}, Sr^{2+}, Zn^{2+}.
Enache et al. (1999)	−(log EC_{50}) = 3.59 − 14.4(σ_p)	0.83	Ba^{2+}, Ca^{2+}, Cd^{2+}, Co^{2+}, Cu^{2+}, Mg^{2+}, Mn^{2+}, Ni^{2+}, Pb^{2+}, Zn^{2+}
Mendes et al. (2010)	Log EC_{50} = −0.77 + 17.52(σ_p)	0.799	Ag^{1+}, Ba^{2+}, Ca^{2+}, Cd^{2+}, Co^{2+}, Cs^{1+}, Cu^{2+}, Fe^{2+}, Hg^{2+}, K^{1+}, Li^{1+}, Mg^{2+}, Mn^{2+}, Na^{1+}, Ni^{2+}, Pb^{2+}, Sr^{2+}, Zn^{2+}

[NRI] assay, Coomassie blue [CB] assay, and cellular growth rate [CGR] assay) to develop QSARs to predict IC_{50} values, that is, the concentration that inhibited 50% of the L-929 cell line of murine fibroblasts (Table 5.20). In addition, Sauvant et al. (1997) developed a QSAR to predict the IC_{50} values for the ciliated protozoa, *Tetrahymena pyriformis* GL, using the doubling time from a *T. pyriformis* GL population assay or DTP (Table 5.20).

Enache et al. (2003) developed 18 QSARs to predict cation toxicity (Table 5.1). Enache et al. (2003) determined the concentration at which there is a 50% decrease in the rate of callus development in the sunflower (*Helianthus annuus* F.1. Sunspot) and found a large number of correlations with relevant physical and chemical descriptors for 16 of their QSARs (Table 5.20).

Hickey (2005) developed linear solvation energy relationships (LSERs) for neutral inorganic species based on LSER equations developed for neutral organic compounds. The general LSER equation was log(Property) = $mV_i/100 + s\pi^* + b\beta_m + a\alpha_m$, where V_i was the intrinsic (van der Waals) molecular volume, π^* was the solute ability to stabilize a neighboring charge or dipole by nonspecific dielectric interactions, and β_m and α_m were the solute ability to accept or donate a hydrogen in a hydrogen bond. The coefficients m, s, b, and a were constants for a particular set of conditions, determined by multiple linear regression of the LSER variable values for a series of chemicals with the measured value for a particular chemical property (Table 5.3). Hickey's (2005) LSERs to predict the water solubility and toxicity of cations are referred to as QSARs in Table 5.4. Hickey (2005) used data from McCloskey et al. (1996), Newman and McCloskey (1996), and deZwart and Sloof (1983) to predict the toxicity of 40 cations to *Vibrio fischeri* in the Microtox® test (Table 5.4). Hickey (2005) used data reported by Bringmann and Kuhn (1977, 1982), Khangarot and Ray (1989), and LeBlanc (1984) to predict the toxicity of 53 cations to *Daphnia magna* and data from Junhke and Ludemann (1978) to predict the toxicity of 32 cations to the Golden orfe *Leuciscus idus melanotus* (Table 5.4). The LSERs/QSARs from Hickey (2005) are listed after those of Walker et al. (2007) in Table 5.20, so that the r^2 adj. values could be used without interruption from Enache et al. (2000) to Walker et al. (2007).

TABLE 5.20

QCARs with Less Numerous Physicochemical Properties Used to Predict Cation Toxicity

Reference	Test Substances	QCARs	r^2	p	n	
Biesinger and Christensen (1972)	Ba^{2+}, Ca^{2+}, Co^{2+}, Cu^{2+}, K^+, Mg^{2+}, Mn^{2+}, Na^+, Ni^{2+}, Sr^{2+}, Zn^{2+}	$pM = 1.6 \log K_{eq} - 1.9$	0.55	<0.01	11	A

Reference	Test substance(s)	QCARs	r^2	s	p	n	
			r^2	MSE	AIC		
McCloskey et al. (1996)	Ag^+, Ba^{2+}, Ca^{2+}, Cd^{2+}, Co^{2+}, Cr^{3+}, Cs^+, Cu^{2+}, Fe^{3+}, Hg^{2+}, K^+, La^{3+}, Li^+, Mg^{2+}, Mn^{2+}, Na^+, Ni^{2+}, Pb^{2+}, Sr^{2+}, Zn^{2+}	$\text{Log EC50} = 5.59 - 1.40(X^2_m r)$	0.64	1.560	77.74	20	A
		$\text{Log EC50} = 2.51 + 0.51(\text{llog } K_{OH})$	0.56	1.716	82.44	20	B
		$\text{Log EC50} = 1.48 - 1.00(X^2_m r) + 0.32(\text{llog } K_{OH})$	0.81	1.161	65.48	20	D
		$\text{Log EC50} = 6.36 - 1.39(X^2_m r) - 0.16(Z^2/r)$	0.70	1.464	75.28	20	F
Newman and McCloskey (1996)	Ca^{2+}, Cd^{2+}, Cu^{2+}, Hg^{2+}, Mg^{2+}, Mn^{2+}, Ni^{2+}, Pb^{2+}, Zn^{2+}	$\text{Log EC50} = -16.65 + 1.87(\text{llog } K_{OH})$	0.93	1.534	35.13	9	B
		$\text{Log EC50} = 5.71 - 1.56(X^2_m r)$	0.34	4.595	66.44	9	A
		$\text{Log EC50} = -0.55 + 1.44(\Delta\beta)$	0.82	2.403	45.26	9	
		$\text{Log EC50} = 12.59 - 0.78(\text{Log-}K_{so}\text{MOH})$	0.75	2.845	49.87	9	
		$\text{Log EC50} = 1.70 - 1.62(X^2_m r) + 0.690(Z^2/r)$	0.35	4.926	66.42	9	F
		$\text{Log EC50} = -5.22 + 1.36(\Delta\beta) + 0.98(Z^2/r)$	0.85	2.381	44.64	9	
Sauvant et al (1997)	Ba^{2+}, Cd^{2+}, Co^{2+}, Cr^{3+}, Cu^{2+}, Fe^{3+}, Ge^{4+}, Hg^{2+}, Mn^{3+}, Nb^{5+}, Pb^{2+}, Sb^{3+}, Sn^{4+}, Ti^{4+}, V^{5+}, Zn^{2+}	RNA: $\text{logIC}_{50} = 2.012 - 0.226\text{IP}(eV)$	0.397		0.01	16	C
		MTT: $\text{logIC}_{50} = 7.024 - 0.796\text{IP}(eV)$	0.340		0.02	16	C
		NRI: $\text{logIC}_{50} = 7.641 - 0.847\text{IP}(eV)$	0.270		0.04	16	C
		CB: $\text{logIC}_{50} = 7.814 - 0.859\text{IP}(eV)$	0.321		0.02	16	C
		CGR: $\text{logIC}_{50} = 5.375 - 0.566\text{IP}(eV)$	0.141		0.15	16	C
		DTP: $\text{logIC}_{50} = 2.988 - 0.303\text{IP}(eV)$	0.278		0.04	16	C

(Continued)

TABLE 5.20 (Continued)

QCARs with Less Numerous Physicochemical Properties Used to Predict Cation Toxicity

Reference	Test Substances	QCARs	r^2	s	p	n	
Tatara et al. (1997)	Ca^{2+}, Cd^{2+}, Cu^{2+}, Hg^{2+}, Mg^{2+}, Mn^{2+}, Ni^{2+}, Pb^{2+}, Zn^{2+}	Log EC50 $= 3.06 - 1.23\left(X_m^2 r\right)$	0.22	17.058		9	A
		Log EC50 $= -1.21 + 1.11(\Delta\beta)$	0.71	6.335		9	
		Log EC50 $= 8.64 - 0.58(\text{Log-}K_{so}\text{MOH})$	0.61	8.507		9	B
		Log EC50 $= -13.83 + 1.47(\text{llog } K_{OH})$	0.83	3.702		9	F
		Log EC50 $= -3.13 - 0.86\left(X_m^2 r\right) + 1.06(Z^2/r)$	0.25	19.008		9	
		Log EC50 $= -5.54 + 1.03(\Delta\beta) + 0.90(Z^2/r)$	0.75	6.459		9	
Tatara et al. (1998)	Ca^{2+}, Cd^{2+}, Co^{2+}, Cr^{3+}, Cs^+, Cu^{2+}, Fe^{3+}, K^+, La^{3+}, Li^+, Mg^{2+}, Mn^{2+}, Na^+, Ni^{2+}, Pb^{2+}, Sr^{2+}, Zn^{2+}.	Log EC50 $= 2.59 - 0.64\left(X_m^2 r\right)$	0.54	0.694	42.04	17	A
		Log EC50 $= -1.43 + 0.27(\text{llog } K_{OH})$	0.67	0.501	36.50	17	B
		Log EC50 $= 3.22 - 0.60\left(X_m^2 r\right) - 0.14(Z^2/r)$	0.75	0.397	33.32	17	F
		Log EC50 $= 0.05 - 0.39\left(X_m^2 r\right) + 0.20(\text{llog } K_{OH})$	0.83	0.280	27.64	17	D

Reference	Test Substance(s)	QCARs	r^2 adj.	s	p	n	
Enache et al. (1999)	Ba^{2+}, Ca^{2+}, Cd^{2+}, Co^{2+}, Mg^{2+}, Mn^{2+}, Ni^{2+}, Pb^{2+}, Zn^{2+}	$-(\log EC_{50}) = 2.67 + 0.000012(\Delta H_o)$		0.19	0.000	10	
		$-(\log EC_{50}) = 2.07 + 0.00876(\Delta H_{aq\ ion.})$		0.21	0.000	10	
Enache et al. (2000)	Cd^{2+}, Co^{2+}, Cr^{3+}, Cr^{6+}, Fe^3, Hg^+, Hg^{2+}, Mn^{2+}, Ni^2, V^{3+}, Zn^{2+}	$-(\log Rrg_{50}) = -1.55 + 0.0442\ AN + 0.246\ \Delta IP$	0.65	0.67	0.006	11	
		$-(\log Rrg_{50}) = -1.28 + 0.0165\ AW + 0.242\ \Delta IP$	0.65	0.66	0.006	11	
Enache et al. (2003)	Ca^{2+}, Cd^{2+}, Co^{2+}, Cu^{2+}, Li^+, Mg^{2+}, Mn^{2+}, Na^{1+}, Ni^{2+}, Zn^{2+}, La^{3+}	$-(\log EC_{50}) = 3.74 - 0.0167\ \Delta H_h$	0.66	0.74	0.000	12	
		$-(\log EC_{50}) = 0.867 + 0.000296\ \rho$	0.70	0.69	0.000	12	
		$-(\log EC_{50}) = -0.145 + 0.182\ IP$	0.67	0.73	0.000	12	C
		$-(\log EC_{50}) = 7.93 - 0.502\ \text{llog } K_{OH}$	0.77	0.61	0.000	12	B
		$-(\log EC_{50}) = 0.430 + 0.165\ \log K(EDTA)$	0.74	0.64	0.000	12	
		$-(\log EC_{50}) = 0.212 + 1.37\ X_m^2 r$	0.65	0.75	0.000	12	A
		$-(\log EC_{50}) = -1.04 + 0.445\ \Delta IP$	0.66	0.74	0.000	12	

	Ions	Equation				A
	Ca²⁺, Cd²⁺, Co²⁺, Cu²⁺, K¹⁺, Li¹⁺, Mg²⁺, Mn²⁺, Na¹⁺, Ni²⁺, Zn²⁺	$-(\log EC_{50}) = -0.149 + 1.48\ X_m^2 r$	0.88	0.43	0.000	11
		$-(\log EC_{50}) = 2.01 - 3.64\ IR$	0.81	0.38	0.000	11
		$-(\log EC_{50}) = 1.87 - 0.0222\ AR$	0.80	0.40	0.000	11
	Ca²⁺, Co²⁺, Cu²⁺, K¹⁺, Mg²⁺, Mn²⁺, Na¹⁺, Ni²⁺, Zn²⁺	$-(\log EC_{50}) = -3.83 + 0.552\ \log K(ATP)$	0.82	0.46	0.000	9
	Ca²⁺, Co²⁺, Cu²⁺, Mg²⁺, Mn²⁺, Ni²⁺, Zn²⁺	$-(\log EC_{50}) = -3.30 + 2.33\ \log K(AMP)$	0.90	0.36	0.001	7
		$-(\log EC_{50}) = 1.23 + 1.47\ Y'$	0.92	0.41	0.000	7
		$-(\log EC_{50}) = -3.14 + 2.35\ \log K(AMP\text{-}2)$	0.93	0.31	0.000	7
		$-(\log EC_{50}) = -3.55 + 2.59\ \log K(AMP\text{-}3)$	0.95	0.25	0.000	7
		$-(\log EC_{50}) = -0.475 + 2.19\ \log K(Ac)$	0.93	0.32	0.000	7
	Ca²⁺, Cd²⁺, Cu²⁺, Mg²⁺, Ni²⁺, Zn²⁺	$-(\log EC_{50}) = 1.03 + 0.820\ \log K(NH_3)$	0.99	0.14	0.000	6
		$-(\log EC_{50}) = -2.85 + 0.369\ \log K(EDTA)$	0.94	0.29	0.001	6
Walker et al. 2007	Ag¹⁺, Al³⁺, Ba²⁺, Ca²⁺, Cd²⁺, Co²⁺, Cu²⁺, Fe³⁺, K¹⁺, La³⁺, Li¹⁺, Mg²⁺, Mn²⁺, Na¹⁺, Ni²⁺, Pb²⁺, Zn²⁻	$-(\log EC_{50}) = 0.756 + 0.000140\ \rho$	0.72	0.32	0.000	17
		$-(\log EC_{50}) = 0.534 + 0.000157\ \rho - 0.00166\ \Delta H_s$	0.73	0.32	0.000	17
		$-(\log EC_{50}) = 0.636 + 0.000209\ \rho - 0.00367\ \Delta H_s - 0.242\ \log K_1\ (\text{sulphate})$	0.81	0.265	0.000	17
		$-(\log EC_{50}) = 1.17 - 0.254\ \log M_{soil}$	0.69	0.34	0.000	17
		$-(\log EC_{50}) = 0.916 - 0.337\ \log M_{soil} + 0.000014\ \text{mg } X/\text{kg soil}$	0.81	0.27	0.000	17
		$-(\log EC_{50}) = 1.04 - 0.302\ \log M_{soil} + 0.000013\ \text{mg } X/\text{kg soil} - 0.000025\ \text{Land plants } X/\text{kg soil}$	0.83	0.25	0.000	17

(Continued)

TABLE 5.20 (Continued)
QCARs with Less Numerous Physicochemical Properties Used to Predict Cation Toxicity

Reference	Test Substances	QCARs	r^2	s	p	n	
Hickey (2005)	Cations from McCloskey et al. (1996), Newman and McCloskey (1996) and deZwart and Sloof (1983)	$\log(EC50)=7.49 - 7.39V_f/100 - 1.38\,\pi^* + 3.70\beta - 1.66\alpha$	0.97	0.319		40	
	Cations from Bringmann and Kuhn (1977,1982), Khangarot and Ray (1989), and LeBlanc (1984)	$\log(EC50)=4.18 - 4.73V_f/100 - 1.67\,\pi^* + 1.48\beta - 0.93\alpha$	0.95	0.221		53	
	Cations from Junhke and Ludemann (1978)	$\log(LC50)=2.90 - 5.71\,V_f/100 - 0.92\,\pi^* + 4.36\beta - 1.27\alpha$	0.94	0.246		32	

Reference	Test Substances	QCARs	r^2	SE	p	n	
van Kolck et al. (2008)	Ag^+, Cd^{2+}, Cu^{2+}, Hg^{2+}, Zn^{2+} toxicity to *M. edulis*	$\text{Log } LC_{50} = 2.8 - 0.7X_m^2 r$	0.79	0.28	0.04	5	A
		$\text{Log } LC_{50} = 1.1 + 0.4 \log|K_{OH}|$	0.05	0.81	0.71	5	B
		$\text{Log } LC_{50} = -0.19 + 0.25 Z^2/r$	0.42	0.63	0.24	5	E
	Cd^{2+}, Cr^{3+}, Cu^{2+}, Hg^{2+}, Pb^{2+}, Zn^{2+} toxicity to *P. viridis*	$\text{Log } LC_{50} = 3.6 - 0.8\, X_m^2 r$	0.74	0.47	0.03	6	A
		$\text{Log } LC_{50} = 0.99 + 0.05 \log|K_{OH}|$	0.03	0.91	0.74	6	B
		$\text{Log } LC_{50} = 0.6 + 0.13 Z^2/r$	0.41	0.71	0.17	6	E

Reference	Test Substances	QCARs	r^2	s	F	n
Lepădatu et al. (2009)	Ag^{1+}, Al^{3+}, Ba^{2+}, Cd^{2+}, Li^{1+}, Pb^{2+}, Zn^{2+}	$-\log EC_{50}= 1.88 - 1.65 \text{ HELH}$	0.078	0.613	0.42	7
		$-\log EC_{50}= 2.74 - 0.246 \text{ HEO}$	0.502	0.450	5.03	7
		$-\log EC_{50}= 1.64 + 0.081 \text{ HEM}$	0.521	0.442	5.43	7
		$-\log EC_{50}= 2.74 - 0.049 \text{ O-EO}$	0.503	0.450	5.05	7
		$-\log EC_{50}= 1.64 + 0.016 \text{ O-EM}$	0.522	0.441	5.45	7
		$-\log EC_{50}= 2.44 - 0.033 \text{ U-ELH}$	0.231	0.560	1.5	7
		$-\log EC_{50}= 3.20 - 0.272 \text{ U-EO}$	0.581	0.413	6.94	7
		$-\log EC_{50}= 0.979 + 0.031 \text{ U-EM}$	0.285	0.285	19.98	7

Test Substances: Co^{2+}, Cu^{2+}, Fe^{3+}, Mn^{2+}, Ni^{2+}

QCARs	r^2	s	F	n	
$-\log EC_{50} = 1.38 + 1.55\ HELH$	0.73	0.188	5.75	5	
$-\log EC_{50} = 1.81 - 0.021\ HEO$	0.01	0.362	0.0	5	
$-\log EC_{50} = -0.39 + 0.152\ HEM$	0.555	0.241	3.75	5	
$-\log EC_{50} = 1.75 + 0.009\ O\text{-}EO$	0.017	0.359	0.05	5	
$-\log EC_{50} = 1.49 + 0.004\ O\text{-}EM$	0.042	0.355	0.13	5	
$-\log EC_{50} = 1.00 + 0.033\ U\text{-}ELH$	0.754	0.180	9.18	5	
$-\log EC_{50} = 0.975 + 0.124\ U\text{-}EO$	0.503	0.255	3.04	5	
$-\log EC_{50} = 2.06 - 0.012\ U\text{-}EM$	0.657	0.212	5.75	5	

Reference Roy et al. (2009)

Test Substances: Ca^{2+}, Co^{2+}, Cr^{3+}, Cu^{2+}, Fe^{3+}, Mg^{2+}, Mn^{2+}, Na^+, Ni^{2+}, Zn^{2+}

QCARs	r^2	s	F	n
$Calc.LC_{50} = -0.0181\ AN + 0.5272$	0.923	0.054		10

Test Substances: Ca^{2+}, Co^{2+}, Cr^{3+}, Fe^{3+}, K^+, Li^+, Mg^{2+}, Mn^{2+}, Na^+, Ni^{2+}

QCARs	r^2	s	F	n
$Calc.LC_{50} = -7.1 \times 10^{-04}\ \omega + 0.2987$	0.793	4.337		10

Reference Sacan et al. (2009)

Test Substances: Ag^+, Cd^{2+}, Co^{2+}, Cr^{3+}, Cu^{2+}, Hg^{2+}, Ni^{2+}, Zn^{2+}

QCARs	r^2	SE	F	n	
$pTO_2 = -1.18(\pm0.14)\ Z + 3.47(\pm0.29)$	0.92	0.20	67.40	8	
$pTO_2 = 0.07(\pm0.02)\ E_{LUMO(g)} + 2.61(\pm0.38)$	0.75	0.36	17.51	8	
$pTO_2 = 0.08(\pm0.01)\ E_{HOMO(g)} + 3.74(\pm0.47)$	0.85	0.27	33.32	8	
$pTO_2 = -0.10(\pm0.02)\ IP + 2.87(\pm0.32)$	0.85	0.27	34.25	8	C
$pTO_2 = -0.14(\pm0.04)\ Z^2/r + 1.92(\pm0.25)$	0.72	0.37	15.67	8	E
$pTO_2 = 0.06(\pm0.01)\ E_{PSS} + 2.28(\pm0.29)$	0.77	0.34	20.34	8	
$pTO_2 = 0.08(\pm0.01)\ \mu_{(g)} + 3.39(\pm0.32)$	0.90	0.22	54.12	8	
$pTO_2 = 0.22(\pm0.06)\ \mu_{(aq)} + 2.96(\pm0.52)$	0.70	0.39	13.76	8	
$pTCO_2 = -1.17(\pm0.21)\ Z + 3.26(\pm0.43)$	0.84	0.30	31.25	8	
$pTCO_2 = 0.06(\pm0.02)\ E_{LUMO(g)} + 2.36(\pm0.47)$	0.63	0.45	10.36	8	
$pTCO_2 = 0.08(\pm0.02)\ E_{HOMO(g)} + 3.62(\pm0.51)$	0.83	0.31	28.89	8	
$pTCO_2 = -0.09(\pm0.02)\ IP + 2.63(\pm0.43)$	0.74	0.38	17.13	8	C
$pTCO_2 = -0.14(\pm0.04)\ Z^2/r + 1.70(\pm0.30)$	0.61	0.46	9.38	8	E

(Continued)

TABLE 5.20 (Continued)
QCARs with Less Numerous Physicochemical Properties Used to Predict Cation Toxicity

Reference	Test Substances	QCARs	adj. r^2	p	AIC	n	E
	Ag⁺, Cd²⁺, Co²⁺, Cr³⁺, Cu²⁺, Ni²⁺, Zn²⁺	$pTCO_2 = 0.05(\pm0.01)\,E_{PSS} + 2.07(\pm0.35)$	0.69	0.26	13.10	8	
		$pTCO_2 = 0.08(\pm0.02)\,\mu_{(g)} + 3.18(\pm0.45)$	0.82	0.31	27.11	8	
		$pTCO_2 = 0.22(\pm0.07)\,\mu_{(aq)} + 2.72(\pm0.60)$	0.62	0.45	9.70	8	
		$pTO_2\text{-labile} = 0.27(\pm0.05)^a\,\mu_{aq} + 3.70(\pm0.47)$	0.84	0.34	26.26	7	E
		$pTO_2\text{-labile} = 0.06(\pm0.01)\,E_{PSS} + 2.71(\pm0.27)$	0.86	0.32	30.52	7	
		$pTO_2\text{-labile} = -0.15(\pm0.05)\,Z^2/r + 2.29(\pm0.33)$	0.69	0.47	11.02	7	
		$pTO_2\text{-labile} = 0.08(\pm0.02)E_{HOMO(g)} + 3.93(\pm0.80)$	0.68	0.48	10.61	7	
		$pTO_2\text{-labile} = 0.07(\pm0.02)E_{LUMO(g)} + 3.06(\pm0.45)$	0.76	0.41	16.01	7	
		$pTO_2\text{-labile} = 0.09(\pm0.02)\,\mu_{(g)} + 3.75(\pm0.52)$	0.82	0.36	22.65	7	
	Cd²⁺, Co²⁺, Cu²⁺, Ni²⁺, Zn²⁺	$pTO_2\text{-labile} = 0.23(\pm0.04)\,E_{PSS} + 5.82(\pm0.89)$	0.89	0.17	25.03	5	
	Ag⁺, Cd²⁺, Co²⁺, Cr³⁺, Cu²⁺, Ni²⁺, Zn²⁺	$pTCO_2\text{-labile} = 0.26(\pm0.07)\,\mu_{aq} + 3.41(\pm0.58)$	0.76	0.42	15.43	7	E
		$pTCO_2\text{-labile} = 0.06(\pm0.01)\,E_{PSS} + 2.48(\pm0.33)$	0.79	0.39	19.23	7	
		$pTCO_2\text{-labile} = -0.15(\pm0.05)\,Z^2/r + 2.06(\pm0.37)$	0.61	0.53	7.68	7	
		$pTCO_2\text{-labile} = 0.08(\pm0.02)E_{HOMO(g)} + 3.74(\pm0.82)$	0.67	0.49	10.60	7	
		$pTCO_2\text{-labile} = 0.07(\pm0.02)E_{LUMO(g)} + 2.80(\pm0.52)$	0.68	0.48	10.74	7	
		$pTCO_2\text{-labile} = 0.08(\pm0.02)\,\mu_{(g)} + 3.50(\pm0.59)$	0.76	0.41	16.28	7	
	Cd²⁺, Co²⁺, Cu²⁺, Ni²⁺, Zn²⁺	$pTCO_2\text{-labile} = 0.23(\pm0.07)\,E_{PSS} + 5.68(\pm1.45)$	0.76	0.27	9.69	5	
	Ag⁺, Cd²⁺, Co²⁺, Cr³⁺, Cu²⁺, Hg²⁺, Ni²⁺, Zn²⁺	$pTO_2 = 0.08\,(\pm0.01)^a\,E_{HOMO(g)} + 0.75\,(\pm0.24)\,\Delta E_0 + 3.16\,(\pm0.35)$	0.95	0.17	46.30	8	
		$pTO_2 = 0.037\,(\pm0.009)\,E_{PSS} + 0.002\,(\pm0.00)\,ESE + 1.06\,(\pm0.39)$	0.93	0.20	34.44	8	
		$pTO_2 = -0.07\,(\pm0.01)\,IP + 0.04\,(\pm0.01)\,ESE + 1.66\,(\pm0.46)$	0.95	0.18	44.28	8	

	Equation				
	$pTO_2 = 0.07\ (\pm0.01)\ E_{HOMO(g)} + 0.18\ (\pm0.06)\ E_{LUMO}\ C_{max} + 3.64\ (\pm0.29)$	0.95	0.17	47.46	8
	$pTO_2 = 0.08\ (\pm0.01)\ E_{HOMO(g)} + 0.13\ (\pm0.04)\ E_{LUMO}\ C_{min} + 4.03\ (\pm0.32)$	0.95	0.18	44.69	8
	$pTO_2 = 0.10\ (\pm0.03)\ \log K_{OH} + 0.57\ (\pm0.09)\ X_m^2 r - 1.33\ (\pm0.34)$	0.92	0.22	27.82	8
	$pTO_2 = 0.12\ (\pm0.05)\ E_{HOMO(aq)} + 0.34\ (\pm0.10)\ E_{LUMO}\ C_{max} + 3.21\ (\pm0.59)$	0.78	0.36	9.08	8
	$pTCO_2 = 0.14\ (\pm0.05)\ E_{HOMO(aq)} + 0.31\ (\pm0.12)\ E_{LUMO}\ C_{max} + 3.14\ (\pm0.67)$	0.74	0.41	7.13	8
Mendes et al. (2010) $Ag^{1+}, Ba^{2+}, Ca^{2+}, Cd^{2+}, Co^{2+}, Cs^{1+}, Cu^{2+}, Fe^{2+}, Hg^{2+}, K^{1+}, Li^{1+}, Mg^{2+}, Mn^{2+}, Na^{1+}, Ni^{2+}, Pb^{2+}, Sr^{2+}, Zn^{2+}$	$\text{Log } EC_{50} = 3.57 - 0.07(\text{AN})$	0.708	0.0007	−47.32	18	
	$\text{Log } EC_{50} = -1.91 + 2.65(\text{AR})$	0.462	0.0128	−40.59	18	
	$\text{Log } EC_{50} = 4.93 - 0.37(\Delta\text{IP})$	0.639	0.0019	−44.97	18	
	$\text{Log } EC_{50} = -3.10 + 0.46(\log|K_{OH}|)$	0.757	0.003	−49.33	18	B
	$\text{Log } EC_{50} = 0.21 + 2.15r$	0.011	0.3187	−33.9	18	
	$\text{Log } EC_{50} = 3.05 - 0.40(\text{AN}/\Delta\text{IP})$	0.219	0.0828	−36.5	18	
	$\text{Log } EC_{50} = 4.24 - 1.27\left(X_m^2 r\right)$	0.911	0.0001	−60.42	18	A
	$\text{Log } EC_{50} = 3.52 - 0.38(Z^2/r)$	0.405	0.0209	−39.49	18	E
	$\text{Log } EC_{50} = 3.69 - 1.18(X_m^2 r) + 0.04(\log|K_{OH}|)$	0.901	0.0001	−55.31	18	D

Log of the equilibrium constant (log K_{eq}) of the metal-ATP complex
Covalent index $\left(X_m^2 r\right)$
Log of the first hydrolysis constant (|log K_{OH}|)
Ionic index (Z^2/r)
Log of the stability constant for the metal fluoride; log of the stability constant for the metal chloride ($\Delta\beta$)
Log metal hydroxide solubility, (Log-K_{so}MOH)
Ionization potential (IP)

(Continued)

TABLE 5.20 (*Continued*)

QCARs with Less Numerous Physicochemical Properties Used to Predict Cation Toxicity

The heat of formation of inorganic oxides (ΔH_o)

The heat of formation of aqueous ion ($\Delta H_{aq \, ion}$)

Atomic number (AN)

Difference in ionization potentials between ion oxidation number (OX) and (OX - 1) (ΔIP)

Atomic weight (AW)

Enthalpy of hydration (ΔH_h)

Density (ρ)

Ionic Radius (IR)

Atomic Radius (AR)

Stability constant of metal ion complexes with EDTA (log K (EDTA))

Formation constants with ATP (log K (ATP))

Stability constants for AMP with divalent metal ions (log K (AMP))

Misono and Saito (1970) softness parameter of (Y')

Stability constant of metal ion complexes with AMP-2 (log K (AMP-2))

Stability constant of metal ion complexes with AMP-3 (log K (AMP-3))

Observed stability constants of metal ion-acetato complexes (log K (Ac))

Calculated constant of metal ion complexes with NH$_3$ (log K (NH$_3$))

Enthalpy of formation of sulfides (ΔH_s)

Stability constant of metal ion complexes with sulfate (log K (sulfate))

Selected average of metal concentration in soil (molar concentration at 10% moisture, log M$_{soil}$)

Medians of elemental composition of soils (mg X/kg soil)

Calculated mean of the elemental content in land plants (Land plants)

Intrinsic (van der Waals) molecular volume (V_i)

Solute ability to stabilize a neighboring charge or dipole by non-specific dielectric interactions (π^*)

Solute ability to accept or donate hydrogen in a hydrogen bond (β_m and α_m)

Highest Occupied Molecular Orbital (HOMO)

Lowest Unoccupied Molecular Orbital LUMO

HELH is the electronegativity of HOMO state for ELH (hydrogen)

HEO is the electronegativity of HOMO state for EO (oxygen)

HEM is the electronegativity of HOMO state for EM (metal)

O-EO is the electronegativity of OMO (occupied molecular orbital) state for EO (oxygen)

O-EM is the electronegativity of OMO state for EM (metal)

U-ELH is the electronegativity of UMO (unoccupied molecular orbital) state for ELH (hydrogen)

U-EO is the electronegativity of UMO state for EM (oxygen)

U-EM is the electronegativity of UMO state for EM (metal)

Electrophilicity (ω)

Chemical potential of the gaseous phase ($\mu_{(g)}$)

Chemical potential of the aqueous phase ($\mu_{(aq)}$)

Electronic spatial extent (ESE)

Energy of the polarized solute solvent (E_{PSS})

Aqueous phase energy of HOMO energy ($E_{HOMO(aq)}$)

Gaseous phase energy of HOMO energy ($E_{HOMO(g)}$)

Gaseous phase energy of LUMO energy ($E_{LUMO(g)}$)

Maximum value for LUMO energy of the most important cation ($E_{LUMO}C_{max}$)

Cationic charge (Z)

Crystal ionic radius (r)

Walker et al. (2007) developed 6 QSARs to predict the toxicity of 17 cations to sunflower seeds (F.1. *Helianthus annuus* Sunspot) from the metal's physical properties and natural occurrence levels (Table 5.4). The QSARs predicted EC_{50} values based on the cation concentration producing a 50% inhibition of radicle growth one day after emergence. The QSAR developed with density of the elements (ρ), enthalpy of formation of metal sulfides (ΔH_s), and the stability constants of metal ions with sulfate (log K_1 [sulphate]) produced the highest adjusted r^2 (Table 5.20). For natural occurrence levels, the QSAR developed with metal concentrations in soil (log M_{soil}), the median elemental composition of soils (mg X/kg soil), and the calculated mean of the elemental content in land plants (Land Plants) produced the highest adjusted r^2 (Table 5.20).

As discussed in Section 5.2.5, Van Kolck et al. (2008) developed 4 QSARs to predict the 96-hour LC_{50} values of 5 cations to the mussel *Mytilis edulis* and 4 QSARs to predict the 96-hour LC_{50} values of 6 cations to the mussel *Perna viridis* (Table 5.17). Six of these QSARs included 3 of the less numerous physical properties used to predict cation toxicity, viz., covalent index $\left(X_m^2 r\right)$, absolute value of the logarithm of the first hydrolysis constant ($|\log K_{OH}|$), and ionic index (Z^2/r). The QSARs developed with the covalent index $\left(X_m^2 r\right)$ produced the highest r^2 value (Table 5.20).

Kinraide (2009) developed softness and toxicity scales by compiling data from 8 previously published softness scales and 10 previously published toxicity scales (Table 5.3). A QSAR was developed that predicted toxicity based on softness and ion charge (Table 5.4). For this QSAR ($T_{Con\ obs} = a\sigma_{Con\ comp} + b\sigma_{Con\ comp}Z + cZ$), $r^2 = 0.923$, $a = 2.16$, $b = -0.521$, $c = 0.0778$, observed toxicity $= T_{Con\ obs}$, computed softness $= \sigma Con$ comp, and ionic charge $= Z$.

Lepădatu et al. (2009) applied molecular fingerprint descriptors representing the electronegativities of the highest occupied molecular orbital (HOMO) and lowest unoccupied molecular orbital (LUMO) quantum molecular states to 12 cations from Walker et al. (2007) with *s*, *p*, and *d^N* valence shells that inhibited the growth of sunflower. Coefficients of determination for descriptors representing HOMO and LUMO for all 12 cations were consistently lower than coefficients of determination for descriptors representing HOMO and LUMO for just the cations with *s*, *p* valence shells (Ag^{1+}, Al^{3+}, Ba^{2+}, Cd^{2+}, Li^{1+}, Pb^{2+}, Zn^{2+}). A similar, but less consistent, comparison also existed for cations with *d^N* valence shells (Co^{2+}, Cu^{2+}, Fe^{3+}, Mn^{2+}, Ni^{2+}). Based on those comparisons, descriptors representing HOMO and LUMO were used to develop separate QSARs for cations with *s*, *p* valence shells and cations with *d^N* valence shells (Table 5.4). For the 7 cations with *s*, *p* valence shells, the highest r^2 value was obtained for the QSAR that used the U-EO (electronegativity of UMO state for EM [oxygen]) descriptor (Table 5.20). For the 5 cations with *d^N* valence shells, the highest r^2 values were obtained for the QSARs that used the HELH (electronegativity of HOMO state for ELH [hydrogen]) and U-ELH (electronegativity of UMO [unoccupied molecular orbital] state for ELH [hydrogen]) descriptors (Table 5.20).

Roy et al. (2009) used the uncommon physicochemical property, electrophilicity (ω), and a training set of 10 cations and experimental toxicity data for the soil nematode, *Caenorhabditis elegans*, to develop 2 QSARs to predict the toxicity of As^{3+} and As^{5+} ions (Table 5.4). The r^2 value decreased when atomic number replaced electrophilicity (ω) in the QSARs (Table 5.20). However, only the electrophilicity (ω)-based QSAR was able to distinguish between As^{3+} and As^{5+} toxicity

Sacan et al. (2009) developed 38 QSARs to predict the toxicity of cations to nitrifying bacteria (Table 5.4). They measured oxygen consumption and carbon dioxide production from nitrifying bacteria in the presence of unspeciated cation concentrations and speciated (labile) cation concentrations. Labile concentrations included the free metal concentrations plus the concentration of the metal–anion complexes having weak stability constants (log K). A 50% reduction in oxygen consumption and carbon dioxide production was calculated for each metal and the corresponding IC$_{50}$ values were converted to logarithmic toxicity units as pTO$_2$ and pTCO$_2$ (Table 5.4). pTO$_2$ and pTCO$_2$ QSARs were developed for 8 unspeciated cations using 8 physicochemical properties (Table 5.20). The pTO$_2$ QSARs for 8 unspeciated cations with the highest r^2 values, lowest SE values, and highest F values were those developed with the cationic charge (Z) and the chemical potential of the gaseous phase ($\mu_{(g)}$) (Table 5.20). The pTCO$_2$ QSARs for 8 unspeciated cations with the highest r^2 values, lowest SE values, and highest F values were those developed with Z, gaseous phase energy of HOMO energy ($E_{HOMO(g)}$), and the chemical potential of the gaseous phase ($\mu_{(g)}$) (Table 5.20). Since the labile toxicity of Hg could not be calculated, it was not included in the 7 cations used to develop pTO$_2$ –labile and pTCO$_2$ –labile QSARs (Table 5.20). The pTO$_2$ –labile QSARs with the highest r^2 values, lowest SE values, and highest F values were those developed with the chemical potential of the aqueous phase ($\mu_{(aq)}$), $\mu_{(g)}$ and the energy of the polarized solute solvent (E_{PSS}) (Table 5.20). As a result of eliminating Ag^{+3} and Cr^{+3} (the most and least toxic cations to nitrifiers), the pTO$_2$ –labile QSAR for only 5 divalent cations had a higher r^2 value, lower SE value, and higher F value than those QSARs developed for 7 cations (Table 5.20). The pTCO$_2$ –labile QSARs for 7 cations and 5 cations with the highest r^2 values, lowest SE values, and highest F values were those developed with the E_{PSS} (Table 5.20). Using 2 physicochemical properties to develop QSARs for 8 unspeciated cations almost always consistently produced higher r^2 value, lower SE value, and higher F value than those QSARs that used 1 physicochemical property (Table 5.20).

As noted above, Mendes et al. (2010) developed QSARs for predicting cation toxicity using standard reduction-oxidation potential, electronegativity, and the Pearson and Mawby (1967) softness parameter (Table 5.12). However, Mendes et al. (2010) developed 9 QSARs for predicting cation toxicity using some less numerous physicochemical properties (Table 5.20). The QSARs developed with the covalent index $\left(X_m^2 r\right)$ or $X_m^2 r$ and the absolute value of the logarithm of the first hydrolysis constant (|log K_{OH}|) had the highest r^2 values, lowest p values, and lowest AIC values (Table 5.20).

Su et al. (2010) developed 2 QSARs to predict the combined toxicity of phenols and Pb to the bioluminescent bacterium *Vibrio fischeri* (formerly *Photobacterium phosphoreum*) (Table 5.4). However, since the 2 QSARs were not developed to predict cation toxicity, they were not included in Table 5.20.

Among the 128 QSARs in Table 5.20 that used less numerous physicochemical properties used to predict cation toxicity, there were some physicochemical properties that were used more than a few times by different teams of investigators or in different laboratories. These physicochemical properties are labeled A–F in Table 5.20. The covalent index $(X_m^2 r)$, labeled A in Table 5.20 was used in 4 different QSARs by Newman's group (McCloskey et al. 1996; Newman and McCloskey 1996; Tatara et al. 1997; Tatara et al. 1998; Newman et al. 1998] and the QSAR with the highest r^2 value was developed by McCloskey et al. (1996) for 20 cations. The covalent index $(X_m^2 r)$ was also used by Enache et al. (2003), Van Kolck et al. (2008), and Mendes et al. (2010) to develop QSARs (Table 5.20). The QSARs developed by Enache et al. (2003) for 11 cations and Mendes et al. (2010) for 18 cations had the highest r^2 values (Table 5.20). The log of the first hydrolysis constant ($|\log K_{OH}|$) labeled B in Table 5.20 was used in 4 different QSARs by Newman's group and the QSAR with the highest r^2 value was developed by McCloskey et al. (1996) for 9 cations. The log of the first hydrolysis constant ($|\log K_{OH}|$) was also used by Enache et al. (2003), van Kolck et al. (2008), and Mendes et al. (2010) to develop QSARs (Table 5.20). The QSARs developed by Enache et al. (2003) for 12 cations and Mendes et al. (2010) for 18 cations had the highest r^2 values (Table 5.20). Ionization potential (IP), labeled C in Table 5.20, was used by 3 different teams of investigators (Sauvant et al. 1997; Enache et al. 2003; and Sacan et al. 2009) to develop QSARs. The IP QSAR developed by Sauvant et al. (1997) to predict the toxicity of 8 cations on oxygen consumption by nitrifying bacteria had the highest r^2 values (Table 5.20). QSARs developed with $X_m^2 r$ and $|\log K_{OH}|$ are labeled D in Table 5.20. These QSARs were used only by Newman's group (McCloskey et al. 1996; Tatara et al. 1998) and Mendes et al. (2010). All 3 QSARs included a large number of cations (17–20) and had high r^2 values, but the Mendes et al. (2010) QSAR had the lowest p value. Van Kolck et al. (2008), Sacan et al. (2009), and Mendes et al. (2010) developed QSARs with the ionic index (Z^2/r), labeled E in Table 5.20. The r^2 values of the van Kolck et al. (2008) and Mendes et al. (2010) QSARs were lower than the r^2 values for the Sacan et al. (2009) QSARs (Table 5.20). The only other QSARs that used less numerous physicochemical properties used to predict cation toxicity were those developed by Newman's group (McCloskey et al. 1996; Newman and McCloskey 1996; Tatara et al. 1997; Tatara et al. 1998) for $X_m^2 r$ and Z^2/r (Table 5.20). These QSARs are labeled F in Table 5.20 and had higher r^2 values for QSARs for predicting toxicity for a high (17–20) number of cations (Table 5.20).

5.4 NONPHYSICOCHEMICAL PROPERTIES USED TO PREDICT CATION TOXICITY

Six non-physico-chemical properties have been used to predict cation toxicity (Table 5.21). These include serum concentrations, calmodulin activity, freshwater concentrations, abundance in the earth's crust, soil concentrations, and elemental composition of soils and plants.

Joseph and Meltzer (1909) established that the toxicity of the alkaline earth metals (magnesium, calcium, potassium, and sodium) to dogs was in inverse proportion to the amounts in the dog's serum. Cox and Harrison (1983) compared Williams et al. (1982) mouse LD_{50} cation toxicity values with the cation's ability to mimic Ca^{2+} in stimulating the intracellular Ca^{2+} receptor protein, calmodulin. Williams et al.

TABLE 5.21

Nonphysicochemical Properties Used to Predict Cation Toxicity

Property	Test Systems	Test Substance(s)	Reference
Serum concentrations	Intravenous determination of dog's lethal dose	Ca^{2+}, Mg^{2+}, Na^{1+}, K^{1+}	Joseph and Meltzer (1909)
Calmodulin activity	Mouse LD_{50} values from Williams et al. (1982)	Ba^{2+}, Be^{2+}, Cd^{2+}, Co^{2+}, Cu^{2+}, Hg^{2+}, Mn^{2+}, Ni^{2+}, Pb^{2+}, Pd^{2+}, Pt^{2+}, Sr^{2+}, Zn^{2+}	Cox and Harrison (1983)
Calmodulin activity	Chinese hamster ovary cells of Tan et al. (1984)	Be^{2+}, Cd^{2+}, Co^{2+}, Cu^{2+}, Hg^{2+}, Mn^{2+}, Ni^{2+}, Pd^{2+}, Sr^{2+}, Zn^{2+}	Williams et al. (1987)
Fresh water concentrations	Stickleback and planarian data from Jones (1939, 1940) and *Daphnia magna* data from Biesinger and Christensen (1972)	Al^{3+}, As^{3+}, Ca^{2+}, Cd^{2+}, Co^{2+}, Cr^{3+}, Cu^{2+}, Fe^{2+}, Hg^{2+}, K^{1+}, Mg^{2+}, Mn^{2+}, Na^{1+}, Ni^{2+}, Pb^{2+}, Pt^{2+}, Zn^{2+}	Svanberg and Lithner (1978)
Fresh water concentrations	*Daphnia magna* data from Biesinger and Christensen (1972)	Al^{3+}, As^{3+}, Ca^{2+}, Cd^{2+}, Co^{2+}, Cr^{3+}, Cu^{2+}, Fe^{2+}, Hg^{2+}, K^{1+}, Mg^{2+}, Mn^{2+}, Na^{1+}, Ni^{2+}, Pb^{2+}, Zn^{2+}	Lithner (1989)
Abundance in the earth's crust	*Daphnia magna* data from Biesinger and Christensen (1972)	Al^{3+}, As^{3+}, Ca^{2+}, Cd^{2+}, Co^{2+}, Cr^{3+}, Cu^{2+}, Fe^{2+}, Hg^{2+}, K^{1+}, Mg^{2+}, Mn^{2+}, Na^{1+}, Ni^{2+}, Pb^{2+}, Zn^{2+}	Lithner (1989)
Abundance in the earth's crust	White lupin (*Lupinus albus*) (concentration in which radicles of seedlings in test solution were able to survive for 24 hours).	$AgNO_3$, $BaCl_2$, $BeSO_4$, $CaCl_2$, $CdCl_2$, $CoSO_4$, $CsCl$, $FeSO_4$, $HgCl_2$, KCl, $LiCl$, $MgCl_2$, $NaCl$, $NiSO_4$, $RbCl$, $SrCl_2$, $ZnCl_2$	True (1930)
Soil concentrations and soil or plant elemental composition	F.1. *Helianthus annuus* "Sunspot" plant (concentration giving 50% inhibition of radicle growth (EC_{50}) one day after emergence)	Ag^{1+}, Al^{3+}, Ba^{2+}, Ca^{2+}, Cd^{2+}, Co^{2+}, Cu^{2+}, Fe^{3+}, K^{1+}, La^{3+}, Li^{1+}, Mg^{2+}, Mn^{2+}, Na^{1+}, Ni^{2+}, Pb^{2+}, Zn^{2+}	Walker et al. (2007)

(1987) used the Chinese hamster ovary cell cloning efficiency (CE_{50}) data of Tan et al. (1984) and the inhibition in calmodulin activity (IC_{50}) data of Cox and Harrison (1983) to demonstrate that IC_{50} values could be used in predict cation toxicity better than the Pearson and Mawby (1967) softness parameter (σ_p). The 13 cations used by Cox and Harrison (1983) and 10 cations used by Williams et al. (1987) are listed in Table 5.21.

Svanberg and Lithner (1978) used Biesinger and Christensen (1972) *Daphnia magna* toxicity data, Jones (1939) stickleback *Gasterosteus aculeatus* L. toxicity data, and Jones (1940) planarian *Polycelis nigra* data to demonstrate that the relative toxicity of 17 cations was essentially inversely related to freshwater concentrations. The 17 cations used by Svanberg and Lithner (1978) are listed in Table 5.21. Lithner (1989) concluded that the abundance of elements in freshwater was a better cation toxicity predictor than the abundance of the elements in the earth's crust, as postulated previously by True (1930). The 16 cations used by Lithner (1989) and the 17 metal salts used by True (1930) are listed in Table 5.21. The Walker et al. (2007) studies are discussed in Section 5.3; the QSARs are listed in Table 5.20 (see the last three QSARs listed under Walker et al. [2007] in Table 5.20) and the 17 cations are listed in Tables 5.20 and 5.21.

5.5   PHYSICOCHEMICAL PROPERTIES USED TO PREDICT CATION BINDING

Kinraide and Yermiyahu (2007) developed a scale of binding strength to inorganic and organic ligands for 49 cations (Table 5.3). Binding strength correlated well with cation toxicity to wheat root elongation, growth of sunflower callus tissue, survival of the nematode *Caenorhabditis elegans*, and bioluminescence response of *Vibrio fischeri*. The "QSAR" that determines the value of ion activities at the plasma membrane surface ($a_{PM,M}$) that will ensure inhibition of root elongation is listed in Table 5.4. The following relationships were also developed by Kinraide and Yermiyahu (2007):

$$\log K_{\text{Scale, M}} = a + Z(b + cPE)$$

$$\text{Scale values} = -1.68 + Z(1.22 + 0.444PE); \; R^2 = 0.969$$

$$\text{Scale of toxicities} = a + b(\text{Atomic mass}) + cPE, \; R^2 = 0.625 - 0.896$$

$$Z = \text{ionic charge}, \; PE = \text{Pauling electronegativity}$$

$$\log K_{\text{Scale, M}} = \text{values in the scale of relative binding strengths}$$

$$\log a_{PM,M} = 1.52 - 0.209(\log K_{\text{Scale, M}})^2 - 0.538/HI, \text{ where } HI = \text{hardness index}$$

Zhou et al. (2011) evaluated 50 conditional binding constants (K) for biotic ligands from 27 studies and calculated mean ligand-specific conditional binding constants (K). They developed $\log K$ QSARs for five species using the Pearson and Mawby softness parameter (σ_p), the covalent index ($X_m^2 r$), the absolute value of the logarithm of the first hydrolysis constant ($|\log K_{OH}|$), and the ionic index Z^2/r) (Table 5.4). The $\log K$ QSARs for the water flea (*Daphnia magna*) were developed from 14 studies with 7 cations. The $\log K$ QSARs for *Daphnia magna* that used σ_p had the highest r^2 value and the lowest p value (Table 5.22). The $\log K$ QSARs for the fathead minnow (*Pimephales promelas*) were developed from 11 studies with 7 cations. The $\log K$ QSARs for *Pimephales promelas* that used σ_p also had the highest r^2 value and the lowest p value, even though the 7 cations for *Pimephales promelas* were different from those used for *Daphnia magna* (Table 5.22). The $\log K$-based QSARs for the rainbow trout (*Oncorhynchus mykiss*) were developed from 14 studies with 7 cations. The $\log K$-based QSARs for *Oncorhynchus mykiss* that used σ_p also had the highest r^2 value and the lowest p value (Table 5.22). The QSARs for barley (*Hordeum vulgare*) were developed from 6 studies with 6 cations. The $\log K$-based QSARs for *Hordeum vulgare* that used $|\log K_{OH}|$ had the highest r^2 value and the lowest p value (Table 5.22). Similar QSARs for the earthworm (*Eisenia fetida*) were developed from 6 studies with 6 cations. The $\log K$ QSARs for the *Eisenia fetida* that used $|\log K_{OH}|$ had the highest r^2 value and the lowest p value (Table 5.22).

TABLE 5.22
Zhou et al. (2011) Binding Constant QCARs

Species	Cations	QCAR	r^2	p
Water flea (*Daphnia magna*)	Ag^{1+}, Ca^{2+}, Cd^{2+}, Cu^{2+}, Mg^{2+}, Na^{1+}, Zn^{2+}	$\text{Log } K = 11.69 - 43.9\sigma_p$	0.89	0.0001
		$\text{Log } K = 2.36 + 1.71\ X_m^2 r$	0.73	0.0001
		$\text{Log } K = 13.99 - 0.72 \text{ Log } K_{OH}$	0.41	0.01
		$\text{Log} K = 5.88 - 0.11\ Z^2/r$	0	0.96
Fathead Minnow (*Pimephales promelas*)	Ag^{1+}, Ca^{2+}, Cd^{2+}, Cu^{2+}, Na^{1+}, Ni^{2+}, Zn^{2+}	$\text{Log } K = 10.57 - 37.62\sigma_p$	084	0.0001
		$\text{Log } K = 2.18 + 1.54\ X_m^2 r$	0.66	0.002
		$\text{Log} K = 12.6 - 0.66 \text{ Log } K_{OH}$	0.53	0.01
		$\text{Log} K = 3.96 + 0.33\ Z^2/r$	0.09	0.37
Rainbow Trout (*Oncorhynchus mykiss*)	Ca^{2+}, Cd^{2+}, Co^{2+}, Cu^{2+}, Mg^{2+}, Pb^{2+}, Zn^{2+}	$\text{Log } K = 11.08 - 42.73\ \sigma_p$	0.82	0.0001
		$\text{Log } K = 3.9 + 0.69\ X_m^2 r$	0.40	0.04
		$\text{Log } K = 10.2 - 0.49 \text{ Log } K_{OH}$	0.41	0.01
		$\text{Log } K = 3.69 + 0.37\ Z^2/r$	0.02	0.57
Barley (*Hordeum vulgare*)	Ca^{2+}, Co^{2+}, Cu^{2+}, K^{1+}, Mg^{2+}, Ni^{2+},	$\text{Log } K = 0.51 + 1.91\ X_m^2 r$	0.61	0.06
		$\text{Log } K = 9.41 - 34.59\ \sigma_p$	0.60	0.07
		$\text{Log } K = 12.44 - 0.76 \text{ Log } K_{OH}$	0.74	002
		$\text{Log } K = 1.52 + 0.55\ Z^2/r$	0.33	0.29
Earthworm (*Eisenia fetida*)	Ca^{2+}, Cd^{2+}, Co^{2+}, Cu^{2+}, Mg^{2+}, Ni^{2+},	$\text{Log } K = 0.82 + 1.68\ X_m^2 r$	0.60	0.07
		$\text{Log } K = 7.78 - 25.36\ \sigma_p$	0.35	0.22
		$\text{Log } K = 11.61 - 0.68 \text{ Log } K_{OH}$	0.82	0.01
		$\text{Log } K = 1.35 + 0.63\ Z^2/r$	0.21	0.35

Source: Data from D.M. Zhou, L.Z. Li, W.J.G.M. Peijnenburg, D.R. Ownby, A.J. Hendriks, P. Wang, and D.D. Li, "A QICAR Approach for Quantifying Binding Constants for Metal-Ligand Complexes." *Ecotox. Environ. Safety* 74 (2011):1036–1042.

5.6   PHYSICOCHEMICAL PROPERTIES USED TO PREDICT CATION BIOCONCENTRATION

Van Kolck et al. (2008) developed 4 QSARs to predict the bioconcentration factors (BCF) of cations to the mussel *Mytilus edulis*, and 4 QSARs to predict the BCFs of cations to the mussel *Perna viridis* (Table 5.4). The BCFs for *Mytilus edulis* were developed for 8 cations and the QSARs with highest r^2 values were obtained using the ionic index (Z^2/r) and the covalent index $\left(X_m^2 r\right)$ (Table 5.18). The BCFs for *Perna viridis* were developed for 7 cations and the QSARs with highest r^2 values were obtained using the Pearson and Mawby softness parameter (σ_p) and the covalent index $\left(X_m^2 r\right)$ (Table 5.18).

5.7   PHYSICOCHEMICAL PROPERTIES USED TO PREDICT CATION BIOSORPTION

Can and Jianlong (2007), Chen and Wang (2007), and Zamil et al. (2009) developed 54 QSARs to predict the maximum biosorption capacity (q_{max}) of cations to either the yeast *Saccharomyces cerevisiae* or the bacterium *Staphylococcus saprophyticus* BMSZ711 (Table 5.4). Can and Jianlong (2007) used the t test to evaluate the contribution of each physicochemical property to the QSAR and set a level of significance as $\alpha = 0.05$. The QSARs that did not meet this level of significance are indicated with the superscript "b" in Table 5.23 and are not discussed further. Can and Jianlong (2007) (labeled C&J in Table 5.23) used 7 physicochemical properties to develop QSARs for predicting the biosorption capacity of 6 borderline cations (Cd^{2+}, Co^{2+}, Cr^{3+}, Cu^{2+}, Ni^{2+}, and Zn^{2+}) and 2 hard cations (Cs^+, Sr^{2+}) (Table 5.23). For these 8 cations, the QSARs developed with the ionic index (Z^2/r) and the absolute value of the logarithm of the first hydrolysis constant (|log K_{OH}|) were the most statistically significant (Table 5.23). Can and Jianlong (2007) removed Ni^{2+} from the 8 cations listed above and developed 10 QSARs with $\alpha = 0.05$ to predict their biosorption capacity. While the QSAR developed with ionization potential (IP) and atomic number (AN)/ the difference of the ionization potential in volts between its oxidation number (OX) and the next lower one (OX − 1) (ΔIP) was the most robust, those developed with Z^2/r, and ionic potential (Z/r) were also highly statistically significant (Table 5.23). Can and Jianlong (2007) developed 5 QSARs to predict the biosorption capacity of 6 borderline cations (Cd^{2+}, Co^{2+}, Cr^{3+}, Cu^{2+}, Ni^{2+}, and Zn^{2+}) and 2 soft cations (Ag^+, Pb^{2+}), but none were very statistically significant (Table 5.23). Can and Jianlong (2007) also developed 7 QSARs to predict the biosorption capacity of 6 borderline cations. Only the QSAR developed with |log K_{OH}| was highly statistically significant (Table 5.23).

Chen and Wang (2007) (labeled C&W in Table 5.23) developed 3 QSARs to predict the q_{max} of 10 cations to *Saccharomyces cerevisiae*; the QSAR developed with the covalent index $\left(X_m^2 r\right)$ was the most statistically significant (Table 5.23). Chen and Wang (2007) also developed 6 QSARs to predict the biosorption capacity of 8 cations and again the QSAR developed with $X_m^2 r$ was the most statistically significant (Table 5.23). Only the Chen and Wang (2007) QSAR based on $X_m^2 r$ and AN/ΔIP did not meet the level of significance as $\alpha = 0.05$ (Table 5.23).

TABLE 5.23

QCARs for Predicting Cation Biosorption to the Yeast, *Saccharomyces cerevisiae* and QCARs for Predicting Cation Biosorption to the Bacterium, *Staphylococcus saprophyticus* BMSZ711

Ref.[a]	QCARs	Cations	R^2adj.	SE	F	P	AIC	MAPE
C&J	qmax = $-0.013 + 0.077$ (OX)	Cd^{2+}, Co^{2+}, Cr^{3+}, Cs^+, Cu^{2+}, Ni^{2+}, Sr^{2+}, Zn^{2+}	0.71	0.026	17.76	0.005	-4.25	15.94
a C&J	qmax = $0.27 - 0.013$ (logK$_{OH}$)	Cd^{2+}, Co^{2+}, Cr^{3+}, Cs^+, Cu^{2+}, Ni^{2+}, Sr^{2+}, Zn^{2+}	0.80	0.021	29.21	0.002	-4.64	12.01
b C&J	qmax = $0.079 + 0.011$ (Z^2/r)	Cd^{2+}, Co^{2+}, Cr^{3+}, Cs^+, Cu^{2+}, Ni^{2+}, Sr^{2+}, Zn^{2+}	0.85	0.019	39.21	0.001	-4.89	10.2
c C&J	qmax = $0.047 + 0.006$ (IP)	Cd^{2+}, Co^{2+}, Cr^{3+}, Cs^+, Cu^{2+}, Ni^{2+}, Sr^{2+}, Zn^{2+}	0.77	0.023	24.42	0.003	-4.49	11.8
C&J	qmax = $0.069 + 0.07$ (Z/AR^2)	Cd^{2+}, Co^{2+}, Cr^{3+}, Cs^+, Cu^{2+}, Ni^{2+}, Sr^{2+}, Zn^{2+}	0.57	0.031	10.2	0.019	-3.86	15.92
d C&J	qmax = $0.053 + 0.035$(Z/r)	Cd^{2+}, Co^{2+}, Cr^{3-}, Cs^+, Cu^{2+}, Ni^{2+}, Sr^{2+}, Zn^{2+}	0.73	0.025	20.09	0.004	-4.34	13.01
e C&J	qmax = $0.044 + 0.069$(Z/AR)	Cd^{2+}, Co^{2+}, Cr^{3+}, Cs^+, Cu^{2+}, Ni^{2+}, Sr^{2+}, Zn^{2+}	0.64	0.029	13.47	0.01	-4.05	14.52
C&J	qmax = $0.10 + 0.02$ (Z^2/r)[b]	Cd^{2+}, Co^{2+}, Cr^{3+}, Cs^+, Cu^{2+}, Ni^{2+}, Sr^{2+}, Zn^{2+}	0	0.048	0.945	0.369	-3.02	17.46
a C&J	qmax = $-0.008 + 0.077$ (OX)	Cd^+, Co^{2+}, Cr^{3+}, Cs^+, Cu^{2+}, Sr^{2+}, Zn^{2+}	0.78	0.023	21.95	0.005	-4.44	13.42
a C&J	qmax = $0.28 - 0.013$ (LogK$_{OH}$)	Cd^+, Co^{2+}, Cr^{3+}, Cs^+, Cu^{2+}, Sr^{2+}, Zn^{2+}	0.88	0.017	45.27	0.001	-5.06	9.57
b C&J	qmax = $0.084 + 0.011$(Z^2/r)	Cd^+, Co^{2+}, Cr^{3+}, Cs^+, Cu^{2+}, Sr^{2+}, Zn^{2+}	0.95	0.011	112.23	0	-5.91	5.56
f C&J	qmax = $1.90 + 0.008$ (AN/ΔIP)	Cd^+, Co^{2+}, Cr^{3+}, Cs^+, Cu^{2+}, Sr^{2+}, Zn^{2+}	0.41	0.038	5.11	0.07	-3.46	15.95
c C&J	qmax = $0.050 + 0.006$ (IP)	Cd^+, Co^{2+}, Cr^{3+}, Cs^+, Cu^{2+}, Sr^{2+}, Zn^{2+}	0.89	0.016	51.55	0.001	-5.18	8.42

(Continued)

TABLE 5.23 (Continued)

QCARs for Predicting Cation Biosorption to the Yeast, *Saccharomyces cerevisiae* and QCARs for Predicting Cation Biosorption to the Bacterium, *Staphylococcus saprophyticus* BMSZ711

Ref.[a]	QCARs	Cations	R²adj.	SE	F	P	AIC	MAPE
b C&J	$qmax = 0.080 + 0.02(Z^2/r)$	$Cd^{2+}, Co^{2+}, Cr^{3+}, Cs^+, Cu^{2+}, Sr^{2+}, Zn^{2+}$	0.93	0.013	80.53	0	−5.59	6.17
C&J	$qmax = 0.069 + 0.078(Z/AR^2)$	$Cd^{2+}, Co^{2+}, Cr^{3+}, Cs^+, Cu^{2+}, Sr^{2+}, Zn^{2+}$	0.77	0.024	20.94	0.006	−4.40	11.02
d C&J	$qmax = 0.055 + 0.037(Z/r)$	$Cd^{2+}, Co^{2+}, Cr^{3+}, Cs^+, Cu^{2+}, Sr^{2+}, Zn^{2+}$	0.91	0.015	58.71	0.001	−5.3	8.14
e C&J	$qmax = 0.045 + 0.073(Z/AR)$	$Cd^{2+}, Co^{2+}, Cr^{3+}, Cs^+, Cu^{2+}, Sr^{2+}, Zn^{2+}$	0.81	0.022	25.77	0.004	−4.57	10.84
C&J	$qmax = 0.102 + 0.023(Z^{*2}/r)$[b]	$Cd^{2+}, Co^{2+}, Cr^{3+}, Cs^+, Cu^{2+}, Sr^{2+}, Zn^{2+}$	0.04	0.049	1.22	0.319	−2.97	17.24
C&J	$qmax = 0.145 - 0.004 (LogK_{OH}) + 0.008(Z^2/r)$[b]	$Cd^{2+}, Co^{2+}, Cr^{3+}, Cs^+, Cu^{2+}, Sr^{2+}, Zn^{2+}$	0.96	0.01	69.17	0.001	−6.04	4.31
C&J	$qmax = 0.086 + 0.011(Z^2 r) - 0.0003\ (AN/\Delta IP)$[b]	$Cd^{2+}, Co^{2+}, Cr^{3+}, Cs^+, Cu^{2+}, Sr^{2+}, Zn^{2+}$	0.94	0.012	45.17	0.002	−5.63	5.66
C&J	$qmax = -0.03 + 0.008\ (IP) + 0.006\ (AN/\Delta IP)$	$Cd^{2+}, Co^{2+}, Cr^{3+}, Cs^+, Cu^{2+}, Sr^{2+}, Zn^{2+}$	0.97	0.009	90.18	0	−6.3	4.13
C&J	$qmax = 0.03 - 0.018(LogK_{OH}) + 0.005\ (AN/\Delta IP)$[a]	$Cd^{2+}, Co^{2+}, Cr^{3+}, Cs^+, Cu^{2+}, Sr^{2+}, Zn^{2+}$	0.92	0.014	36.22	0.003	−5.42	6.71
h C&J	$qmax = 0.037 + 0.004\ (AN)$	$Ag^+, Cd^{2+}, Co^{2+}, Cr^{3+}, Cu^{2+}, Ni^{2+}, Pb^{2+}, Zn^{2+}$	0.46	0.089	6.94	0.039	−1.79	32.83
C&J	$qmax = -0.16 + 0.44r$	$Ag^+, Cd^{2+}, Co^{2+}, Cr^{3+}, Cu^{2+}, Ni^{2+}, Pb^{2+}, Zn^{2+}$	0.53	0.082	9.04	0.0024	−1.94	31.36
g C&J	$qmax = 0.017 + 0.064\left(X_m^2 r\right)$	$Ag^+, Cd^{2+}, Co^{2+}, Cr^{3+}, Cu^{2+}, Ni^{2+}, Pb^{2+}, Zn^{2+}$	0.59	0.077	11.21	0.015	−2.08	30.15
f C&J	$qmax = 0.087 + 0.028(AN/\Delta IP)$	$Ag^+, Cd^{2+}, Co^{2+}, Cr^{3+}, Cu^{2+}, Ni^{2+}, Pb^{2+}, Zn^{2+}$	0.43	0.091	6.38	0.045	−1.75	35.98
C&J	$qmax = -0.71 + 0.68(AR)$	$Ag^+, Cd^{2+}, Co^{2+}, Cr^{3+}, Cu^{2+}, Ni^{2+}, Pb^{2+}, Zn^{2+}$	0.42	0.092	6.04	0.049	−1.72	33.92
i C&J	$qmax = 0.47 - 14.08(AR/AW)$[b]	$Ag^+, Cd^{2+}, Co^{2+}, Cr^{3+}, Cu^{2+}, Ni^{2+}, Pb^{2+}, Zn^{2+}$	0.34	0.098	4.58	0.076	−1.59	38.93
a C&J	$qmax = 0.33 - 0.020\ (logK_{OH})$	$Cd^{2+}, Co^{2+}, Cr^{3+}, Cu^{2+}, Ni^{2+}, Zn^{2+}$	0.94	0.012	77.4	0.001	−5.74	5.6

g	C&J	$qmax = 0.38 - 0.93(X^2_m r)$[a]	$Cd^+, Co^{2+}, Cr^{3+}, Cu^{2+}, Ni^{2+}, Zn^{2+}$	0.51	0.034	6.24	0.067	-3.67	16.51
b	C&J	$qmax = 0.076 + 0.011 (Z^2/r)$	$Cd^{2+}, Co^{2+}, Cr^{3+}, Cu^{2+}, Ni^{2+}, Zn^{2+}$	0.78	0.023	18.85	0.012	-4.47	11.75
c	C&J	$qmax = -0.018 + 0.008(IP)$	$Cd^{2+}, Co^{2+}, Cr^{3+}, Cu^{2+}, Ni^{2+}, Zn^{2+}$	0.88	0.017	36.66	0.004	-5.03	7.89
	C&J	$qmax = 0.065 + 0.022(Z/r^2)$	$Cd^{2+}, Co^{2+}, Cr^{3+}, Cu^{2+}, Ni^{2+}, Zn^{2+}$	0.63	0.029	9.5	0.037	-3.94	15.34
	C&J	$qmax = 0.014 + 0.11(Z/AR)$[b]	$Cd^{2+}, Co^{2+}, Cr^{3+}, Cu^{2+}, Ni^{2+}, Zn^{2+}$	0.51	0.034	6.24	0.067	-3.66	18.24
d	C&J	$qmax = 0.021 + 0.045(Z/r)$	$Cd^{2+}, Co^{2+}, Cr^{3+}, Cu^{2+}, Ni^{2+}, Zn^{2+}$	0.69	0.027	12.26	0.025	-4.13	14.15
e	C&J	$qmax = -0.03 + 0.11(Z/AR)$	$Cd^{2+}, Co^{2+}, Cr^{3+}, Cu^{2+}, Ni^{2+}, Zn^{2+}$	0.67	0.028	11.02	0.029	-4.05	15.11
	C&J	$qmax = 0.28 - 0.050(Z^2/r)$[a]	$Cd^{2+}, Co^{2+}, Cr^{3+}, Cu^{2-}, Ni^{2+}, Zn^{2+}$	-0.07	0.05	0.67	0.46	-2.88	22.03
g	C&W	$qmax = 0.029 - 0.061 (X^2_m r)$	$Ag^+, Cd^{2+}, Co^{2+}, Cr^{3+}Cs^+, Cu^{2+}, Ni^{2+}, Pb^{2+}, Sr^{2+}, Zn^{2+}$	0.67	0.067	19.04	0.002	-2.38	27.36
	C&W	$qmax = -0.127 + 0.023(Z^*)$	$Ag^+, Cd^{2+}, Co^{2-}, Cr^{3+}Cs^+, Cu^{2+}, Ni^{2+}, Pb^{2+}, Sr^{2+}, Zn^{2+}$	0.34	0.095	5.61	0.045	-1.70	39.97
i	C&W	$qmax = 0.47 - 14.9(AR/AW)$	$Ag^+, Cd^{2+}, Co^{2+}, Cr^{3+}Cs^+, Cu^{2+}, Ni^{2+}, Pb^{2+}, Sr^{2+}, Zn^{2+}$	0.39	0.091	6.84	0.030	-1.78	37.82
h	C&W	$qmax = -0.026 + 0.005 (AN)$	$Cd^{2+}, Co^{2+}, Cr^{3+}, Cu^{2+}, Ni^{2+}, Pb^{2+}, Sr^{2+}, Zn^{2+}$	0.79	0.049	24.01	0.004	-2.96	24.20
g	C&W	$qmax = 0.029 + 0.061 (X^2_m r)$	$Cd^{2+}, Co^{2+}, Cr^{3+}, Cu^{2+}, Ni^{2+}, Pb^{2+}, Sr^{2+}, Zn^{2+}$	0.87	0.039	39.77	0.001	-3.39	22.30
f	C&W	$qmax = 0.039 + 0.026 (AN/\Delta IP)$	$Cd^{2+}, Co^{2+}, Cr^{3+}, Cu^{2+}, Ni^{2+}, Pb^{2+}, Sr^{2+}, Zn^{2+}$	0.50	0.076	7.07	0.045	-2.08	32.30
	C&W	$qmax = 0.002 + 0.002 (AW)$	$Cd^{2+}, Co^{2+}, Cr^{3+}, Cu^{2+}, Ni^{2+}, Pb^{2+}, Sr^{2+}, Zn^{2+}$	0.81	0.047	25.94	0.004	-3.03	23.14
i	C&W	$qmax = 0.457 - 15.56(AR/AW)$	$Cd^{2+}, Co^{2+}, Cr^{3+}, Cu^{2+}, Ni^{2+}, Pb^{2+}, Sr^{2+}, Zn^{2+}$	0.64	0.064	11.87	0.018	-2.42	23.42
	C&W	$qmax = -0.024 + 0.028 (Z^*)$	$Cd^{2+}, Co^{2+}, Cr^{3+}, Cu^{2+}, Ni^{2+}, Pb^{2+}, Sr^{2+}, Zn^{2+}$	0.58	0.069	9.35	0.028	-2.26	30.38

(Continued)

TABLE 5.23 (Continued)
QCARs for Predicting Cation Biosorption to the Yeast, *Saccharomyces cerevisiae* and QCARs for Predicting Cation Biosorption to the Bacterium, *Staphylococcus saprophyticus* BMSZ711

Ref.[a]	QCARs	Cations	R²adj.	SE	F	P	AIC	MAPE
C&W	$qmax = -0.018 + 0.05(X_m^2 r) - 0.01\ (AN/\Delta IP)$[b]	$Cd^{2+}, Co^{2+}, Cr^{3+}, Cu^{2+}, Ni^{2+}, Pb^{2+}, Sr^{2+}, Zn^{2+}$	0.91	0.032	31.66	0.004	−3.74	15.70
g Z	$qmax = 0.091 + 0.111\ (X_m^2 r)$	$Cd^{2+}, Cr^{3+}, Co^{2+}, Cu^{2+}, Hg^{2+}, Pb^{2+}, Ni^{2+}, K^+, Zn^{2+}$	0.69	0.114	18.70	0.004	−4.14	25.23
Z	$qmax = 0.191 + 0.002\ (AW)$	$Cd^{2+}, Cr^{3+}, Co^{2+}, Cu^{2+}, Hg^{2+}, Pb^{2+}, Ni^{2+}, K^+, Zn^{2+}$	0.44	0.153	7.38	0.029	−3.56	29.15
h Z	$qmax = 0.159 + 0.006\ (AN)$	$Cd^{2+}, Cr^{3+}, Co^{2+}, Cu^{2+}, Hg^{2+}, Pb^{2+}, Ni^{2+}, K^+, Zn^{2+}$	0.44	0.153	7.33	0.030	−3.56	28.88
Z	$qmax = -0.194 + 0.342\ (X_m)$	$Cd^{2+}, Cr^{3+}, Co^{2+}, Cu^{2+}, Hg^{2+}, Pb^{2+}, Ni^{2+}, K^+, Zn^{2+}$	0.39	0.160	6.14	0.042	−3.47	32.27
f Z	$qmax = 0.209 + 0.041\ (AN/\Delta IP)$	$Cd^{2+}, Cr^{3+}, Co^{2+}, Cu^{2+}, Hg^{2+}, Pb^{2+}, Ni^{2+}, K^+, Zn^{2+}$	0.34	0.167	5.09	0.059	−3.39	36.98
Z	$qmax = 1.100 - 1.412\ (AR) + 1.534\ (IR)$	$Cd^{2+}, Cr^{3+}, Co^{2+}, Cu^{2+}, Hg^{2+}, Pb^{2+}, Ni^{2+}, K^+, Zn^{2+}$	0.63	0.125	7.81	0.021	−3.90	19.32
Z	$qmax = -0.046 + 0.739\ (IR) - 0.309\ (\Delta E_0)$	$Cd^{2+}, Cr^{3+}, Co^{2+}, Cu^{2+}, Hg^{2+}, Pb^{2+}, Ni^{2+}, K^+, Zn^{2+}$	0.70	0.111	10.48	0.011	−4.12	21.63

Source: Data from C. Can and W. Jianlong, "Correlating Metal Ionic Characteristics with Biosorption Capacity Using QSAR Model." *Chemosphere* 69 (2007):1610–1616; C. Chen and J. Wang. "Influence of Metal Ionic Characteristics on their Biosorption Capacity by *Saccharomyces cerevisiae*." *Appl. Microbiol. Biot.* 74 (2007):911–917; S.S. Zamil, S. Ahmad, M.H. Choi, J.Y. Park, and S.C. Yoon, "Correlating Metal Ionic Characteristics with Biosorption Capacity of *Staphylococcus saprophyticus* BMSZ711 Using QICAR Model." *Bioresource Technol.* 100 (2009):1895–1902.

[a] C&J = Can and Jianlong; C&W = Chen and Wang; Z = Zamil et al.

[b] Physicochemical property was not significant on q_{max} ($\alpha = 0.05$).

Zamil et al. (2009) (labeled Z in Table 5.23) developed 7 QSARs to predict the maximum biosorption capacity (q_{max}) of cations to the bacterium, *Staphylococcus saprophyticus* BMSZ711. The QSARs developed with X_m^2r and a combination of ionic radius (IR) and standard reduction-oxidation potential, also known as the absolute value of the electrochemical potential between the ion and its first stable reduced state (ΔE_0), were the most statistically significant (Table 5.23).

Several physicochemical properties were used 3 or more times to develop QSARs for predicting the biosorption capacity of cations (labeled a–j in column 1 of Table 5.23). The absolute value of the logarithm of the first hydrolysis constant (|log K_{OH}|), ionic index (Z^2/r), and ionization potential (IP) (labeled a, b, and c in column 1 of Table 5.23) were used by Can and Jianlong (2007) to develop 3 QSARs each, and all were statistically significant. The ionic potential (Z/r) was used by Can and Jianlong (2007) to develop 3 QSARs (labeled d in column 1 of Table 5.23) and the one used to predict the biosorption capacity of 7 cations with a r^2 of 0.91 was the most statistically significant. The ion charge/atomic radius (Z/AR) was used by Can and Jianlong (2007) to develop 3 QSARs (labeled e in column 1 of Table 5.23) and the one with an r^2 of 0.81 was the most statistically significant. The atomic number (AN)/the difference of the ionization potential in volts between its oxidation number (OX) and the next lower one (OX − 1) (ΔIP), was used by Can and Jianlong (2007), Chen and Wang (2007), and Zamil et al. (2009) to develop 4 QSARs for predicting the biosorption capacity of cations (labeled f in column 1 of Table 5.23), but none were highly statistically significant. Can and Jianlong (2007), Chen and Wang (2007), and Zamil et al. (2009) used the covalent index $\left(X_m^2r \right)$ to develop 4 QSARs for predicting the biosorption capacity of cations (labeled g in column 1 of Table 5.23) and the one with an r^2 of 0.87 was the most statistically significant. AN and AR/AW were used by Can and Jianlong (2007), Chen and Wang (2007), and Zamil et al. (2009) to develop 3 QSARs each for predicting the biosorption capacity of cations (labeled h and i in column 1 of Table 5.23), but none were highly statistically significant.

REFERENCES

Babich, H., J.A. Puerner, and E. Borenfreund. 1986. In vitro cytotoxicity of metal to bluegill (BF-2) cells. *Arch Environ. Contam. Toxicol.* 15:31–37.

Biesinger, K.L., and G.M. Christensen. 1972. Effects of various metals on survival growth reproduction and metabolism of *Daphnia magna. J. Fish. Res. Board Can.* 29:1691–1700.

Bringmann, V.-G., and R. Kuhn. 1977. Befunde der Schadwirkung wassergefardender Stoffe gegen Daphnia magna. *Z. Wasser Abwasser-Forschung* 10:161–166.

Bringmann, V.-G., and R. Kuhn. 1982. Ergebnisse der Schadwirkung wassergefahrdender Stoffe gegen Daphnia magna in einem weiterenwickelten standardisierten Testverfahren. *Z. Wasser Abwasser-Forschung* 15,1–6.

Can, C., and W. Jianlong. 2007. Correlating metal ionic characteristics with biosorption capacity using QSAR model. *Chemosphere* 69:1610–1616.

Chen, C., and J. Wang. 2007. Influence of metal ionic characteristics on their biosorption capacity by *Saccharomyces cerevisiae. Appl. Microbiol. Biot.* 74:911–917.

Cox, J.L., and Harrison, S.D. Jr., 1983. Correlation of metal toxicity with in vitro calmodulin inhibition. *Biochem. Biophys. Research. Commun.* 115:106–111.

Danielli, J.F., and J.T. Davies. 1951. Reactions at interfaces in relation to biological problems. *Advan. Enzymol.* 11:35–89.

deZwart, D., and W. Sloof. 1983. The Microtox as an alternative assay in the acute toxicity assessment of water pollutants. *Aquatic Tox.* 4, 129–138.

Dounce, A.L., and T.H. Lan. 1949. The action of uranium on enzymes and proteins. In *National Nuclear Energy Series*, Div. VI, Vol. 1, edited by A.L. Dounce, 759–888. New York: McGraw-Hill.

Enache, M., P. Palit, J.C. Dearden, and N.W. Lepp. 1999. Evaluation of cation toxicities to sunflower: A study related to the assessment of environmental hazards. *Proceedings of Extended Abstracts, 5th International Conference on the Biogeochemistry of Trace Elements*, Vol. 2. Vienna, Austria, July 11–15, 1999, 1120–1121.

Enache, M., P. Palit, J.C. Dearden, and N.W. Lepp. 2000. Correlation of physico-chemical parameters with toxicity of metal ions to plants. *Pest Manag Sci* 56:821–824.

Enache, M., J.C. Dearden, and J.D. Walker. 2003. QSAR analysis of metal ion toxicity data in sunflower callus cultures (*Helianthus annuus*, Sunspot). *QSAR Comb. Sci.* 22:234–240.

Grushkin, B. 1956. Some effects of metals on living systems. MS thesis, University of Texas, Austin, TX, USA.

Hara, T., and Y. Sonoda. 1979. Comparison of the toxicity of heavy metals to cabbage growth. *Plant Soil,* 51:127–133.

Hickey, J.P. 2005. Estimation of inorganic species aquatic toxicity. In *Techniques in Aquatic Toxicology: Volume 2*, chap. 13, edited by G.K. Ostrander, 617–629. Boca Raton, FL: CRC Press/Lewis.

Jacobson, K.B., J.E. Turner, N.T. Christie, and R.K. Owenby. 1983. Toxic and biochemical effects of divalent metal ions in *Drosophila*: Correlations to effects in mice and to chemical softness parameters. *Sci. Total Environ.* 28:355–366.

Jones, J.R.E. 1939. The relation between the electrolytic solution pressures of the metals and their toxicity to the stickleback (*Gasterosteus aculeatus* L.). *J. Exp. Biol.* 16:425–437.

Jones, J.R.E. 1940. A further study of the relation between toxicity and solution pressure with *Polycelis nigra* as test animal. *J. Exp. Biol.* 17:408–415.

Jones, M.M., and W.K. Vaughn. 1978. HSAB theory and acute metal ion toxicity and detoxification processes. *J. Inorg. Nucl. Chem.* 40:2081–2088.

Joseph, D.R., and S.J. Meltzer. 1909. The comparative toxicity of the chlorides of magnesium, calcium, potassium and sodium. *J. Pharmacol.* 1:1–26.

Kaiser, K.L.E. 1980. Correlation and prediction of metal toxicity to aquatic biota. *Can. J. Fish Aquat. Sci.* 37:211–218.

Kaiser, K.L.E. 1985. Correlation of metal ion toxicities to mice. *Sci. Total Environ.* 46:113–119.

Khangarot, B.S., and S. Das. 2009. Acute toxicity of metals and reference toxicants to a freshwater ostracod, *Cypris subglobosa* Sowerby, 1840 and correlation to EC50 values of other test models. *J. Hazardous Materials* 172:641–649.

Khangarot, B.S., and P.K. Ray. 1989. Investigation of correlation between physicochemical properties of metals and their toxicity to the water flea *Daphnia magna* Strauss. *Ecotox. Environ. Safety* 18:109–120.

Kinraide, T.B. 2009. Improved scales for metal ion softness and toxicity. *Environ. Toxicol. Chem.* 28:525–533.

Kinraide, T.B., and U. Yermiyahu. 2007. A scale of metal ion binding strengths correlating with ionic charge: Pauling electronegativity, toxicity, and other physiological effects. *J. Inorg. Biochem.* 101:1201–1213.

LeBlanc, G. 1984. Interspecies relationships in acute toxicity of chemicals to aquatic organisms. *Environ. Tox. & Chem.* 3 47–60.

Lepădatu, C., M. Enache, and J.D. Walker. 2009. Toward a more realistic QSAR approach to predicting metal toxicity. *QSAR Combin. Sci.* 28:520–525.

Lewis, D.F.V., M. Dobrota, M.G. Taylor, and D.V. Parke. 1999. Metal toxicity in two rodent species and redox potential: Evaluation of quantitative structure-activity relationships. *Environ. Toxicol. Chem.* 18:2199–2204.

Lithner, G. 1989. Some fundamental relationships between metal toxicity in freshwater physico-chemical properties and background levels. *Sci. Total. Environ.* 87/88:365–380.

Magwood S., and S. George. 1996. *In-vitro* alternatives to whole animal testing-comparative cytotoxicity studies of divalent metals in established cell-lines derived from tropical and temperate water fish species in a neutral red assay. *Marine Environ. Res.* 42:37–40.

Mathews, A.P. 1904. The relation between solution tension atomic volume and the physiological action of the elements. *Am. J. Physiol.* 10:290–323.

McCloskey, J.T., M.C. Newman, and S.B. Clark. 1996. Predicting the relative toxicity of metal-ions using ion characteristics: Microtox® bioluminescence assay. *Environ. Toxicol. Chem.* 15:1730–1737.

McGuigan, H. 1904. The relation between the decomposition-tension of salts and their antifermentative properties. *Am. J. Physiol.* 10:444–451.

Mendes, L.F., E.L. Bastos, and C.V. Stevani. 2010. Prediction of metal cation toxicity to the bioluminescent fungus *Gerronema viridilucens*. *Environ. Toxicol. Chem.* 29:2177–2181.

Misono, M., and Y. Saito. 1970. Evaluation of softness from the stability constants of metal-ion complexes. *Bull. Chem. Soc. Japan* 43: 3680–3684.

Newman, M.C., and J.T. McCloskey. 1996. Predicting relative toxicity and interactions of divalent metal ions: Microtox bioluminescence assay. *Environ. Tox. Chem.* 15:275–281.

Newman, M.C., J.T. McCloskey, and C.P. Tatara. 1998. Using metal-ligand binding characteristics to predict metal toxicity: Quantitative ion character-activity relationships (QICARs). *Environ. Health Persp.* 106:1419–1425.

Pearson, R.G., and F.J. Mawby. 1967. The nature of metal-halogen bonds. In *Halogen Chemistry,* Vol. 3., edited by V. Gutmann, 55–84. London: Academic Press.

Pauling, L. 1932. The nature of the chemical bond. IV. The energy of single bonds and the relative electronegativity of atoms. *J. Am. Chem Soc.* 54(9):3570–3582.

Roy, D.R., S. Giri, and P.K. Chattaraj. 2009. Arsenic toxicity: An atom counting and electrophilicity-based protocol. *Molecular Diversity* 13:551–556.

Sacan, M.T., F. Cecen, M.D. Erturk, and N. Semerci. 2009. Modelling the relative toxicity of metals on respiration of nitrifiers using ion characteristics. *SAR QSAR Environ. Res.* 20:727–740.

Sauvant, M.P., D. Pepin, J. Bohatier, C.A. Groliere, and J. Guillot. 1997. Toxicity assessment of 16 inorganic environmental pollutants by six bioassays. *Ecotoxicol. Environ. Safety* 37:131–140.

Schmidt, E.G. 1928. The inactivation of unease. *J. Biol. Chem.* 73:53–61.

Segner, H., D. Lenz, W. Hanke, and G. Schüürmann. 1994. Cytotoxicity of metals toward rainbow trout R1 cell-line. *Environ. Toxicol. Water Qual.* 9:273–279.

Seifriz, W. 1949. Toxicity and the chemical properties of ions. *Science* 110:193–196.

Shaw, W.H.R. 1954a. The inhibition of urease by various metal ions. *J. Am. Chem. Soc.* 76:2160–2163.

Shaw, W.H.R. 1954b. Toxicity of cations toward living systems. *Science* 120:361–363.

Shaw, W.H.R, and B. Grushkin. 1957. The toxicity of metal ions to aquatic organisms. *Arch. Biochem. Biophys.* 67:447–452.

Somers, E. 1959. Fungitoxicity of metal ions. *Nature* 184:475–476.

Su, L.M., Y.H. Zhao, X. Yuan, C.F. Mu, N. Wang, and J.C. Yan. 2010. Evaluation of combined toxicity of phenols and lead to *Photobacterium phosphoreum* and quantitative structure-activity relationships. *Bull. Environ. Contam. Toxicol.* 84:311–314.

Svanberg, O., and G. Lithner. 1978. Evaluation of harmful effects of heavy metals on aquatic organisms, seminar on heavy metals. Technological methods for the limitation of discharges under the convention of the Protection of the Marine Environment of the Baltic Sea Area, Copenhagen. NBL Rep 75. Swedish Environmental Protection Board.

Tan, E-L., M.W. Williams, R.L. Schenley, S.W. Perdue, T.L. Hayden, J.E. Turner, and A.W. Hsie. 1984. The toxicity of sixteen metallic compounds in Chinese hamster ovary cells. *Toxicol. Appl. Pharmacol.* 74:330–336.

Tatara, C.P., M.C. Newman, J.T. McCloskey, and P.L. Williams. 1997. Predicting relative metal toxicity with ion characteristics: *Caenorhabditis elegans* LC50. *Aquatic Toxicol.* 39:279–290.

Tatara, C.P., M.C. Newman, J.T. McCloskey, and P.L. Williams. 1998. Use of ion characteristics to predict relative toxicity of mono-, di-, and trivalent metal ions: *Caenorhabditis elegans* LC50. *Aquatic Toxicol.* 42: 255–269.

True, R.H. 1930. The toxicity of molecules and ions, *Proc. Am. Phil. Soc* 69:231–245.

Turner, J.E., E.H. Lee, K.B. Jacobson, N.T. Christie, M.W. Williams, and J.D. Hoeschele. 1983. Investigation of correlations between chemical parameters of metal ions and acute toxicity in mice and *Drosophila*. *Sci. Total Environ.* 28:343–354.

Turner, J.E., M.W. Williams, K.B. Jacobson, and B.E. Hingerty. 1985. Correlations of acute toxicity of metal ions and the covalent/ionic character of their bonds. In *QSAR in Toxicology and Xenobiochemistry*, edited by M. Tichý, 171–178. Amsterdam: Elsevier.

Van Kolck, M., M.A.J. Huijbregts, K. Veltman, and A.J. Hendriks. 2008. Estimating bioconcentration factors, lethal concentrations and critical body residues of metals in the mollusks *Perna viridis* and *Mytilus edulis* using ion characteristics. *Environ. Toxicol. Chem.* 27:272–276. Supplementary Information available at http://onlinelibrary.wiley.com/store/10.1897/07–224R.1/asset/supinfo/10.1897_07–224.S1.pdf?v=1&s=fda547dc5689332031992621f5e3833819a8e96f.

Venugopal, B., and T.D. Luckey. 1978. *Metal Toxicity in Mammals, Vol. 2, Chemical Toxicity of Metals and Metalloids.* New York: Plenum.

Walker, J.D., M. Enache, and J.C. Dearden. 2003. Quantitative cationic activity relationships for predicting toxicity of metals. *Environ. Toxicol. Chem.* 22:1916–1935.

Walker, J.D., M. Enache, and J.C. Dearden. 2007. Quantitative cationic activity relationships for predicting toxicity of metal ions from physicochemical properties and natural occurrence levels. *QSAR Combin. Sci.* 26:522–527.

Williams, M.W., J.D. Hoeschele, J.E. Turner, K.B. Jacobson, N.T. Christie, C.L. Paton, L.H. Smith, H.R. Witschi, and E.H. Lee. 1982. Chemical softness and acute metal toxicity in mice and *Drosophila*. *Toxicol. Appl. Pharmacol.* 63:461–469.

Williams, M.W., and J.E. Turner. 1981. Comments on softness parameters and metal ion toxicity. *J. Inorg. Nucl. Chem.* 43:1689–1691.

Williams, M.W., J.E. Turner, and A.W. Hsie. 1987. Calmodulin inhibition: A possible predictor of metal-ion toxicity. In *QSAR in Environmental Toxicology* II, edited by K.L.E. Kaiser and D. Reidel, 401–405. Netherlands: Dordrecht.

Woodruff, L.L., and H.H. Bunzel. 1909. The relative toxicity of various salts and acids toward *Paramecium*. *Am. J. Physiol.* 25:190–194.

Workentine, M.L., J.J. Harrison, P.U. Stenroos, H, Ceri, and R.J. Turner. 2008. *Pseudomonas fluorescens*' view of the periodic table. *Environ. Microbiol.* 10(1):238–250.

Zamil, S.S., S. Ahmad, M.H. Choi, J.Y. Park, and S.C. Yoon. 2009. Correlating metal ionic characteristics with biosorption capacity of *Staphylococcus saprophyticus* BMSZ711 using QICAR model. *Bioresource Technol.* 100:1895–1902.

Zhou, D.M., L.Z. Li, W.J.G.M. Peijnenburg, D.R. Ownby, A.J. Hendriks, P. Wang, and D.D. Li. 2011. A QICAR approach for quantifying binding constants for metal-ligands complexes. *Ecotox. Environ. Safety* 74:1036–1042.

APPENDIX: DEFINITIONS FOR CHAPTER 5 ACRONYMS

Statistical Definitions	Acronyms
Standard error	SE
Coefficient of determination	r^2
Variance ratio (F test)	F
Probability of a chance correlation	p
Mean absolute percent error between observed and predicted values in the prediction	MAPE
Akaike's Information Criterion	AIC
Mean square error	MSE
Adjusted r square	r^2 adj.
Physicochemical Property Definitions	
Standard electrode potential	$E°$
Standard reduction-oxidation potential	ΔE_0
Absolute value of the electrochemical potential between the ion and its first stable reduced state	ΔE_0
Negative logarithm of solubility product equilibrium constant of the corresponding metal sulphide	pK_{sp}
Ion charge	Z
Effective ion charge	Z*
Ionic potential	Z/r
Log of the equilibrium constant of the metal-ATP complex	$\log K_{eq}$
Covalent index	$X_m^2 r$
Log of the first hydrolysis constant	$|\log K_{OH}|$
Ionic index	Z^2/r
Log of the stability constant for the metal fluoride; log of the stability constant for the metal chloride	$\Delta\beta$
Log metal hydroxide solubility	$Log-K_{so}MOH$
Ionization potential	IP
Allred-Rochow electronegativity	X_{AR}
Pauling's electronegativity	X
Mulliken Electronegativity	X_m
Pearson and Mawby softness parameter	σ_p
The heat of formation of inorganic oxides	ΔH_o
The heat of formation of aqueous ion	$\Delta H_{aq\ ion}$
Atomic number	AN
Difference in ionization potentials between ion oxidation number (OX) and (OX − 1)	ΔIP
Atomic weight	AW
Enthalpy of hydration	ΔH_h
Density	ρ
Ionic Radius	IR
Atomic Radius	AR
Stability constant of metal ion complexes with EDTA	log K (EDTA)
Formation constants with ATP	log K (ATP)
Stability constants for AMP with divalent metal ions	log K (AMP)
Misono and Saito (1970) softness parameter	Y'

Stability constant of metal ion complexes with AMP-2	log K (AMP-2)
Stability constant of metal ion complexes with AMP-3	log K (AMP-3)
Observed stability constants of metal ion-acetato complexes	log K (Ac)
Calculated constant of metal ion complexes with NH_3	log K (NH_3)
Enthalpy of formation of sulfides	ΔH_s
Stability constant of metal ion complexes with sulfate	log K (sulfate)
Average molar metal concentration in soil at 10% moisture	log M_{soil}
Medians of elemental composition of soils	mg X/kg soil
Calculated mean of the elemental content in land plants	Land plants
Intrinsic (van der Waals) molecular volume	V_i
Solute ability to stabilize a neighboring charge or dipole by non-specific dielectric interactions	π^*
Solute ability to accept or donate hydrogen in a hydrogen bond	β_m and α_m
Highest Occupied Molecular Orbital	HOMO
Lowest Unoccupied Molecular Orbital	LUMO
Electronegativity of HOMO state for ELH (hydrogen)	HELH
Electronegativity of HOMO state for EO (oxygen)	HEO
Electronegativity of HOMO state for EM (metal)	HEM
Electronegativity of OMO (occupied molecular orbital) state for EO (oxygen)	O-EO
Electronegativity of OMO state for EM (metal)	O-EM
Electronegativity of UMO (unoccupied molecular orbital) state for ELH (hydrogen)	U-ELH
Electronegativity of UMO state for EM (oxygen)	U-EO
Electronegativity of UMO state for EM (metal)	U-EM
Electrophilicity	ω
Chemical potential of the gaseous phase	$\mu_{(g)}$
Chemical potential of the aqueous phase	$\mu_{(aq)}$
Electronic spatial extent	ESE
Energy of the polarized solute solvent	E_{PSS}
Aqueous phase energy of HOMO energy	$E_{HOMO(aq)}$
Gaseous phase energy of HOMO energy	$E_{HOMO(g)}$
Gaseous phase energy of LUMO energy	$E_{LUMO(g)}$
Maximum value for LUMO energy of the most important cation	$E_{LUMO}C_{max}$
Cationic charge	Z
Crystal ionic radius	r
Oxidation number	OX

Other Definitions

Quantitative Cationic Activity Relationship	QCAR
Quantitative Ion Character Activity Relationships	QICAR
Linear Solvation Energy Relationship	LSER
RNA synthesis rate assay	RNA
MTT reduction assay	MTT
Neutral red incorporation assay	NRI
Coomassie blue assay	CB

6 QSARs versus BLM

6.1 INTRODUCTION

In Chapter 1 it was noted that the biotic ligand model (BLM) estimates equilibrium concentrations or activities of dissolved metals species that can compete for and adsorb to ligands on biological surfaces. Chapter 5 provided detailed information about quantitative structure-activity relationships (QSARs) for metal ions. The purpose of this chapter is to distinguish between QSARs and the BLM for metals.

6.2 BLM

The technical basis for the BLM was described by DiToro et al. (2001). Its application to acute aquatic copper toxicity was described by Santore et al. (2001). Paquin et al. (2002) provided a historical overview of the BLM, while the final section of this chapter provides a list of 175 references describing BLM studies published from 2001 to 2011.

The BLM concept is a permutation of the free ion activity model (FIAM) (e.g., see Campbell 1995). As a primary distinction, the FIAM predicts metal uptake by aquatic organisms by assuming that metal internalization is related to the free metal ion activity in the bulk solution; however, the BLM assumes that the association of metals with biotic ligands of the organism determines the toxic effects (Zeng et al., 2009). Additional differences between the FIAM and BLM were discussed by Hassler et al. (2004).

The aim of the BLM is to explain and predict the effects that metal speciation in aqueous media has upon metal toxicity to aquatic organisms. The term *biotic ligand* refers to a specific biological receptor of an organism that interacts with the metal ions in aqueous media. This interaction is similar to those involving any other type of chemical ligand. It is considered in certain organisms such as fish that the biotic ligand would be Na^+ and K^+ channel proteins of gills, which have a role in regulating the ionic composition of blood. Therefore, in more explicit terms, the biotic ligand is a biological receptor.

The BLM for acute toxicity to aquatic organisms is based on the hypothesis that the mortality through intoxication occurs if the metal–biotic ligand complex reaches or exceeds a critical concentration. This threshold is directly linked to the quantity of solvated metal ions available in water. It follows that metal bioavailability in water can be influenced by different chemical factors, such as (a) the agonist or antagonist competition with other cations for interaction with the biotic ligand, (b) the action of anions present in water, and (c) the action of some organic chemicals that can form metal complexes, and in so doing, prevent or influence metal binding to the biotic ligand.

6.3   QSARS VERSUS THE BLM

In QSARs, the biological activity is also produced by the interaction between the ligand and the biological receptor of an organism. The only difference is that the *ligand* in QSARs is a chemical species, and the biological receptor is the *biotic ligand*.

The BLM is very similar to QSARs, in that they both use a series of physico-chemical and biological parameters that would characterize and describe:

- the transformation of the ions on their path,
- the solvation in water of the metals,
- the chemical speciation,
- the sedimentation and accumulation in an aqueous environment, and
- the interaction with the biotic ligand in the organisms present in an aqueous environment.

However, in a QSAR, these physicochemical and biological parameters are called *descriptors*, and their number is infinitely larger because descriptors of different nature are used (e.g., physical, chemical, geometrical, quanto-molecular, informational natures).

The BLM and QSARs are distinct in at least three ways:

1. The BLM correlates the biological activity (toxicity) for a multitude of metal ions (a set of metal ions) and other chemical species that exist simultaneously in an aquatic environment, and describes their fate and their interaction with the biotic ligand (which is equivalent to the biological receptor in a QSAR) with the use of a multitude of experimental physicochemical parameters such as formation constants, pH, and temperature.
2. QSARs have a statistical component where each chemical species (ligand) in a chemical's class is represented by a series of descriptors (the physico-chemical parameters in the BLM) that are the same for all chemical species in the class.
3. In QSAR studies, the regression equation that is established gives information on the mechanism of interaction, ligand-biological receptor, or other physicochemical processes that influence biological activity. The linear regression correlates the descriptors and the biological response as follows:

$$A = a_0 + a_1 X_1 + a_2 X_2 + ...,$$

where A = biological activity, a_0, a_1, ... are the regression coefficients, and X_1, X_2, ... are the structural descriptors.

The prediction capability of a QSAR depends on the statistical features of the regression output (regression coefficient r^2, the a_i coefficients).

REFERENCES

Adams, W.J., R. Blust, U. Borgmann, et al. 2011. Utility of tissue residues for predicting effects of metals on aquatic organisms. *Integr. Environ. Assess. Manag.* 7:75–98.

Antunes, P.M., E.J. Berkelaar, D. Boyle, et al. 2006. The biotic ligand model for plants and metals: Technical challenges for field application. *Environ. Toxicol. Chem.* 25: 875–882.

Antunes, P.M., B.A. Hale, and A.C. Ryan. 2007. Toxicity versus accumulation for barley plants exposed to copper in the presence of metal buffers: Progress towards development of a terrestrial biotic ligand model. *Environ. Toxicol .Chem.* 26:2282–2289.

Antunes, P.M., and N.J. Kreager. 2009. Development of the terrestrial biotic ligand model for predicting nickel toxicity to barley (*Hordeum vulgare*): Ion effects at low pH. *Environ. Toxicol. Chem.* 28:1704–1710.

Arnold, W.R., R.L. Diamond, and D.S. Smith. 2010a. Acute and chronic toxicity of copper to the euryhaline rotifer, *Brachionus plicatilis* ("L" strain). *Arch. Environ. Contam. Toxicol.* 60(2):250–260.

Arnold, W.R., R.L. Diamond, and D.S. Smith. 2010b. The effects of salinity, pH, and dissolved organic matter on acute copper toxicity to the rotifer, *Brachionus plicatilis* ("L" strain). *Arch. Environ. Contam. Toxicol.* 59(2): 225–234.

Arnold, W.R., R.C. Santore, and J.S. Cotsifas. 2005. Predicting copper toxicity in estuarine and marine waters using the Biotic Ligand Model. *Mar. Pollut. Bull.* 50:1634–1640.

Besser, J.M., C.A. Mebane, D.R. Mount, et al. 2007. Sensitivity of mottled sculpins (*Cottus bairdi*) and rainbow trout (*Onchorhynchus mykiss*) to acute and chronic toxicity of cadmium, copper, and zinc. *Environ. Toxicol. Chem.* 26:1657–1665.

Bianchini, A., R.C. Playle, C.M. Wood, et al. 2005. Mechanism of acute silver toxicity in marine invertebrates. *Aquat. Toxicol.* 72:67–82.

Bianchini, A., and C.M. Wood. 2008. Does sulfide or water hardness protect against chronic silver toxicity in Daphnia magna? A critical assessment of the acute-to-chronic toxicity ratio for silver. *Ecotoxicol. Environ. Saf.* 71:32–40.

Bielmyer, G.K., M. Grosell, P.R. Paquin, et al. 2007. Validation study of the acute biotic ligand model for silver. *Environ. Toxicol. Chem.* 26:2241–2246.

Bielmyer, G.K., S.J. Klaine, J.R. Tomasso, et al. 2004. Changes in water quality after addition of sea salts to fresh water: Implications during toxicity testing. *Chemosphere* 57:1707–1711.

Birceanu, O., M.J. Chowdhury, P.L. Gillis, et al. 2008. Modes of metal toxicity and impaired branchial ionoregulation in rainbow trout exposed to mixtures of Pb and Cd in soft water. *Aquat. Toxicol.* 89:222–231.

Blanchard, J., and M. Grosell. 2005. Effects of salinity on copper accumulation in the common killifish (*Fundulus heteroclitus*). *Environ. Toxicol. Chem.* 24:1403–1413.

Borgmann, U., W.P. Norwood, and D.G. Dixon. 2004. Re-evaluation of metal bioaccumulation and chronic toxicity in *Hyalella azteca* using saturation curves and the biotic ligand model. *Environ. Pollut.* 131:469–484.

Borgmann, U., M. Nowierski, and D.G. Dixon. 2005. Effect of major ions on the toxicity of copper to *Hyalella azteca* and implications for the biotic ligand model. *Aquat. Toxicol.* 73:268–287.

Bossuyt, B.T., K.A. De Schamphelaere, and C.R. Janssen. 2004. Using the biotic ligand model for predicting the acute sensitivity of cladoceran-dominated communities to copper in natural surface waters. *Environ. Sci. Technol.* 38:5030–5037.

Bravin, M.N., A.M. Michaud, B. Larabi, et al. 2010. RHIZOtest: A plant-based biotest to account for rhizosphere processes when assessing copper bioavailability. *Environ. Pollut.* 158:3330–3337.

Bringolf, R.B., B.A. Morris, C.J. Boese, et al. 2006. Influence of dissolved organic matter on acute toxicity of zinc to larval fathead minnows (*Pimephales promelas*). *Arch. Environ. Contam. Toxicol.* 51:438–444.

Brix, K.V., J. Keithly, R.C. Santore, et al. 2010. Ecological risk assessment of zinc from storm-water runoff to an aquatic ecosystem. *Sci. Total Environ.* 408:1824–1832.

Brooks, B.W., J.K. Stanley, J.C. White, et al. 2004. Laboratory and field responses to cadmium: An experimental study in effluent-dominated stream mesocosms. *Environ. Toxicol. Chem.* 23:1057–1064.

Campbell, P.G.C. 1995. Interactions between trace metals and aquatic organisms: A critique of the free-ion activity model. In *Metal Speciation and Bioavailability in Aquatic Systems*, Vol. 1, edited by A. Tessier, D.R. Turner, 45–102. New York: John Wiley.

Chen, B.C., W.Y. Chen, and C.M. Liao. 2009. A biotic ligand model-based toxicodynamic approach to predict arsenic toxicity to tilapia gills in cultural ponds. *Ecotoxicology* 18:377–383.

Chen, W.Y., and C.M. Liao. 2010. Dynamic features of ecophysiological response of fresh-water clam to arsenic revealed by BLM-based toxicological model. *Ecotoxicology* 19:1074–1083.

Chen, W.Y., J.W. Tsai, Y.R. Ju, et al. 2010. Systems-level modeling the effects of arsenic exposure with sequential pulsed and fluctuating patterns for tilapia and freshwater clam. *Environ. Pollut.* 158:1494–1505.

Chen, Z., L. Zhu, and K.J. Wilkinson. 2010. Validation of the biotic ligand model in metal mixtures: Bioaccumulation of lead and copper. *Environ. Sci. Technol.* 44:3580–3586.

Chen, Z.Z., L. Zhu, K. Yao, et al. 2009. [Interaction between calcium and lead affects the toxicity to embryo of zebrafish (*Danio rerio*)]. *Huan Jing Ke Xue* 30:1205–1209.

Choi, O., T.E. Clevenger, B. Deng, et al. 2009. Role of sulfide and ligand strength in controlling nanosilver toxicity. *Water Res.* 43:1879–1886.

Clifford, M., and J.C. McGeer. 2009. Development of a biotic ligand model for the acute toxic-ity of zinc to *Daphnia pulex* in soft waters. *Aquat. Toxicol.* 91:26–32.

Clifford, M., and J.C. McGeer. 2010. Development of a biotic ligand model to predict the acute toxicity of cadmium to *Daphnia pulex*. *Aquat. Toxicol.* 98:1–7.

Comber, S.D., G. Merrington, L. Sturdy, et al. 2008. Copper and zinc water quality standards under the EU Water Framework Directive: The use of a tiered approach to estimate the levels of failure. *Sci. Total Environ.* 403:12–22.

Constantino, C., M. Scrimshaw, S. Comber, et al. 2010. An evaluation of biotic ligand mod-els predicting acute copper toxicity to *Daphnia magna* in wastewater effluent. *Environ. Toxicol. Chem.* 30(4): 852–860.

Craig, P.M., C.M. Wood, and G.B. McClelland. 2010. Water chemistry alters gene expression and physiological end points of chronic waterborne copper exposure in zebrafish, *Danio rerio*. *Environ. Sci. Technol.* 44:2156–2162.

Croteau, M.N., and S.N. Luoma. 2007. Characterizing dissolved Cu and Cd uptake in terms of the biotic ligand and biodynamics using enriched stable isotopes. *Environ. Sci. Technol.* 41:3140–3145.

De Schamphelaere, K.A., B.T. Bossuyt, and C.R. Janssen. 2007. Variability of the protective effect of sodium on the acute toxicity of copper to freshwater cladocerans. *Environ. Toxicol. Chem.* 26:535–542.

De Schamphelaere, K.A., D.G. Heijerick, and C.R. Janssen. 2006. Cross-phylum comparison of a chronic biotic ligand model to predict chronic toxicity of copper to a freshwater rotifer, *Brachionus calyciflorus* (Pallas). *Ecotoxicol. Environ. Saf.* 63:189–195.

De Schamphelaere, K.A., and C.R. Janssen. 2004a. Development and field validation of a biotic ligand model predicting chronic copper toxicity to *Daphnia magna*. *Environ. Toxicol. Chem.* 23:1365–1375.

De Schamphelaere, K.A., and C.R. Janssen. 2004b. Bioavailability and chronic toxic-ity of zinc to juvenile rainbow trout (*Oncorhynchus mykiss*): Comparison with other fish species and development of a biotic ligand model. *Environ. Sci. Technol.* 38:6201–6209.

De Schamphelaere, K.A., and C.R. Janssen. 2004c. Effects of chronic dietary copper exposure on growth and reproduction of *Daphnia magna*. *Environ. Toxicol. Chem.* 23:2038–2047.

De Schamphelaere, K.A., and C.R. Janssen. 2010. Cross-phylum extrapolation of the *Daphnia magna* chronic biotic ligand model for zinc to the snail *Lymnaea stagnalis* and the rotifer *Brachionus calyciflorus*. *Sci.Total Environ.* 408:5414–5422.

De Schamphelaere, K.A., S. Lofts, and C.R. Janssen. 2005. Bioavailability models for predicting acute and chronic toxicity of zinc to algae, daphnids, and fish in natural surface waters. *Environ. Toxicol. Chem.* 24:1190–1197.

De Schamphelaere, K.A., J.L. Stauber, K.L. Wilde, et al. 2005. Toward a biotic ligand model for freshwater green algae: Surface-bound and internal copper are better predictors of toxicity than free Cu2+-ion activity when pH is varied. *Environ. Sci. Technol.* 39:2067–2072.

De Schamphelaere, K.A., F.M. Vasconcelos, F.M. Tack, et al. 2004. Effect of dissolved organic matter source on acute copper toxicity to *Daphnia magna*. *Environ. Sci. Technol.* 23:1248–1255.

Deforest, D.K., R.W. Gensemer, E.J. Van Genderen, et al. 2010. Protectiveness of water quality criteria for copper in Western United States waters relative to predicted olfactory responses in juvenile Pacific salmon. *Integr. Environ. Assess. Manag.* 7(3):336–347.

Deleebeeck, N.M., K.A. De Schamphelaere, D.G. Heijerick, et al. 2008. The acute toxicity of nickel to *Daphnia magna*: Predictive capacity of bioavailability models in artificial and natural waters. *Ecotoxicol. Environ. Saf.* 70:67–78.

Deleebeeck, N.M., K.A. De Schamphelaere, and C.R. Janssen. 2008. A novel method for predicting chronic nickel bioavailability and toxicity to *Daphnia magna* in artificial and natural waters. *Environ. Toxicol. Chem.* 27:2097–2107.

Deleebeeck, N.M., K.A. De Schamphelaere, and C.R. Janssen. 2009. Effects of Mg(2+) and H(+) on the toxicity of Ni(2+) to the unicellular green alga *Pseudokirchneriella subcapitata*: Model development and validation with surface waters. *Sci. Total Environ.* 407:1901–1914.

Di Toro, D.M., H.E. Allen, H.L. Bergman, J.S. Meyer, P.R. Paquin, and R.C. Santore. 2001. Biotic ligand model of the acute toxicity of metals. 1. Technical basis. *Environ. Toxicol. Chem.* 20:2383–2396.

Di Toro, D.M., J.A. McGrath, D.J. Hansen, et al. 2005. Predicting sediment metal toxicity using a sediment biotic ligand model: Methodology and initial application. *Environ. Toxicol. Chem.* 24:2410–2427.

Fortin, C., F.H. Denison, and J. Garnier-Laplace. 2007. Metal-phytoplankton interactions: Modeling the effect of competing ions (H+, Ca2+, and Mg2+) on uranium uptake. *Environ. Toxicol. Chem.* 26:242–248.

Fortin, C., L. Dutel, and J. Garnier-Laplace. 2004. Uranium complexation and uptake by a green alga in relation to chemical speciation: The importance of the free uranyl ion. *Environ. Toxicol. Chem.* 23:974–981.

Francois, L., C. Fortin, and P.G. Campbell. 2007. pH modulates transport rates of manganese and cadmium in the green alga *Chlamydomonas reinhardtii* through non-competitive interactions: Implications for an algal BLM. *Aquat. Toxicol.* 84:123–132.

Franklin, N., G. McClelland, and C.M. Wood. 2005. A biotic ligand model approach to copper toxicity in tropical freshwater zebrafish (*Danio rerio*). *Canadian Technical Report of Fisheries and Aquatic Sciences* 2617, 34.

Franklin, N.M., C.N. Glover, J.A. Nicol, et al. 2005. Calcium/cadmium interactions at uptake surfaces in rainbow trout: Waterborne versus dietary routes of exposure. *Environ. Toxicol. Chem.* 24:2954–2964.

Galvez, F., N.M. Franklin, R.B.Tuttle, et al. 2007. Interactions of waterborne and dietary cadmium on the expression of calcium transporters in the gills of rainbow trout: Influence of dietary calcium supplementation. *Aquat. Toxicol.* 84:208–214.

Gandhi, N., M.L. Diamond, D. van de Meent, et al. 2010. New method for calculating comparative toxicity potential of cationic metals in freshwater: Application to copper, nickel, and zinc. *Environ. Sci. Technol.* 44:5195–5201.

Gillis, P.L., J.C. McGeer, G.L. Mackie, et al. 2010. The effect of natural dissolved organic carbon on the acute toxicity of copper to larval freshwater mussels (glochidia). *Environ. Toxicol. Chem.* 29:2519–2528.

Gillis, P.L., R.J. Mitchell, A.N. Schwalb, et al. 2008. Sensitivity of the glochidia (larvae) of freshwater mussels to copper: Assessing the effect of water hardness and dissolved organic carbon on the sensitivity of endangered species. *Aquat. Toxicol.* 88:137–145.

Glover, C.N., and C.M. Wood. 2004. Physiological interactions of silver and humic substances in *Daphnia magna*: Effects on reproduction and silver accumulation following an acute silver challenge. *Comp. Biochem. Physiol. C Toxicol. Pharmacol.* 139:273–280.

Gorski, P.R., D.E. Armstrong, J.P. Hurley, et al. 2006. Speciation of aqueous methylmercury influences uptake by a freshwater alga (*Selenastrum capricornutum*). *Environ. Toxicol. Chem.* 25:534–540.

Gravenmier, J.J., D.W. Johnston, R.C. Santore, et al. 2005. Acute toxicity of copper to the threespine stickleback, *Gasterosteus aculeatus*. *Environ. Toxicol.* 20:150–159.

Hassler, C.S., R. Behra, and K.J. Wilkinson. 2005. Impact of zinc acclimation on bioaccumulation and homeostasis in *Chlorella kesslerii*. *Aquat. Toxicol.* 74:139–149.

Hassler, C.S., R.D. Chafin, M.B. Klinger, et al. 2007. Application of the biotic ligand model to explain potassium interaction with thallium uptake and toxicity to plankton. *Environ. Toxicol. Chem.* 26:1139–1145.

Hassler, C.S., V.I. Slaveykova, and K.J. Wilkinson. 2004. Some fundamental (and often overlooked) considerations underlying the free ion activity and biotic ligand models. *Environ. Toxicol. Chem.* 23:283–291.

Hatano, A., and R. Shoji. 2008. Toxicity of copper and cadmium in combinations to duckweed analyzed by the biotic ligand model. *Environ. Toxicol.* 23:372–378.

Hatano, A., and R. Shoji. 2010. A new model for predicting time course toxicity of heavy metals based on Biotic Ligand Model (BLM). *Comp. Biochem. Physiol. C Toxicol. Pharmacol.* 151:25–32.

Hayashi, T.I., and N. Kashiwagi. 2010. A Bayesian approach to probabilistic ecological risk assessment: Risk comparison of nine toxic substances in Tokyo surface waters. *Environ. Sci. Pollut. Res. Int.* 18(3):365–375.

Heijerick, D.G., B.T. Bossuyt, K.A. De Schamphelaere, et al. 2005. Effect of varying physicochemistry of European surface waters on the copper toxicity to the green alga *Pseudokirchneriella subcapitata*. *Ecotoxicol.* 14:661–670.

Heijerick, D.G., K.A. De Schamphelaere, P.A. Van Sprang, et al. 2005. Development of a chronic zinc biotic ligand model for *Daphnia magna*. *Ecotoxicol. Environ. Saf.* 62:1–10.

Hiriart-Baer, V.P., C. Fortin, D.Y. Lee, et al. 2006. Toxicity of silver to two freshwater algae, *Chlamydomonas reinhardtii* and *Pseudokirchneriella sub-capitata*, grown under continuous culture conditions: Influence of thiosulphate. *Aquat .Toxicol.* 78:136–148.

Hoang, T.C., J.R. Tomasso, and S.J. Klaine. 2004. Influence of water quality and age on nickel toxicity to fathead minnows (*Pimephales promelas*). *Environ. Toxicol. Chem.* 23:86–92.

Hoang, T.C., J.R. Tomasso, and S.J. Klaine. 2007. An integrated model describing the toxic responses of *Daphnia magna* to pulsed exposures of three metals. *Environ. Toxicol. Chem.* 26:132–138.

Jou, L.J., W.Y. Chen, and C.M. Liao. 2009. Online detection of waterborne bioavailable copper by valve daily rhythms in freshwater clam *Corbicula fluminea*. *Environ. Monit. Assess.* 155:257–272.

Kalis, E.J., E.J. Temminghoff, and L. Weng, et al. 2006. Effects of humic acid and competing cations on metal uptake by *Lolium perenne*. *Environ. Toxicol. Chem.* 25:702–711.

Kamo, M., and T. Nagai. 2008. An application of the biotic ligand model to predict the toxic effects of metal mixtures. *Environ. Toxicol. Chem.* 27:1479–1487.

Keithly, J., J.A. Brooker, and D.K. DeForest, et al. 2004. Acute and chronic toxicity of nickel to a cladoceran (*Ceriodaphnia dubia*) and an amphipod (*Hyalella azteca*). *Environ. Toxicol Chem.* 23:691–696.

Kinraide, T.B. 2006. Plasma membrane surface potential (psiPM) as a determinant of ion bio-availability: A critical analysis of new and published toxicological studies and a simplified method for the computation of plant psiPM. *Environ. Toxicol. Chem.* 25:3188–3198.

Klinck, J., M. Dunbar, S. Brown, et al. 2005. Influence of water chemistry and natural organic matter on active and passive uptake of inorganic mercury by gills of rainbow trout (*Oncorhynchus mykiss*). *Aquat. Toxicol.* 72:161–175.

Kolts, J.M., M.L. Brooks, B.D. Cantrell, et al. 2008. Dissolved fraction of standard laboratory cladoceran food alters toxicity of waterborne silver to *Ceriodaphnia dubia*. *Environ. Toxicol. Chem.* 27:1426–1434.

Koster, M., A. de Groot, M.G. Vijver, et al. 2006. Copper in the terrestrial environment: Verification of a laboratory-derived terrestrial biotic ligand model to predict earthworm mortality with toxicity observed in field soils. *Soil Biology and Biochemistry.* 38:7.

Kozlova, T., C.M. Wood, and J.C. McGeer. 2009. The effect of water chemistry on the acute toxicity of nickel to the cladoceran *Daphnia pulex* and the development of a biotic ligand model. *Aquat. Toxicol.* 91:221–228.

Lamelas, C., and V.I. Slaveykova. 2007. Comparison of Cd(II), Cu(II), and Pb(II) biouptake by green algae in the presence of humic acid. *Environ. Sci. Technol.* 41:4172–4178.

Lauer, M.M., and A. Bianchini. 2010. Chronic copper toxicity in the estuarine copepod *Acartia tonsa* at different salinities. *Environ. Toxicol. Chem.* 29:2297–2303.

Lee, D.Y., C. Fortin, and P.G. Campbell. 2004. Influence of chloride on silver uptake by two green algae, *Pseudokirchneriella subcapitata* and *Chlorella pyrenoidosa*. *Environ. Toxicol. Chem.* 23:1012–1018.

Lee, D.Y., C. Fortin, and P.G. Campbell. 2005. Contrasting effects of chloride on the toxicity of silver to two green algae, *Pseudokirchneriella subcapitata* and *Chlamydomonas reinhardtii*. *Aquat. Toxicol.* 75:127–135.

Li, B.,X. Zhang, X. Wang, et al. 2009. Refining a biotic ligand model for nickel toxicity to barley root elongation in solution culture. *Ecotoxicol. Environ. Saf.* 72:1760–1766.

Li, L.Z., D.M. Zhou, X.S. Luo, et al. 2008. Effect of major cations and pH on the acute toxicity of cadmium to the earthworm *Eisenia fetida*: Implications for the biotic ligand model approach. *Arch. Environ. Contam. Toxicol.* 55:70–77.

Li, L.Z., D.M. Zhou, P. Wang, et al. 2009. Effect of cation competition on cadmium uptake from solution by the earthworm *Eisenia fetida*. *Environ. Toxicol. Chem.* 28:1732–1738.

Li, X.F., J.W. Sun, Y.Z. Huang, et al. 2010. Copper toxicity thresholds in Chinese soils based on substrate-induced nitrification assay. *Environ. Toxicol. Chem.* 29:294–300.

Liao, C.M., L.J. Jou, C.M. Lin, et al. 2007. Predicting acute copper toxicity to valve closure behavior in the freshwater clam *Corbicula fluminea* supports the biotic ligand model. *Environ. Toxicol.* 22:295–307.

Liao, C.M., Y.R. Ju, and W.Y. Chen. 2010. Subcellular partitioning links BLM-based toxicokinetics for assessing cadmium toxicity to rainbow trout. *Environ. Toxicol.* 26(6):600–609.

Liao, C.M., C.M. Lin, L.J. Jou, et al. 2007. Linking valve closure behavior and sodium transport mechanism in freshwater clam *Corbicula fluminea* in response to copper. *Environ. Pollut.* 147:656–667.

Linbo, T.L., D.H. Baldwin, J.K. McIntyre, et al. 2009. Effects of water hardness, alkalinity, and dissolved organic carbon on the toxicity of copper to the lateral line of developing fish. *Environ. Toxicol. Chem.* 28:1455–1461.

Lock, K., K.A. De Schamphelaere, S. Becaus, et al. 2006. Development and validation of an acute biotic ligand model (BLM) predicting cobalt toxicity in soil to the potworm *Enchytraeus albidus*. *Soil Biology and Biochemistry* 38:1924–1932.

Lock, K., K.A. De Schamphelaere, S. Becaus, et al. 2007. Development and validation of a terrestrial biotic ligand model predicting the effect of cobalt on root growth of barley (*Hordeum vulgare*). *Environ. Pollut.* 147:626–633.

Lock, K., H. Van Eeckhout, K.A. De Schamphelaere, et al. 2007. Development of a biotic ligand model (BLM) predicting nickel toxicity to barley (*Hordeum vulgare*). *Chemosphere* 66:1346–1352.

Lopez-Chuken, U.J., and S.D. Young. 2010. Modelling sulphate-enhanced cadmium uptake by *Zea mays* from nutrient solution under conditions of constant free Cd2+ ion activity. *J. Environ. Sci.* (China) 22:1080–1085.

Lopez-Chuken, U.J., S.D. Young, and J.L. Guzman-Mar. 2010. Evaluating a "biotic ligand model" applied to chloride-enhanced Cd uptake by *Brassica juncea* from nutrient solution at constant Cd2+ activity. *Environ. Technol.* 31:307–318.

Luo, J., H. Zhang, F.J. Zhao, et al. 2010. Distinguishing diffusional and plant control of Cd and Ni uptake by hyperaccumulator and nonhyperaccumulator plants. *Environ. Sci. Technol.* 44:6636–6641.

Luo, X.S., L.Z. Li, and D.M. Zhou. 2007. Development of a terrestrial biotic ligand model (t-BLM): Alleviation of the rhizotoxicity of copper to wheat by magnesium. *Shengtai Duli Xuebao* 2:41–48.

Luo, X.S., L.Z. Li, and D.M. Zhou. 2008. Effect of cations on copper toxicity to wheat root: implications for the biotic ligand model. *Chemosphere* 73:401–406.

Mager, E.M., K.V. Brix, R.M. Gerdes, et al. 2010. Effects of water chemistry on the chronic toxicity of lead to the cladoceran, *Ceriodaphnia dubia*. *Ecotoxicol. Environ. Saf.* 74(3):238–243.

Mager, E.M., A.J. Esbaugh, K.V. Brix, et al. 2011. Influences of water chemistry on the acute toxicity of lead to *Pimephales promelas* and *Ceriodaphnia dubia*. *Comp. Biochem. Physiol. C Toxicol. Pharmacol.* 153:82–90.

March, F.A., F.J. Dwyer, T. Augspurger, et al. 2007. An evaluation of freshwater mussel toxicity data in the derivation of water quality guidance and standards for copper. *Environ. Toxicol. Chem.* 26: 2066–2074.

Markich, S.J., A.R. King, and S.P. Wilson. 2006. Non-effect of water hardness on the accumulation and toxicity of copper in a freshwater macrophyte (*Ceratophyllum demersum*): How useful are hardness-modified copper guidelines for protecting freshwater biota? *Chemosphere* 65:1791–1800.

Martins Cde, M., I.F. Barcarolli, E.J. de Menezes, et al. 2011. Acute toxicity, accumulation and tissue distribution of copper in the blue crab *Callinectes sapidus* acclimated to different salinities: In vivo and in vitro studies. *Aquat. Toxicol.* 101:88–99.

Mertens, J., F. Degryse, D. Springael, et al. 2007. Zinc toxicity to nitrification in soil and soilless culture can be predicted with the same biotic ligand model. *Environ. Sci. Technol.* 41:2992–2997.

Meyer, J.S., and W.J. Adams. 2010. Relationship between biotic ligand model-based water quality criteria and avoidance and olfactory responses to copper by fish. *Environ. Toxicol. Chem.* 29:2096–2103.

Meyer, J.S., C.J. Boese, and J.M. Morris. 2007. Use of the biotic ligand model to predict pulse-exposure toxicity of copper to fathead minnows (*Pimephales promelas*). *Aquat. Toxicol.* 84:268–278.

Miao, A.J., and W.X. Wang. 2007. Predicting copper toxicity with its intracellular or subcellular concentration and the thiol synthesis in a marine diatom. *Environ. Sci. Technol.* 41:1777–1782.

Morgan, T.P., M. Grosell, R.C. Playle, et al. 2004. The time course of silver accumulation in rainbow trout during static exposure to silver nitrate: Physiological regulation or an artifact of the exposure conditions? *Aquat. Toxicol.* 66:55–72.

Morgan, T.P., C.M. Guadagnolo, M. Grosell, et al. 2005. Effects of water hardness on the physiological responses to chronic waterborne silver exposure in early life stages of rainbow trout (*Oncorhynchus mykiss*). *Aquat. Toxicol.* 74:333–350.

Morgan, T.P., and C.M. Wood. 2004. A relationship between gill silver accumulation and acute silver toxicity in the freshwater rainbow trout: Support for the acute silver biotic ligand model. *Environ. Toxicol. Chem.* 23:1261–1267.

Nadella, S.R., J.L. Fitzpatrick, N. Franklin, et al. 2009. Toxicity of dissolved Cu, Zn, Ni and Cd to developing embryos of the blue mussel (*Mytilus trossolus*) and the protective effect of dissolved organic carbon. *Comp. Biochem. Physiol. C Toxicol. Pharmacol.* 149:340–348.

Natale, O.E., C.E. Gomez, and M.V. Leis. 2007. Application of the biotic ligand model for regulatory purposes to selected rivers in Argentina with extreme water-quality characteristics. *Integr. Environ. Assess. Manag.* 3:517–528.

Ng, T.Y., M.J. Chowdhury, and C.M. Wood. 2010. Can the biotic ligand model predict Cu toxicity across a range of pHs in softwater-acclimated rainbow trout? *Environ. Sci. Technol.* 44:6263–6268.

Nichols, J.W., S. Brown, C.M. Wood, et al. 2006. Influence of salinity and organic matter on silver accumulation in Gulf toadfish (*Opsanus beta*). *Aquat. Toxicol.* 78:253–261.

Niyogi, S., R. Kent, and C.M. Wood. 2008. Effects of water chemistry variables on gill binding and acute toxicity of cadmium in rainbow trout (*Oncorhynchus mykiss*): A biotic ligand model (BLM) approach. *Comp. Biochem. Physiol. C Toxicol. Pharmacol.* 148:305–314.

Niyogi, S., and C.M. Wood. 2004. Biotic ligand model, a flexible tool for developing site-specific water quality guidelines for metals. *Environ. Sci. Technol.* 38:6177–6192.

Ore, S., J. Mertens, K.K. Brandt, et al. 2010. Copper toxicity to bioluminescent *Nitrosomonas europaea* in soil is explained by the free metal ion activity in pore water. *Environ. Sci. Technol.* 44:9201–9206.

Paganini, C.L., and A. Bianchini. 2009. Copper accumulation and toxicity in isolated cells from gills and hepatopancreas of the blue crab (*Callinectes sapidus*). *Environ. Toxicol. Chem.* 28:1200–1205.

Paquin, P.R., J.M. Gorsuch, S. Apte, G.E. Bartley, K.C. Bowles, P.G.C. Campbell, C.G. Delos, D.M. Di Toro, R.L. Dwyer, F. Galvez, R.W. Gensemer, G.G. Goss, C. Hogstand, C.R. Janssen, J.C. McGeer, R.B. Naddy, R.C. Playle, R.C. Santore, U. Schneider, W.A. Stubblefield, C.M. Wood, and K.B. Wu. 2002. The biotic ligand model: A historical overview. *Comp. Biochem. Physiol., Part C: Toxicol. Pharmacol.* 133:3–35.

Pedroso, M.S., J.G. Bersano, and A. Bianchini. 2007. Acute silver toxicity in the euryhaline copepod *Acartia tonsa*: Influence of salinity and food. *Environ. Toxicol. Chem.* 26:2158–2165.

Peters, A., G. Merrington, K. de Schamphelaere, et al. 2010. Regulatory consideration of bioavailability for metals: Simplification of input parameters for the chronic copper biotic ligand model. *Integr. Environ. Assess. Manag.* 7(3):437–444.

Pinho, G.L., and A. Bianchini. 2010. Acute copper toxicity in the euryhaline copepod *Acartia tonsa*: Implications for the development of an estuarine and marine biotic ligand model. *Environ. Toxicol. Chem.* 29:1834–1840.

Playle, R.C. 2004. Using multiple metal-gill binding models and the toxic unit concept to help reconcile multiple-metal toxicity results. *Aquat. Toxicol.* 67:359–370.

Reiley, M.C. 2007. Science, policy, and trends of metals risk assessment at EPA: How understanding metals bioavailability has changed metals risk assessment at US EPA. *Aquat. Toxicol.* 84:292–298.

Rosen, G., I. Rivera-Duarte, D.B. Chadwick, et al. 2008. Critical tissue copper residues for marine bivalve (*Mytilus galloprovincialis*) and echinoderm (*Strongylocentrotus purpuratus*) embryonic development: Conceptual, regulatory and environmental implications. *Mar. Environ. Res.* 66:327–336.

Ryan, A.C., J.R. Tomasso, and S.J. Klaine. 2009. Influence of pH, hardness, dissolved organic carbon concentration, and dissolved organic matter source on the acute toxicity of copper to *Daphnia magna* in soft waters: Implications for the biotic ligand model. *Environ. Toxicol. Chem.* 28:1663–1670.

Ryan, A.C., E.J. Van Genderen, J.R. Tomasso, et al. 2004. Influence of natural organic matter source on copper toxicity to larval fathead minnows (*Pimephales promelas*): Implications for the biotic ligand model. *Environ. Toxicol. Chem.* 23:1567–1574.

Santore, R.C., D.M. Di Toro, P.R. Paquin, H.E. Allen, and J.S. Meyer. 2001. Biotic ligand model of the acute toxicity of metals. 2. Application to acute copper toxicity in freshwater fish and *Daphnia*. *Environ. Toxicol. Chem.* 20:2397–2402.

Sappington, K.G., T.S. Bridges, S.P. Bradbury, et al. 2010. Application of the tissue residue approach in ecological risk assessment. *Integr. Environ. Assess. Manag.* 7(1):116–140.

Sappington, K.G., T.S. Bridges, S.P. Bradbury, et al. 2011. Application of the tissue residue approach in ecological risk assessment. *Integr. Environ. Assess. Manag.* 7:116–140.

Sarathy, V., and H.E. Allen. 2005. Copper complexation by dissolved organic matter from surface water and wastewater effluent. *Ecotoxicol. Environ. Saf.* 61:337–344.

Schlekat, C.E., E. Van Genderen, K.A. De Schamphelaere, et al. 2010. Cross-species extrapolation of chronic nickel biotic ligand models. *Sci. Total Environ.* 408:6148–6157.

Schmidt, T.S., W.H. Clements, K.A. Mitchell, et al. 2010. Development of a new toxic-unit model for the bioassessment of metals in streams. *Environ. Toxicol. Chem.* 29:2432–2442.

Schroeder, J.E., U. Borgmann, and D.G. Dixon. 2010. Evaluation of the biotic ligand model to predict long-term toxicity of nickel to *Hyalella azteca*. *Environ. Toxicol. Chem.* 29:2498–2504.

Schwartz, M.L., and B. Vigneault. 2007. Development and validation of a chronic copper biotic ligand model for *Ceriodaphnia dubia*. *Aquat. Toxicol.* 84:247–254.

Sciera, K.L., J.J. Isely, J.R. Tomasso Jr., et al. 2004. Influence of multiple water-quality characteristics on copper toxicity to fathead minnows (*Pimephales promelas*). *Environ. Toxicol. Chem.* 23:2900–2905.

Shoji, R. 2008. Effect of dissolved organic matter source on phytotoxicity to *Lemna aequinoctialis*. *Aquat. Toxicol.* 87:210–214.

Simpson, S.L., and G.E. Batley. 2007. Predicting metal toxicity in sediments: A critique of current approaches. *Integr. Environ. Assess. Manag.* 3:18–31.

Slaveykova, V.I. 2007. Predicting Pb bioavailability to freshwater microalgae in the presence of fulvic acid: Algal cell density as a variable. *Chemosphere* 69:1438–1445.

Slaveykova, V.I., K. Dedieu, N. Parthasarathy, et al. 2008. Effect of competing ions and complexing organic substances on the cadmium uptake by the soil bacterium *Sinorhizobium meliloti*. *Environ. Toxicol. Chem.* 13:1.

Slaveykova, V.I., K. Dedieu, N. Parthasarathy, et al. 2009. Effect of competing ions and complexing organic substances on the cadmium uptake by the soil bacterium *Sinorhizobium meliloti*. Environ. Toxicol. Chem. 28:741–748.

Smolders, E., K. Oorts, P. Van Sprang, et al. 2009. Toxicity of trace metals in soil as affected by soil type and aging after contamination: Using calibrated bioavailability models to set ecological soil standards. *Environ. Toxicol. Chem.* 28:1633–1642.

Steenbergen, N.T., F. Iaccino, M. de Winkel, et al. 2005. Development of a biotic ligand model and a regression model predicting acute copper toxicity to the earthworm *Aporrectodea caliginosa*. *Environ. Sci. Technol.* 39:5694–5702.

Thakali, S., H.E. Allen, D.M. Di Toro, et al. 2006a. A terrestrial biotic ligand model. 1. Development and application to Cu and Ni toxicities to barley root elongation in soils. *Environ. Sci. Technol.* 40:7085–7093.

Thakali, S., H.E. Allen, D.M. Di Toro, et al. 2006b. Terrestrial biotic ligand model. 2. Application to Ni and Cu toxicities to plants, invertebrates, and microbes in soil. *Environ. Sci. Technol.* 40:7094–7100.

Todd, A.S., S. Brinkman, R.E. Wolf, et al. 2009. An enriched stable-isotope approach to determine the gill-zinc binding properties of juvenile rainbow trout (*Oncorhynchus mykiss*) during acute zinc exposures in hard and soft waters. *Environ. Toxicol. Chem.* 28:1233–1243.

Tran, D., J.C. Massabuau, and J. Garnier-Laplacet. 2004. Effect of carbon dioxide on uranium bioaccumulation in the freshwater clam *Corbicula fluminea*. *Environ. Toxicol. Chem.* 23:739–747.

Tsai, J.W., W.Y. Chen, Y.R. Ju, et al. 2009. Bioavailability links mode of action can improve the long-term field risk assessment for tilapia exposed to arsenic. *Environ. Int.* 35:727–736.

Van Genderen, E., W. Adams, R. Cardwell, et al. 2008. An evaluation of the bioavailability and aquatic toxicity attributed to ambient copper concentrations in surface waters from several parts of the world. *Integr. Environ. Assess. Manag.* 4:416–424.

Van Genderen, E., W. Adams, R. Cardwell, et al. 2009. An evaluation of the bioavailability and aquatic toxicity attributed to ambient zinc concentrations in fresh surface waters from several parts of the world. *Integr. Environ. Assess. Manag.* 5:426–434.

Van Genderen, E., R. Gensemer, C. Smith, et al. 2007. Evaluation of the biotic ligand model relative to other site-specific criteria derivation methods for copper in surface waters with elevated hardness. *Aquat. Toxicol.* 84:279–291.

Van Genderen, E.J., and S.J. Klaine. 2008. Demonstration of a landscape-scale approach for predicting acute copper toxicity to larval fathead minnows (*Pimephales promelas*) in surface waters. *Integr. Environ. Assess. Manag.* 4:237–245.

Van Genderen, E.J., J.R. Tomasso, and S.J. Klaine. 2008. Influence of copper exposure on whole-body sodium levels in larval fathead minnows (*Pimephales promelas*). *Environ. Toxicol. Chem.* 27:1442–1449.

Van Gestel, C.A., and J.E. Koolhaas. 2004. Water-extractability, free ion activity, and pH explain cadmium sorption and toxicity to *Folsomia candida* (Collembola) in seven soil-pH combinations. *Environ. Toxicol. Chem.* 23:1822–1833.

Van Sprang, P.A., F.A. Verdonck, F. Van Assche, et al. 2009. Environmental risk assessment of zinc in European freshwaters: A critical appraisal. *Sci. Total Environ.* 407:5373–5391.

Vanengelen, M.R., R.K. Szilagyi, R. Gerlach, et al. 2010. Uranium exerts acute toxicity by binding to pyrroloquinoline quinone cofactor. *Environ. Sci. Technol.* 45(3):937–942.

Veltman, K., M.A. Huijbregts, and A.J. Hendriks. 2010. Integration of biotic ligand models (BLM) and bioaccumulation kinetics into a mechanistic framework for metal uptake in aquatic organisms. *Environ. Sci. Technol.* 44:5022–5028.

Vijver, M.G., A. De Koning, and W.J. Peijnenburg. 2008. Uncertainty of water type-specific hazardous copper concentrations derived with biotic ligand models. *Environ. Toxicol. Chem.* 27:2311–2319.

Vijver, M.G., C.A. Van Gestel, R.P. Lanno, et al. 2004. Internal metal sequestration and its ecotoxicological relevance: A review. *Environ. Sci. Technol.* 38:4705–4712.

Villavicencio, G., P. Urrestarazu, C. Carvajal, et al. 2005. Biotic ligand model prediction of copper toxicity to daphnids in a range of natural waters in Chile. *Environ. Toxicol. Chem.* 24:1287–1299.

Voigt, A., W.H. Hendershot, and G.I. Sunahara. 2006. Rhizotoxicity of cadmium and copper in soil extracts. *Environ. Toxicol. Chem.* 25:692–701.

Wang, M., and W.X. Wang. 2008. Cadmium toxicity in a marine diatom as predicted by the cellular metal sensitive fraction. *Environ. Sci. Technol.* 42:940–946.

Wang, N., C.G. Ingersoll, C.G. Greer, et al. 2007. Chronic toxicity of copper and ammonia to juvenile freshwater mussels (*Unionidae*). *Environ. Toxicol. Chem.* 26:2048–2056.

Wang, N., C.G. Ingersoll, D.K. Hardesty, et al. 2007. Acute toxicity of copper, ammonia, and chlorine to glochidia and juveniles of freshwater mussels (*Unionidae*). *Environ. Toxicol. Chem.* 26:2036–2047.

Wang, N., C.A. Mebane, J.L. Kunz, et al. 2009. Evaluation of acute copper toxicity to juvenile freshwater mussels (fatmucket, *Lampsilis Siliquoidea*) in natural and reconstituted waters. *Environ. Toxicol. Chem.* 2:1.

Wang, P., D. Zhou, T.B. Kinraide, et al. 2008. Cell membrane surface potential (psi0) plays a dominant role in the phytotoxicity of copper and arsenate. *Plant Physiol.* 148:2134–2143.

Wang, P., D.M. Zhou, L.Z. Li, et al. 2010. Evaluating the biotic ligand model for toxicity and the alleviation of toxicity in terms of cell membrane surface potential. *Environ. Toxicol. Chem.* 29:1503–1511.

Wang, X., B. Li, Y. Ma, et al. 2010. Development of a biotic ligand model for acute zinc toxicity to barley root elongation. *Ecotoxicol. Environ. Saf.* 73:1272–1278.

Wang, X., Y. Ma, L. Hua, et al. 2008. Development of biotic ligand model (BLM) predicting copper acute toxicity to barley (*Hordeum vulgare*). *Huanjing Kexue Xuebao.* 28:1704–1712.

Wang, X., Y. Ma, L. Hua, et al. 2009. Identification of hydroxyl copper toxicity to barley (*Hordeum vulgare*) root elongation in solution culture. *Environ. Toxicol. Chem.* 28:662–667.

Welsh, P.G., J. Lipton, C.A. Mebane, et al. 2008. Influence of flow-through and renewal exposures on the toxicity of copper to rainbow trout. *Ecotoxicol. Environ. Saf.* 69:199–208.

Wilde, K.L., J.L. Stauber, S.J. Markich, et al. 2006. The effect of pH on the uptake and toxicity of copper and zinc in a tropical freshwater alga (*Chlorella* sp.). *Arch. Environ. Contam. Toxicol.* 51:174–185.

Worms, I., D.F. Simon, C.S. Hassler, et al. 2006. Bioavailability of trace metals to aquatic microorganisms: Importance of chemical, biological and physical processes on biouptake. *Biochimie* 88:1721–1731.

Worms, I.A., and K.J. Wilkinson. 2007. Ni uptake by a green alga. 2. Validation of equilibrium models for competition effects. *Environ. Sci. Technol.* 41:4264–4270.

Zeng, J., L. Yang, and W.X. Wang. 2009. Cadmium and zinc uptake and toxicity in two strains of *Microcystis aeruginosa* predicted by metal free ion activity and intracellular concentration. *Aquat. Toxicol.* 91:212–220.

Zeng, J., L. Yang, and W.X. Wang. 2010. High sensitivity of cyanobacterium *Microcystis aeruginosa* to copper and the prediction of copper toxicity. *Environ. Toxicol. Chem.* 29:2260–2268.

Zhao, F.J., C.P. Rooney, H. Zhang, et al. 2006. Comparison of soil solution speciation and diffusive gradients in thin-films measurement as an indicator of copper bioavailability to plants. *Environ. Toxicol. Chem.* 25:733–742.

Zhou, B., J. Nichols, R.C. Playle, et al. 2005. An in vitro biotic ligand model (BLM) for silver binding to cultured gill epithelia of freshwater rainbow trout (*Oncorhynchus mykiss*). *Toxicol. Appl. Pharmacol.* 202:25–37.

7 Regulatory Limits and Applications

7.1 REGULATORY LIMITS

The purpose of this chapter is to briefly describe some of the organizations that have established regulatory limits for metals. These include nongovernmental organizations such as the American Conference of Governmental Industrial Hygienists, American Industrial Hygiene Association, International Agency for the Research on Cancer, United Nations Environmental Program, and government organizations including the Department of Transportation, Environment Canada, National Institute for Occupational Safety and Health, National Toxicology Program, Organization for Economic Co-operation and Development, Occupational Safety and Health Administration, and the US Environmental Protection Agency.

7.1.1 AMERICAN CONFERENCE OF GOVERNMENTAL INDUSTRIAL HYGIENISTS (ACGIH)

The ACGIH has established Threshold Limit Values-Time Weighted Averages (TLV-TWA) for more than 25 metals and metal compounds, and Short-Term Exposure Limits (STEL) for beryllium and compounds, organic tin compounds, and soluble and insoluble tungsten (ACGIH 2009). Critical effects for these metals on which the TLVs-TWAs were based included factors such as respiratory effects, cancer, central and peripheral nervous system impairment, and kidney damage (Table 7.1). The ACGIH also classified cobalt and its compounds, and lead and its compounds as animal carcinogens; arsenic and inorganic arsenic compounds, beryllium and compounds, cadmium and cadmium compounds, chromium VI, and insoluble nickel were classified as human carcinogens (Table 7.2).

7.1.2 AMERICAN INDUSTRIAL HYGIENE ASSOCIATION (AIHA)

The AIHA has established an Emergency Response Planning Guideline (ERPG) for one metal, mercury (AIHA 2009). ERPGs are guidelines for responding to potential releases of airborne substances for use in community emergency planning.

7.1.3 DEPARTMENT OF TRANSPORTATION (DOT)

The DOT has classified 45 metal substances as hazardous substances (DOT 2009). Of these, 22 are regulated as radionuclides (identified with an asterisk [*] in Table 7.3). The DOT designates hazardous materials for purposes of transportation

243

TABLE 7.1
Occupational Exposure Limits

Chemical Name	Chemical Abstracts Service Registry Number (CASRN)	American Conference of Governmental Industrial Hygienists (ACGIH 2009)		National Institute for Occupational Safety and Health (NIOSH 2005)		Occupational Safety and Health Administration (OSHA 2006)
		Threshold Limit Values-Time Weighted Average (TLV-TWA)	Critical Effects	Recommended Exposure Limits (RELs)	Target Organs	Permissible Exposure Limits (PELs)
Aluminum, metal	7429-90-5	1 mg/m^3	Pneumoconiosis; lower respiratory tract irritation; and neurotoxicity	15 mg/m^3 (total dust) 15 mg/m^3 (respirable fraction)	Eyes, skin, respiratory system	15 mg/m^3 (total dust) 5 mg/m^3 (respirable fraction)
Antimony and compounds	7440-36-0	0.5 mg/m^3	Skin and upper respiratory tract irritation	0.5 mg/m^3	Eyes, skin, respiratory system, cardiovascular system	0.5 mg/m^3
Arsenic	7440-38-2	0.01 mg/m^3	Lung cancer	0.5 mg/m^3	Carcinogen Liver, kidneys, skin, lungs, lymphatic system	0.01 mg/m^3 (inorganic compounds) 0.5 mg/m^3 (organic compounds)
Barium and compounds	7440-39-3	0.5 mg/m^3	Eye, skin, and gastrointestinal irritation; muscular stimulant	0.5 mg/m^3	—	0.5 mg/m^3 (soluble compounds)

Substance	CAS Number					
Beryllium and compounds	7440-41-7	0.002 mg/m³ STEL = 0.01 mg/m³	Cancer (lung); berylliosis	0.002 mg/m³ Ceiling = 0.005 mg/m³	Carcinogen Eyes, skin, respiratory system	0.002 mg/m³ Ceiling = 0.005 mg/m³ Acceptable maximum peak above the acceptable ceiling concentration for an 8-hr shift = 0.025 mg/m³ (maximum duration of 30 min)
Cadmium and cadmium compounds	7440-43-9	0.002 mg/m³ (respirable fraction) 0.002 mg/m³ (as cadmium)	Kidney damage	0.005 mg/m³	Carcinogen Respiratory system, kidneys, prostate, blood	0.005 mg/m³
Chromium Chromium +3	7440-47-3 16065-83-1	0.5 mg/m³ (metal and Cr III compounds) 0.05 mg/m³ (water-soluble Cr VI compounds) 0.01 mg/m³ (insoluble Cr VI compounds)	Upper respiratory tract and skin irritation (metal and Cr III compounds) Upper respiratory tract; cancer (water-soluble Cr VI compounds) Lung cancer (insoluble Cr VI compounds)	0.05 mg/m3 (chromium (II) and (III) compounds) 1 mg/m³ (chromium (VI) compounds)	Eyes, skin, respiratory system	0.05 mg/m³ (chromium (II) and (III) compounds) 1 mg/m³ (chromium (VI) compounds)
Cobalt and cobalt compounds	7440-48-4	0.02 mg/m³	Asthma; pulmonary function; myocardial effects	0.1 mg/m³	Skin, respiratory system	0.1 mg/m³
Copper	7440-50-8	0.2 mg/m³ (fumes) 1 mg/m³ (dusts and mists)	Irritation (fumes); gastrointestinal; metal fume fever Lung; byssinosis (dusts and mists)	0.1 mg/m³ (fumes) 1 mg/m³ (dusts and mists)	Eyes, skin, respiratory system, liver, kidneys (increased risk with Wilson's disease)	0.1 mg/m³ (fumes) 1 mg/m³ (dusts and mists)

(Continued)

TABLE 7.1 (*Continued*)
Occupational Exposure Limits

| Chemical Name | Chemical Abstracts Service Registry Number (CASRN) | American Conference of Governmental Industrial Hygienists (ACGIH 2009) | | National Institute for Occupational Safety and Health (NIOSH 2005) | | Occupational Safety and Health Administration (OSHA 2006) |
		Threshold Limit Values-Time Weighted Average (TLV-TWA)	Critical Effects	Recommended Exposure Limits (RELs)	Target Organs	Permissible Exposure Limits (PELs)
Indium	7440-74-6	0.01 mg/m^3	Pulmonary edema; pneumonitis; dental erosion; malaise	—	—	—
Lead	7439-92-1	0.05 mg/m^3	Central nervous system impairment; peripheral nervous system impairment; hematologic effects	0.05 mg/m^3	Eyes, gastrointestinal tract, central nervous system, kidneys, blood, gingival tissue	0.05 mg/m^3 (inorganic)
Tetraethyl lead	78-00-2					0.075 mg/m^3
Tetramethyl lead	75-74-1					0.075 mg/m^3
Manganese and inorganic compounds	7439-96-5	0.2 mg/m^3	Central nervous system impairment	Ceiling = 5 mg/m^3	Respiratory system, central nervous system, blood, kidneys	Ceiling = 5 mg/m^3

				Acceptable ceiling concentration		Acceptable ceiling concentration
Mercury	7439-97-6	0.025 mg/m³ (elemental and inorganic forms) 0.1 mg/m³ (aryl compounds) 0.01 mg/m³ (alkyl compounds) Ceiling = 0.03 mg/m³ (alkyl compounds)	Central nervous system impairment; kidney damage (elemental and inorganic forms and aryl compounds) Central nervous system & peripheral nervous system impairment; kidney damage (alkyl compounds)	Acceptable ceiling concentration = 1 mg/10m³	Eyes, skin, respiratory system, central nervous system, kidneys	Acceptable ceiling concentration = 1 mg/10m³
Nickel	7440-02-0	1.5 mg/m³ (elemental, inhalable fraction) 0.1 mg/m³ (soluble inorganic compounds, inhalable fraction) 0.2 mg/m³ (insoluble inorganic compounds, inhalable fraction)	Dermatitis; pneumoconiosis (elemental) Lung damage; nasal cancer (soluble inorganic compounds) Lung cancer (insoluble inorganic compounds)	1 mg/m³	Carcinogen—lung and nasal cancer Nasal cavities, lungs, skin	1 mg/m³
Platinum	7440-06-4	0.002 mg/m³ (soluble salts, as Pt) 1 mg/m³ (metal)	Asthma; upper respiratory tract irritation	0.002 mg/m³	Eyes, skin, respiratory system	0.002 mg/m³
Rhodium	7440-16-6	0.01 mg/m³ (soluble compounds) 1 mg/m³ (metal and insoluble compounds)	Asthma (soluble compounds) Upper respiratory tract irritation (metal) Lower respiratory tract irritation (insoluble)	0.01 mg/m³ (metal fume and insoluble compounds) 0.001 mg/m³ (soluble compounds)	Respiratory system	0.01 mg/m³ (metal fume and insoluble compounds) 0.001 mg/m³ (soluble compounds)

(Continued)

TABLE 7.1 (*Continued*)
Occupational Exposure Limits

Chemical Name	Chemical Abstracts Service Registry Number (CASRN)	American Conference of Governmental Industrial Hygienists (ACGIH 2009)		National Institute for Occupational Safety and Health (NIOSH 2005)		Occupational Safety and Health Administration (OSHA 2006)
		Threshold Limit Values-Time Weighted Average (TLV-TWA)	Critical Effects	Recommended Exposure Limits (RELs)	Target Organs	Permissible Exposure Limits (PELs)
Silver	7440-22-4	0.01 mg/m^3 (soluble compounds, as Ag) 0.1 mg/m^3 (metal)	Argyria	0.01 mg/m^3	Nasal septum, skin, eyes	0.01 mg/m^3
Tantalum	7440-25-7	5 mg/m^3	Upper respiratory tract irritation	5 mg/m^3	Eyes, skin, respiratory system	5 mg/m^3
Thallium and soluble compounds	7440-28-0	0.1 mg/m^3	Alopecia	0.1 mg/m^3	—	0.1 mg/m^3
Tin	7440-31-5	2 mg/m^3 (oxide and inorganic compounds, except tin hydride) 2 mg/m^3 (metal) 0.1 mg/m^3 (organic compounds) STEL = 0.2 mg/m^3 (organic compounds)	Pneumoconiosis; eye and upper respiratory tract irritation; headache; nausea (oxide and inorganic compounds, except tin hydride)	2 mg/m^3 (inorganic) 0.1 mg/m^3 (organic)	Eyes, skin, respiratory system	2 mg/m^3 (inorganic) 0.1 mg/m^3 (organic)

Tungsten	7440-33-7	5 mg/m^3 (metal and insoluble compounds) 1 mg/m^3 (soluble compounds) STEL = 10 mg/m^3 (metal and insoluble compounds) STEL = 3 mg/m^3 (soluble compounds)	Lower respiratory tract irritation (metal and insoluble compounds) Central nervous system impairment; pulmonary fibrosis (soluble compounds)	—	—	—
Yttrium	7440-65-5	1 mg/m^3	Pulmonary fibrosis	1 mg/m^3	Eyes, respiratory system, liver	1 mg/m^3

TABLE 7.2
Carcinogenicity Classification

Chemical Name	CASRN	American Conference of Governmental Industrial Hygienists	International Agency for the Research on Cancer	National Toxicology Program	US EPA Integrated Risk Information System
Aluminum, metal and soluble salts	7429-90-5	A4: not classifiable as a human carcinogen	—	—	—
Arsenic and inorganic arsenic compounds	7440-38-2	A1: confirmed human carcinogen	Group 1: carcinogenic to humans	Known to be a human carcinogen	—
Barium and compounds	7440-39-3	A4: not classifiable as a human carcinogen	—	—	Group D: not classifiable as to human carcinogenicity
Beryllium and compounds	7440-41-7	A1: confirmed human carcinogen	Group 1: carcinogenic to humans	Known to be a human carcinogen	Group B1: probable human carcinogen, based on limited evidence of carcinogenicity in humans Inhalation Unit Risk = 2.4×10^{-3} per $\mu g/m^3$
Cadmium and cadmium compounds	7440-43-9	A2: suspected human carcinogen	Group 1: carcinogenic to humans	Known to be a human carcinogen	Group B1: probable human carcinogen, based on limited evidence of carcinogenicity in humans Inhalation Unit Risk = 1.8×10^{-3} per $\mu g/m^3$

	CAS				
Chromium +3	16065-83-1	A1: confirmed human carcinogen (water-soluble Cr VI compounds) A1: confirmed human carcinogen (insoluble Cr VI compounds) A4: not classifiable as a human carcinogen (metal and Cr III compounds)	Group 3: not classifiable as to its carcinogenicity to humans (chromium, metallic; chromium (III) compounds)	—	Group D: not classifiable as to human carcinogenicity (chromium(III), insoluble salts)
Cobalt and cobalt compounds	7440-48-4	A3: confirmed animal carcinogen with unknown relevance to humans	Group 2B: possibly carcinogenic to humans	—	—
Copper	7440-50-8	—	—	—	Group D: not classifiable as to human carcinogenicity
Lead and lead compounds	7439-92-1	A3: confirmed animal carcinogen with unknown relevance to humans	Group 2B: possibly carcinogenic to humans	Reasonably anticipated to be human carcinogen	Group B2: probable human carcinogen, based on sufficient evidence of carcinogenicity in animals
Manganese	7439-96-5	—	—	—	Group D: not classifiable as to human carcinogenicity
Mercury	7439-97-6	A4: not classifiable as a human carcinogen (elemental and inorganic forms)	Group 3: not classifiable as to its carcinogenicity to humans (mercury and inorganic mercury compounds)	—	Group D: not classifiable as to human carcinogenicity (mercury, elemental)

(Continued)

TABLE 7.2 (*Continued*)
Carcinogenicity Classification

Chemical Name	CASRN	American Conference of Governmental Industrial Hygienists	International Agency for the Research on Cancer	National Toxicology Program	US EPA Integrated Risk Information System
Nickel	7440-02-0	A1: confirmed human carcinogen (insoluble inorganic compounds) A4: not classifiable as a human carcinogen (soluble inorganic compounds) A5: not suspected as a human carcinogen (elemental)	Group 2B: possibly carcinogenic to humans (nickel, metallic and alloys)	Known to be human carcinogens (nickel compounds) Reasonably anticipated to be a human carcinogen (metallic nickel)	—
Rhodium, metal and soluble/insoluble compounds	7440-16-6	A4: not classifiable as a human carcinogen	—	—	—
Silver	7440-22-4	—	—	—	Group D: not classifiable as to human carcinogenicity
Thorium-232 and its decay products	7440-29-1		Group 1: carcinogenic to humans	—	—
Tin	7440-31-5	A4: not classifiable as a human carcinogen (organic compounds)	—	—	—
Zinc, elemental	7440-66-6		—	—	Group D: not classifiable as to human carcinogenicity

TABLE 7.3

National and International Regulations and Guidelines Pertaining to Metals Identified as Hazardous Substances

Chemical Name	CASRN	DOT Hazardous Materials (49 CFR 172.101)	Environment Canada National Pollutant Release Inventory (NPRI) Databases	EPA Designation, Reportable Quantities, and Notifications (40 CFR 302.4)	EPA Toxic Chemical Release Reporting (40 CFR 372.65)	OECD High Production Volume (HPV) Chemicals	UNEP Concise International Chemical Assessment Document (CICADs)	UNEP Environmental Health Criteria (EHC) Monographs	UNEP Health and Safety Guides (HSG)	UNEP International Chemical Safety Cards (ICSCs)	UNEP Poisons Information Monographs (PIMs)
Aluminum	7429-90-5	Yes*	Yes	Yes*	Yes	Yes	No	Yes	No	No	No
Antimony, elemental	7440-36-0	Yes	Yes	Yes	Yes	Yes	No	No	No	Yes	Yes
Arsenic	7440-38-2	Yes	Yes	Yes	Yes	No	No	Yes	Yes	Yes	Yes
Barium	7440-39-3	Yes	No	Yes	Yes	No	Yes	No	Yes	Yes	Yes
Beryllium, elemental	7440-41-7	Yes	No	Yes	Yes	No	Yes	Yes	Yes	Yes	No
Cadmium	7440-43-9	Yes	Yes	Yes	Yes	Yes	No	Yes	No	Yes	Yes
Calcium	7440-70-2	Yes	No	Yes*	No	Yes	No	No	No	Yes	No
Cerium	7440-45-1	Yes	No	Yes*	No	No	No	No	No	No	No
Cesium, elemental	7440-46-2	Yes	No	Yes*	No	No	No	No	No	No	No
Chromium	7440-47-3	Yes	Yes	Yes	Yes	Yes	No	No	No	Yes	No
Chromium, ion (Cr3+)	16065-83-1	No	No	No	No	No	Yes	No	No	No	No
Cobalt	7440-48-4	Yes*	Yes	Yes	Yes	Yes	Yes	No	No	Yes	No
Copper	7440-50-8	Yes	Yes	Yes	Yes	Yes	No	Yes	No	Yes	Yes

(Continued)

TABLE 7.3 (Continued)
National and International Regulations and Guidelines Pertaining to Metals Identified as Hazardous Substances

Chemical Name	CASRN	DOT Hazardous Materials (49 CFR 172.101)	Environment Canada National Pollutant Release Inventory (NPRI) Databases	EPA Designation, Reportable Quantities, and Notifications (40 CFR 302.4)	EPA Toxic Chemical Release Reporting (40 CFR 372.65)	OECD High Production Volume (HPV) Chemicals	UNEP Concise International Chemical Assessment Document (CICADs)	UNEP Environmental Health Criteria (EHC) Monographs	UNEP Health and Safety Guides (HSG)	UNEP International Chemical Safety Cards (ICSCs)	UNEP Poisons Information Monographs (PIMs)
Gadolinium	7440-54-2	Yes*	No	Yes*	No	No	No	No	No	No	No
Gallium	7440-55-3	Yes	No	Yes*	No	No	No	No	No	No	No
Germanium	7440-56-4	Yes*	No	Yes*	No	No	No	No	No	No	No
Gold	7440-57-5	Yes*	No	Yes*	No	No	No	No	No	Yes	No
Indium	7440-74-6	Yes*	No	Yes*	No	No	No	No	No	Yes	No
Iron	7439-89-6	Yes*	No	Yes*	No	Yes	No	No	No	No	No
Lanthanum	7439-91-0	Yes*	No	Yes*	No	No	No	No	No	No	No
Lead	7439-92-1	Yes	Yes	Yes	Yes	Yes	No	Yes	No	Yes	Yes
Lithium, elemental	7439-93-2	Yes	No	No	No	No	No	No	No	Yes	No
Magnesium	7439-95-4	Yes	No	Yes*	No	Yes	No	No	No	Yes	No
Manganese	7439-96-5	Yes*	Yes	Yes	Yes	Yes	Yes	Yes	No	Yes	No
Mercury	7439-97-6	Yes	Yes	Yes	Yes	No	Yes	Yes	No	Yes	No
Nickel	7440-02-0	Yes	Yes	Yes*	Yes	Yes	No	Yes	Yes	Yes	No
Niobium	7440-03-1	Yes*	No	Yes*	No	Yes	No	No	No	No	No
Osmium, elemental	7440-04-2	Yes*	No	Yes*	No	No	No	No	No	No	No
Palladium	7440-05-3	Yes*	No	Yes*	No	No	No	Yes	No	No	No

Element	CAS No.										
Platinum	7440-06-4	Yes*	No	Yes*	No	No	No	Yes	No	No	No
Potassium	7440-09-7	Yes	No	Yes*	No	No	No	No	No	Yes	No
Rhodium	7440-16-6	Yes*	No	Yes*	No	No	No	No	No	Yes	No
Rubidium	7440-17-7	Yes	No	Yes*	No	No	No	No	No	No	No
Ruthenium	7440-18-8	Yes*	No	Yes*	No	No	No	No	No	No	No
Scandium	7440-20-2	Yes*	No	Yes*	No	No	No	No	No	No	No
Silver	7440-22-4	Yes	Yes	Yes	Yes	Yes	Yes	No	No	Yes	No
Sodium	7440-23-5	Yes	No	Yes	No	Yes	No	No	No	Yes	No
Strontium, elemental	7440-24-6	Yes*	No	Yes*	No	No	Yes	No	No	Yes	No
Tantalum	7440-25-7	Yes*	No	Yes*	No	No	No	No	No	Yes	No
Thallium, elemental	7440-28-0	Yes	No	Yes	Yes	No	No	Yes	Yes	Yes	Yes
Thorium	7440-29-1	Yes*	No	Yes*	No	No	No	No	No	Yes	No
Tin	7440-31-5	Yes*	No	No	No	Yes	Yes	Yes	No	Yes	No
Titanium	7440-32-6	Yes	No	Yes*	No	Yes	No	Yes	No	No	No
Tungsten	7440-33-7	No	No	Yes*	No	Yes	No	No	No	Yes	No
Vanadium, elemental	7440-62-2	Yes	Yes	Yes*	Yes	Yes	Yes	Yes	Yes	No	No
Yttrium	7440-65-5	Yes*	No	Yes*	No	No	No	No	No	No	No
Zinc, elemental	7440-66-6	Yes*	Yes	Yes	No	Yes	No	Yes	No	No	Yes

* Regulated as a radionuclide.

and prescribes the requirements for shipping papers, package marking, labeling, and transport vehicle placards applicable to the shipment and transportation of those hazardous materials (DOT 2009).

7.1.4 ENVIRONMENT CANADA

Environment Canada (2009) has 14 metal substances regulated under the National Pollutant Release Inventory (NPRI) (Table 7.3). The NPRI is Canada's legislated, publicly accessible inventory of pollutant releases (to air, water, and land), disposals, and transfers for recycling (Environment Canada 2009).

7.1.5 INTERNATIONAL AGENCY FOR THE RESEARCH ON CANCER (IARC)

IARC has classified arsenic and inorganic arsenic compounds, beryllium and compounds, cadmium and cadmium compounds, and thorium-232 and its daughters as human carcinogens (Table 7.2) In addition, IARC classified cobalt and cobalt compounds, lead and its compounds, and nickel as possible human carcinogens (IARC 2010).

7.1.6 NATIONAL INSTITUTE FOR OCCUPATIONAL SAFETY AND HEALTH (NIOSH)

NIOSH has Recommended Exposure Limits (RELs) for 20 metals and metal compounds (Table 7.1). The REL is a level that NIOSH purports is protective of worker safety and health over a worker's lifetime if used in combination with engineering and work practice controls, exposure and medical monitoring, posting and labeling of hazards, worker training, and personal protective equipment. Target organs for these metals on which the RELs were based included blood, cardiovascular system, central nervous system, eyes, gastrointestinal tract, heart, kidneys, liver, lungs, lymphatic system, prostate, respiratory system, and skin (NIOSH 2005).

7.1.7 NATIONAL TOXICOLOGY PROGRAM (NTP)

The NTP has classified arsenic and inorganic arsenic compounds, beryllium and compounds, cadmium and cadmium compounds, and nickel compounds as known human carcinogens. In addition, NTP classified lead and lead compounds and nickel as reasonably anticipated to be human carcinogens (NTP 2005) (Table 7.2).

7.1.8 ORGANIZATION FOR ECONOMIC CO-OPERATION AND DEVELOPMENT (OECD)

The OECD has identified 20 metal substances as High Production Volume (HPV) chemicals (OECD 2009) (Table 7.3). HPV chemicals include all chemicals reported to be produced or imported at amounts exceeding 1,000 tons per year in at least one OECD member country or in the European Union region. None of the metals identified as HPV chemicals have an OECD Screening Information Data Set (SIDS) (OECD 2007). The SIDS is described in http://www.oecd.org/Hpv/UI/Search.aspx.

7.1.9 OCCUPATIONAL SAFETY AND HEALTH ADMINISTRATION (OSHA)

OSHA has developed Permissible Exposure Limits (PELs) for 20 metals and metal compounds (Table 7.1). PELs are regulatory limits on the amount or concentration of a substance in the air to protect workers against the health effects of exposure to hazardous substances (OSHA 2006). OSHA PELs are based on an 8-hour time-weighted average (TWA) exposure.

7.1.10 UNITED NATIONS ENVIRONMENTAL PROGRAM (UNEP)

The UNEP provides leadership and encourages partnership in caring for the environment by inspiring, informing, and enabling nations and peoples to improve their quality of life without compromising that of future generations (http://www.unep.org/Documents.Multilingual/Default.asp?DocumentID = 43). The UNEP has developed 6 Health and Safety Guides (HSG) (UNEP 1999), 16 Environmental Health Criteria (EHC) Monographs (UNEP 2006), Concise International Chemical Assessment Documents (CICADS) (UNEP 2009a), 8 Poisons Information Monographs (PIMs) (UNEP 2009b), and 27 International Chemical Safety Cards (ICSCs) (UNEP 2009c) (Table 7.3).

7.1.11 US ENVIRONMENTAL PROTECTION AGENCY (US EPA)

The US EPA implemented four laws that are used to set regulatory limits for metals, including the Clean Air Act, the Clean Water Act, the Emergency Planning and Community Right-to-Know Act, and the Comprehensive Environmental Response, Compensation, and Liability Act. These laws and the metals for which they set regulatory limits are briefly discussed below.

7.1.11.1 Clean Air Act

The US EPA Clean Air Act Amendments of 1990 list antimony compounds, arsenic compounds (inorganic including arsine), beryllium compounds, cadmium compounds, chromium compounds, cobalt compounds, lead compounds, manganese compounds, mercury compounds, and nickel compounds as Hazardous Air Pollutants (HAPs) (US EPA 2008a). The *compound* designation is defined to include any unique chemical substance that contains the specified metal as part of that chemical's structure (US EPA 2008a). The Clean Air Act requires the EPA to set National Ambient Air Quality Standards (NAAQS; 40 C.F.R. Part 50) for pollutants considered harmful to public health and the environment (US EPA 2008). Primary standards set limits to protect public health, including the health of especially sensitive populations (such as asthmatics, children, and the elderly), while secondary standards set limits to protect public welfare, including protection against decreased visibility, and damage to animals, crops, vegetation, and buildings. Of the six principal pollutants for which NAAQS have been set, lead is the only metal. The primary and secondary standards for lead are the same, *viz.*, 0.15 µg/m^3 as a 3 monthly rolling average and 1.5 µg/m^3 as a quarterly average, (see http://www.epa.gov/air/criteria.html).

7.1.11.2   Clean Water Act

The National Primary and Secondary Drinking Water Regulations cover the US EPA primary and secondary drinking water regulations for 10 and 6 metals, respectively (Table 7.4). The 10 metals for which there are primary regulations have a Maximum Contaminant Level Goal (MCLG) and a Maximum Contaminant Level (MCL) (US EPA 2009a). Of the 6 metals for which there are a secondary drinking water regulation (SDWR), only copper has a MCLG and an MCL (Table 7.4). The MCLG is the level of a contaminant in drinking water below which there is no known or expected risk to health, that is, a nonenforceable public health goal. The MCL is the highest level of a contaminant that is allowed in drinking water, that is, an enforceable standard. The SDWR is a nonenforceable federal guideline regarding cosmetic effects or aesthetic effects of drinking water. The US EPA established freshwater and saltwater quality criteria for 11 and 9 metals, respectively (Table 7.4). For fresh- and saltwater, a Criteria Maximum Concentration (CMC) was established for 10 and 9 metals, and a Criterion Continuous Concentration (CCC) for 10 and 8 metals, respectively (Table 7.4). Silver has a freshwater and saltwater CMC but no CCC, while iron only has a freshwater CCC (Table 7.4). The CMC is an estimate of the highest concentration of a material in surface water to which an aquatic community can be briefly exposed without resulting in an unacceptable effect. The CCC is an estimate of the highest concentration of a material in surface water to which an aquatic community can be exposed indefinitely without resulting in an unacceptable effect (US EPA 2009b). Under the same water quality guidelines, the US EPA also established criteria for human health consumption of aquatic organisms and water for 8 metals and only aquatic organisms for 6 metals (Table 7.4).

7.1.11.3   Emergency Planning and Community Right-to-Know Act (EPCRA)

The EPCRA was created under title III of the 1986 Superfund Amendments and Reauthorization Act (SARA). One of the provisions of the EPCRA created the Toxics Release Inventory (TRI), which lists 16 metals (Table 7.3). The Toxic Chemical Activity Threshold triggers the TRI reporting requirements under EPCRA Section 313. These requirements apply to all federal facilities with more than 10 employees if an activity threshold is exceeded. There are three activity thresholds. If the facility manufactures a toxic chemical (including creating, importing, or coincidental manufacture), the activity threshold is 25,000 pounds within one calendar year. The activity threshold is the same if the facility processes a toxic chemical by incorporating it into a product. However, if the facility otherwise uses the toxic chemical, the activity threshold drops to 10,000 pounds. More detailed definitions of the terms *manufacture*, *process*, and *otherwise use* are provided in Title 40 C.F.R. Part 372.3.

7.1.11.4   Comprehensive Environmental Response, Compensation, and Liability Act (CERCLA)

The CERCLA, commonly known as the Superfund, was enacted by the US Congress on December 11, 1980. This law created a tax on the chemical and petroleum industries, and provided broad federal authority to respond directly to releases or

TABLE 7.4

Regulations for Metal Contaminants in Water

| Chemical Name | CASRN | National Primary and Secondary Drinking Water Regulations | | | National Recommended Water Quality Criteria | | | | | |
| | | Primary | | Secondary | Freshwater | | Saltwater | | Human Health for Consumption of: | |
		MCLG (mg/L)	MCL (mg/L)	SDWR (mg/L)	CMC (µg/L)	CCC (µg/L)	CMC (µg/L)	CCC (µg/L)	Water + Organism (µg/L)	Organism Only (µg/L)
Aluminum	7429-90-5	—	—	0.05–0.2	750	87	—	—	—	—
Antimony	7440-36-0	0.006	0.006	—	—	—	—	—	5.6	640
Arsenic	7440-38-2	0	0.010	—	340	150	69	36	0.018	0.14
Barium	7440-39-3	2	2	—	—	—	—	—	1000	—
Beryllium	7440-41-7	0.004	0.004	—	—	—	—	—	—	—
Cadmium	7440-43-9	0.005	0.035	—	2.0	0.25	40	8.8	—	—
Chromium (total)	7440-47-3	0.1	0.1	—	—	—	—	—	—	—
Chromium, ion (Cr3+)	16065-83-1	—	—	—	570	74	—	—	—	—
Chromium, ion (Cr6+)	18540-29-9	—	—	—	16	11	1100	50	—	—
Copper	7440-50-8	1.3	1.3	1.0	—	—	4.8	3.1	1300	—
Iron	7439-89-6	—	—	0.3	—	1000	—	—	—	—
Lead	7439-92-1	0	0.015	—	65	2.8	210	8.1	—	—
Manganese	7439-96-5	—	—	0.05	—	—	—	—	50	100
Mercury	7439-97-6	0.002	0.002	—	1.4	0.77	1.8	0.94	—	—
Nickel	7440-02-0	—	—	—	470	52	74	8.2	610	4600
Silver	7440-22-4	—	—	0.10	3.2	—	1.9	—	—	—

(Continued)

TABLE 7.4 (*Continued*)
Regulations for Metal Contaminants in Water

Chemical Name	CASRN	National Primary and Secondary Drinking Water Regulations			National Recommended Water Quality Criteria						
		Primary		Secondary	Freshwater		Saltwater		Human Health for Consumption of:	
		MCLG (mg/L)	MCL (mg/L)	SDWR (mg/L)	CMC (µg/L)	CCC (µg/L)	CMC (µg/L)	CCC (µg/L)	Water + Organism (µg/L)	Organism Only (µg/L)
Thallium	7440-28-0	0.0005	0.002	—	—	—	—	—	0.24	0.47
Zinc	7440-66-6	—	—	5	120	120	90	81	7400	26,000

MCLG = Maximum Contaminant Level Goal; MCL = Maximum Contaminant Level; SDWR = Secondary Drinking Water Regulations; CMC = Criteria Maximum Concentration; CCC = Criterion Continuous Concentration.

threatened releases of hazardous substances that may endanger public health or the environment. The releases of hazardous substances are controlled through the use of Reportable Quantities or RQs. The US EPA has adopted the five RQs of 1, 10, 100, 1,000, and 5,000 pounds for hazardous substances and RQs for radionuclides of 0.001, 0.01, 0.1, 1, 10, 100, and 1,000 Ci. There are RQs for 44 metals, 27 of which are radionuclides that are identified by an asterisk in Table 7.3.

7.2 REGULATORY APPLICATIONS

QSARs have been used by Australian, Canadian, Danish, European, German, Japanese, Netherlands, and United States government organizations to predict physical and chemical properties, environmental fate, ecological effects, and health effects of organic chemicals (Walker et al. 2002). The Organization for Economic Cooperation and Development (OECD), through the European Economic Commission's Technical Guidance Documents, provides guidance on applications of QSARs to predict the aquatic toxicity, logarithm of the octanol–water partition coefficient (log K_{ow}), soil or sediment organic carbon partition coefficient (Koc), atmospheric oxidation, bioconcentration factors (BCFs), and biodegradation of organic chemicals. In the United States, the Toxic Substances Control Act (TSCA) Interagency Testing Committee (ITC), the Agency for Toxic Substances and Disease Registry (ATSDR), the US Environmental Protection Agency (US EPA), and the Food and Drug Administration (FDA) use QSARs to predict aquatic toxicity, chemical or physical properties, environmental fate parameters, and health effects of organic chemicals (Walker 2003). While these organizations have used QSARs for organic chemicals, only Environment Canada, the US EPA, and the ITC have used SAR- or QSAR-related algorithms to predict the toxicity of metal ions. The European Chemicals Agency is considering using QSARs to predict the toxicity of metal ions.

7.2.1 ENVIRONMENT CANADA

Environment Canada has used a QSAR-related technique, linear solvation energy relationships (LSERs), during their regulatory activities for inorganic chemicals. LSERs are described in Chapters 3 and 5, while Environment Canada's activities are described here. Under Sections 73 and 74 of the 1999 Canadian Environmental Protection Act, Environment Canada and Health Canada had to *categorize* and *screen* about 23,000 substances on the Domestic Substances List (DSL) for persistence (P), bioaccumulation (B), and inherently toxic (iT) properties (MacDonald et al. 2002). The DSL was created for the purpose of defining a new substance. The original list was from January 1, 1984 to December 31, 1986 for all substances manufactured or imported in Canada, in Canadian commerce, or used for commercial manufacturing. About 5–10% of the DSL substances are inorganic chemicals. To categorize the inorganic chemicals, Environment Canada formed the Inorganic Working Group (IWG). The mandate of the IWG was to make recommendations as to the best possible means to categorize inorganic substances against P and B and iT criteria, and to screen inorganic substances. The first task was an approach to categorization together with iT criteria. The goal was to ensure that the most harmful substances posing an environmental risk were adequately identified, and those that were not posing an environmental risk were also identified. The IWG noted that persistence and bioaccumulation criteria for organic chemicals were not suitably discriminatory for inorganic chemicals; transformation relative to stability constants may be an appropriate substitute for persistence; biomagnification was not an issue for inorganic chemicals; chronic toxicity was important; an acute aquatic toxicity LC_{50} or $EC_{50} \leq 1$ mg/L and nominal concentrations of inorganic chemicals could be used to determine or predict aquatic toxicity provided the nominal concentrations did not exceed water solubility values. The IWG considered the use of LSERs to estimate the solution properties of inorganic chemicals (IWG 2001). The LSERs developed by Hickey (2005) for estimating water solubility and toxicity of cations are discussed in Section 5.3 and listed in Table 5.4. The most recent information on Environment Canada's assessment of metals was included in a 2009 presentation to the Society of Environmental Chemistry and Toxicology (Gauthier et al. 2009) and provided to a 2011 OECD Workshop on Metals Specificities in Environmental Risk Assessment (http://www.oecd.org/dataoecd/42/13/48719530.pdf). The workshop report was published in January 2012 (http://www.oecd.org/officialdocuments/publicdisplaydocumentpdf/?cote=ENV/JM/MONO(2012)2&docLanguage=En).

7.2.2 EPA

To predict the health effects of metal ions, the US EPA's Office of Chemical Safety and Pollution Prevention uses mechanism-based SAR. A metal ion's bioavailability, as it exists in inorganic and organic metal compounds, is estimated by considering the ion's solubility, oxidation state, dissociation, and reactivity. To predict potential carcinogenicity, the EPA developed the OncoLogic Cancer Expert system, which has a metals/metalloids subsystem that includes major carcinogenic metals (http://www.epa.gov/oppt/sf/pubs/oncologic.htm).

7.2.3 ITC

Since it was established in 1976, the Interagency Testing Committee (ITC) is probably the oldest government organization that has a statutory requirement to apply SARs and QSARs to identify potentially toxic chemicals. The ITC must apply SARs and QSARs before adding chemicals to the TSCA Section 4(e) Priority Testing List and recommending testing of these chemicals to the US EPA Administrator (Walker 1993). As required by TSCA, the ITC must determine: (1) the extent to which the substance or mixture is closely related to a chemical substance or mixture that is known to present an unreasonable risk of injury to health or the environment, and (2) the extent to which testing of the substance or mixture may result in the development of data upon which the effects of the substance or mixture on health or the environment can reasonably be determined or predicted. The ITC used QSARs to predict the sorption of antimony and related metals and the potential toxicity of other metal ions.

7.2.4 EUROPEAN CHEMICALS AGENCY (ECHA)

The European Chemicals Agency's metal-specific tools were presented to a February 2, 2012 workshop (http://echa.europa.eu/documents/10162/17098/01_jl_metal-specific_tool_use_communication_d2_lrws_20120203_en.pdf). While the tools did not include QSARs, appendix R.7.13-2: Environmental risk assessment for metals and metal compounds, discusses the possible use of QSARs for predicting the toxicity of metal ions (http://guidance.echa.europa.eu/docs/guidance_document/information_requirements_r7_13_2_en.pdf). The following quote is from page 41:

> The development of QSAR methods for metals and inorganic metal compounds has not been as actively pursued as for organic substances. However, for some inorganic substances, predicting toxicity from chemical properties may be relevant. In this respect, Quantitative Ion Character-Activity Relationships (QICARs) and Quantitative Cationic-Activity Relationships (QCARs) have recently been developed (Ownby and Newman 2003, Walker et al. 2003). More research efforts are needed in this field, however, in order to develop and validate appropriate models.

7.3 FUTURE CONSIDERATIONS

Table 5.1 listed 97 QCARs for predicting cation toxicities and Table 5.4 listed 183 QCARs for predicting cation bioconcentration potential, biosorption capacity, binding strength, and toxicity. Moreover, Chapter 8 describes how to construct QSARs for metal ions. Between the QSARs listed in Tables 5.1 and 5.4 and the instructions provided in Chapter 8, it should now be possible for regulatory organizations to consider using QSARs to predict the toxicity of metal ions. Similar methods can be used to construct QSARs for bioactive organometallic complexes using certain descriptors presented in Chapter 4.

REFERENCES

ACGIH (American Conference of Governmental Industrial Hygienists). 2009. TLVs® and BEIs®: Based on the Documentation of Threshold Limit Values for Chemical Substances and Physical Agents and Biological Exposure Indices. Cincinnati, OH: ACGIH.

AIHA (American Industrial Hygiene Association). 2009. Emergency Response Planning Guidelines (ERPGs). American Industrial Hygiene Association, Fairfax, VA, http://www.aiha.org/Pages/default.aspx.

DOT (US Department of Transportation). 2009. Hazardous Materials Table, 49 C.F.R. 172.101, http://www.access.gpo.gov/nara/cfr/waisidx_04/49cfr172_04.html (accessed September 23, 2010).

Environment Canada. 2009. National Pollutant Release Inventory (NPRI) Databases. Gatineau, Quebec, http://www.ec.gc.ca/pdb/npri/npri_dat_rep_e.cfm#search.

Gauthier J., M. Eggleton, Y. Couillard, J. Hill, O. Marois, and A. Gosselin. 2009. Environment Canada's assessment activities on metals under the Chemical Management Plan. Presentation to the Society of Environmental Chemistry and Toxicology. New Orleans, LA.

Hickey, J.P. 2005. Estimation of inorganic species aquatic toxicity. In: Techniques in Aquatic Toxicology: Volume 2. G.K. Ostrander (Ed). CRC Press/Lewis Publishers, Boca Raton, FL.Ch. 34. pp. 617–629.

IARC (International Agency for the Research on Cancer). 2010. Agents Reviewed by the IARC Monographs: Volumes 1–88. Lyon, France: IARC, http://monographs. iarc.fr/.

Inorganic Working Group. 2001. Categorization of inorganic substances on the Domestic Substances List (DSL): Findings and recommendations of the Inorganic Working Group (IWG). IWG Secretariat, Environment Canada, Hull, Quebec K1A 0H3.

MacDonald, D., R. Breton, R. Sutcliffe, and J. Walker. 2002. Uses and limitations of quantitative structure-activity relationships (QSARs) to categorize substances on the Canadian Domestic Substances List as persistent and/or bioaccumulative, and inherently toxic to non-human organisms. *SAR QSAR Environ. Res.* 13:43–55.

NIOSH (National Institute for Occupational Safety and Health). 2005. *NIOSH Pocket Guide to Chemical Hazards* (NPG). Atlanta, GA: Centers for Disease Control and Prevention, http://www.cdc.gov/niosh/npg/.

NTP (National Toxicology Program). 2005. *Report on Carcinogens*, 11th edition. Washington DC: US Department of Health and Human Services, Public Health Service, http://ntp-server.niehs.nih.gov/.

OECD (Organisation for Economic Co-operation and Development). 2007. OECD Screening Information Data Set Project (SIDS) high production volume chemicals, http://www. inchem.org/pages/sids.html.

OECD (Organisation for Economic Co-operation and Development). 2009. *High Production Volume (HPV) Chemicals.* Paris, France, http://www.oecd.org/dataoecd/55/38/33883530. pdf.

OSHA. 2006. Permissible Exposure Limits (PELs). Code of Federal Regulations 29 CFR 1910.1000. U.S. Department of Labor: Washington, DC. Available online from: http://www.osha.gov/pls/oshaweb/owadisp.show_document?p_table=standards&p_ id=9992.

Ownby, D.R., and M.C. Newman. 2003. Advances in quantitative ion character-activity relationships (QICARS): Using metal–ligand binding characteristics to predict metal toxicity. *QSAR Comb. Sci.* 22:241–246.

UNEP (United Nations Environment Programme). 1999. Health and Safety Guides (HSG). Chemical safety information from intergovernmental organizations. Nairobi, Kenya, http://www.inchem.org/pages/hsg.html.

UNEP (United Nations Environment Programme). 2006. Environmental Health Criteria (EHC) monographs. Chemical safety information from intergovernmental organizations. Nairobi, Kenya, http://www.inchem.org/pages/ehc.html.

UNEP (United Nations Environment Programme). 2009a. Concise International Chemical Assessment Document (CICADS). Chemical safety information from intergovernmental organizations. Nairobi, Kenya, http://www.inchem.org/pages/cicads.html.

UNEP (United Nations Environment Programme). 2009b. Poisons Information Monographs (PIMs). Chemical safety information from intergovernmental organizations. Nairobi, Kenya, http://www.inchem.org/pages/pims.html (accessed September 23, 2010).

UNEP (United Nations Environment Programme). 2009c. International Chemical Safety Cards (ICSCs). Chemical safety information from intergovernmental organizations. Nairobi, Kenya, http://www.inchem.org/pages/icsc.html.

US EPA (US Environmental Protection Agency). 2008. Hazardous Air Pollutants (HAPs). Washington, DC: Office of Air Quality Planning and Standards, US Environmental Protection Agency, http://www.epa.gov/lawsregs/search/40cfr.html.

US EPA (US Environmental Protection Agency). 2009a. National Primary Drinking Water Regulations. Washington, DC: Office of Ground Water and Drinking Water, US Environmental Protection Agency, EPA 816-F-02-013, http://www.epa.gov/safewater/ mcl.html.

US EPA (US Environmental Protection Agency). 2009b. National Recommended Water Quality Criteria. Priority Pollutants. Washington, DC: Office of Water, Office of Science and Technology, US Environmental Protection Agency, EPA-822-R-02-047, http://www.epa.gov/waterscience/criteria/wqcriteria.html.

Walker, J.D. 1993. The TSCA Interagency Testing Committee, 1977 to 1992: Creation, structure, functions and contributions. In *Environmental Toxicology and Risk Assessment*, Vol. 2, edited by J.W. Gorsuch, F.J. Dwyer, C.G. Ingersoll, and T.W. La Pointe, ASTM STP 1216, 451–509. Philadelphia, PA: American Society for Testing and Materials.

Walker, J.D. 2003. Applications of QSARs in toxicology: A US government perspective. *J. Molecular Structure—Theochem*. 622:167–184.

Walker, J.D., L. Carlsen, E. Hulzebos, and B. Simon-Hettich. 2002. Global government applications of analogues, SARs and QCARs to predict aquatic toxicity, chemical or physical properties: Environmental fate parameters and health effects of organic chemicals. *SAR QSAR Environ. Res.* 13:607–619.

Walker, J.T., M. Enache, and J.C. Dearden. 2003. Quantitative cationic-activity relationships for predicting toxicity of metals. *Environ. Toxicol. Chem.* 22:1916–1935.

8 Constructing QSARs for Metal Ions

8.1 SELECTION AND TRANSFORMATIONS OF EXPLANATORY VARIABLES

Defining the mode of action to be modeled is the first step in QSAR generation (McKinney et al. 2000). Chapter 1 covers broadly the various modes of action. The class of relevant toxicants and candidate explanatory variables (descriptors) are identified once the mode of action is defined. Previous chapters describe the potential explanatory variables for metal ions. The final step in QSAR development is the generation of quantitative means for selecting and relating the explanatory variable(s) to the effect of interest. The purpose of this chapter is to provide essential details about this last step.

What are the issues that require consideration while generating a tool to relate metal ion qualities and bioactivity? First, a method must be selected for determining which, and perhaps how many, explanatory variables to use. Second, the best approach must be applied to fit the most appropriate model to these data. What is best will depend on the intended use of the model and the data set qualities. Third, some validation method is applied that specifies how useful the resulting model is relative to the intended predictions of bioactivity for a metal ion not used to generate the model. Again, usefulness will depend on the intended application of resulting predictions. Finally, unique issues must be addressed for building predictive models for metal ion mixtures. The objective of this chapter is to explore these four activities: selecting the best explanatory variables, fitting the appropriate model, assessing predictive value of a model, and modeling metal ion mixtures.

8.2 SELECTION AND ADJUSTMENT OF INDEPENDENT VARIABLES

How does one determine which and how many explanatory variables to include in a model? There is no substitute for a sound understanding of the subject: the statistical methods described below should augment, not supplant, a sound understanding of the chemical, physical, and biological processes that translate metal exposure to bioactivity. How many candidate explanatory variables to consider might depend on the group of cations for which predictions are being made. A few candidate models with a single explanatory variable each might be the focus for some sets of metal ions, such as a group of divalent class (b) and intermediate metal ions. Models incorporating more than one explanatory variable might be thoughtfully explored for more diverse sets, such as ones composed of class (a), intermediate, and class (b) metal ions with differing charges. Several metal–ligand binding trends might be anticipated to influence metal ion bioactivity for such a heterogeneous set of cations.

8.3 QUANTITATIVE ION CHARACTERISTIC-ACTIVITY RELATIONSHIP (QICAR) MODELS

8.3.1 Intermetal Ion Trends

Models predicting bioactivity of single metal ions based on binding characteristics can be generated for particular subsets of metals of interest or for metal ions in general. No general rules are needed if intent alone determines the subset, that is, relative toxicity of trivalent lanthanides used by the computer display industry. Wolterbeck and Verburg (2001) recommend a periodic table corner calibration of elements if the intent is to produce a more general metal model for a particular effect to a particular biological entity. A series of metals with minimum and maximum values for relevant properties are selected from the periodic table "corners." For example, they select boron, cesium, germanium, lithium, selenium, and uranium for general model calibration. This theme of selecting metals ranging along relevant binding axes is also the natural choice for more narrowly focused studies such as that of the lanthanides just mentioned. In so doing, the distribution of metal ions along the final scale(s) used for the explanatory variable(s) should be as uniform as reasonable to facilitate later regression fitting.

Selecting the most useful model from among a group of candidates is not as simple as fitting a regression and picking the model with the highest coefficient of determination (r^2) value,

$$r^2 = \frac{b^2 \sum_{i=1}^{n} (X_i - \bar{X})^2}{\sum_{i=1}^{n} (Y_i - \bar{Y})^2} \tag{8.1}$$

where, for a simple model with one explanatory variable, b = slope, X_i and Y_i = the ith X and Y of n data pairs, and $\bar{X}$ and $\bar{Y}$ are averages of the X and Y observations. The coefficient of determination increases as additional explanatory parameters are added to a model, so it will not provide directly useful insight for identifying the model whose parameters contain the most information for making predictions per estimated explanatory variable.

One general scheme for assessing model adequacy and then selecting from among candidate models is provided here. The model adequacy assessment has three initial components: application of subject knowledge to identify candidate models, conventional regression methods, and residual analysis. Additional crucial components involved with gauging predictive adequacy will be discussed later. Approaches to selecting the best model from among candidates can take several forms, including the Minimum Akaike Information Criterion Estimation (MAICE) and Mallows's C_p approaches. Each of these approaches will be described and illustrated. During discussions, it is important that the statistical details not distract the reader from the central theme that subject knowledge should be the touchstone for decisions at all steps.

Modeling begins by satisfying the basic conditions for regression analysis. It is assumed at the onset that the explanatory variable values are independent of each other. Also, Type I regression techniques formally require no error in the explanatory variable(s), although this requirement is relaxed often to the general premise that the explanatory variable has an immaterial amount of error relative to that of the response variable. Another approach, such as functional regression, might be required if this premise was unjustified. Next, the regression residuals (the difference between the observed response variable value and its predicted value) are assumed to be normally distributed. Finally, the sample variance around the regression line is assumed to be independent of the magnitude of the explanatory variable(s), that is, homoscedasticity is assumed. The last two requirements can be satisfied in some cases by transforming one or more of the variables prior to regression fitting. A common instance in this chapter is the logarithmic transformation of the response variable, such as the log of EC_{50}. This transformation can resolve nonnormality of residuals and heteroscedasticity issues while also conforming to the general toxicological paradigm that response is more often related linearly to the logarithm of dose than to the arithmetic dose (Finney 1942, 1947).

A combination of univariate statistics and plots allow exploration of a candidate model relative to these last two requirements. Bacterial bioluminescence 15-minute EC_{50} data for 20 metal ions (McCloskey et al. 1996, Appendix 8.1) and the recently developed softness index (Kinraide 2009) can be applied to illustrate this approach. The following statistical analysis system (SAS) code implements analyses with normality plots and tests of regression residuals (Figure 8.1 top). It also plots predicted and observed data (Figure 8.1 middle) and regression residuals versus the explanatory variable, σ_{con} (Figure 8.1 bottom).

```
PROC GLM;
  MODEL TOTLEC = SOFTCON;
  OUTPUT OUT = LINEAR2 PREDICTED = PRED2 RESIDUAL = RES2;
  RUN;
PROC UNIVARIATE NORMAL PLOT;
  VAR RES2;
  RUN;
SYMBOL1 V = dot COLOR = black; SYMBOL2 V = star COLOR = black;
SYMBOL3 V = dot COLOR = black;
PROC GPLOT;
  PLOT TOTLEC*SOFTCON PRED2*SOFTCON/OVERLAY HAXIS = -1.5 to
  1.5 by 0.5;
  PLOT RES2*SOFTCON/VREF = 0 HAXIS = -1.5 to 1.5 by 0.5;
RUN;
```

The middle plot in Figure 8.1 shows the values predicted (asterisks) with the model shown in the upper right corner and the original observations (solid dots). The observations are distributed uniformly along the axis of the explanatory variable with no obvious gaps. This minimizes the chance of a few extreme observations having more influence on the model fitting than others. The coefficient of determination, r^2,

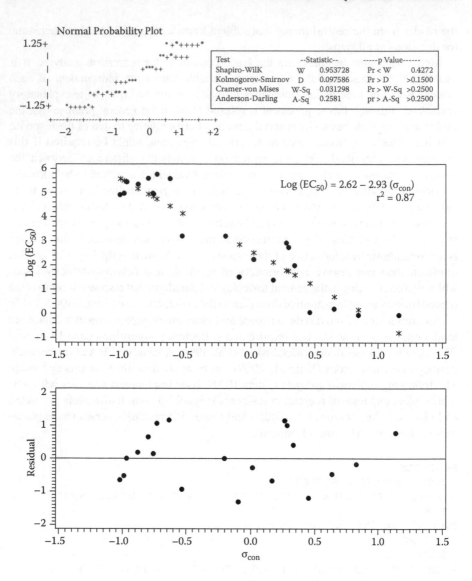

FIGURE 8.1 Bacterial bioluminescence inhibition 15-minute EC_{50} data for 20 metal ions (McCloskey et al. 1996) modeled with Kinraide's softness index (Kinraide 2009). The SAS code listed in Appendix 8.1 produced the normality plots and tests for the regression residuals (top) and also plots of predicted and observed data (middle) and plots of regression residuals versus the explanatory variable, σ_{con} (bottom). (Data from J.T. McCloskey, M.C. Newman, and S.B. Clark. "Predicting the Relative Toxicity of Metal Ions Using Ion Characteristics: Microtox® Bioluminescence Assay. *Environ. Toxicol. Chem.* 15 (1996):1730–1737; and T.B. Kinraide, Improved Scales for Metal Ion Softness and Toxicity. *Environ. Toxicol. Chem.* 28 (2009):525–533.)

of 0.87 indicates that the model accounts for approximately 87% of the variation in the response variable, with only 13% of the variability remaining unexplained. The residuals appear to be randomly distributed around the predicted values (bottom panel). The residuals show no pattern if plotted against the explanatory variable with a reference line indicating the state of perfect prediction at any point (residual = 0). There was a random distribution of residuals with no obvious trends left unexplained by the model along the range of values for the explanatory variable. This will be explored more closely later. Further, there was no trend in the amount of variation in the residuals along the abscissa, providing no reason to doubt the assumption of homoscedasticity. In a normal probability plot (top of Figure 8.1) with the distribution of points expected for a normal distribution (asterisks) and positions of the residuals (+ signs), the residuals conform to the assumed normal distribution. Four tests for normality of the regression residuals also provide no evidence of deviation from this assumption (top right of Figure 8.1).

The next task required in QSAR development is selection of the best model. Several approaches are used and range from statistically uninformed judgment of the researcher, MAICE, and Mallows's C_p method. Model selection guided solely by the researcher's informal judgment can produce the best model, but consistency of good judgment is enhanced by application of more formal methods. For example, model selection from among candidate models based only on the smallest χ^2 value will not always produce the best model,

$$\chi^2 = \sum_{i=1}^{n} \frac{\left(Y_i - \hat{Y}_i\right)^2}{\hat{Y}_i}. \tag{8.2}$$

Use of coefficients of determination or χ^2 values *for two models of similar complexity* might be adequate if combined with subject knowledge and the residual plots just described. As an example, such use would be appropriate if the above SAS code fitting the [ln EC_{50i}, σ_{con}] data were also modified to assess the alternative model generated with the more conventional softness index, σ_p. The r^2 for σ_{con} was 0.87 and that for σ_p was 0.81, lending support to Kinraide's (2009) argument that σ_{con} will perform better than the conventional σ_p during model generation. But the model with the most information per fitted explanatory variable cannot be identified with these otherwise useful goodness-of-fit statistics. The r^2 will increase with each addition of an explanatory variable, but the incremental improvement in fit might carry the cost of increased variance in parameter estimates (Hocking 1976). A straightforward change can be made to Equation (8.1) to generate an adjusted r^2 that incorporates the number of explanatory parameters and model degrees of freedom (Hocking 1976; Walker et al. 2003),

$$r^2_{Adjusted} = 1 - \frac{(n-1)\left(1-r^2\right)}{n-p} \tag{8.3}$$

where n = the number of observations, r^2 = coefficient of determination, and p = the number of estimated parameters.

Alternatively, criteria can be estimated for each model based on the principle of parsimony, that is, all else being equal, select the simplest model. The Akaike Information Criterion (AIC) is one of the most widely used information criterion that combines the model error sum of squares and the number of parameters in the model,

$$AIC = n \: ln\left(\frac{\sum_{i=1}^{n}\left(Y_i - \hat{Y}_i\right)^2}{n}\right) + 2p \tag{8.4}$$

The MAICE approach involves computing AIC values for each candidate model and then selecting the model with the lowest AIC value. The model with the lowest AIC value contains the most information per estimated parameter. Similar criteria, such as the Sawa or Schwarz Bayesian information criteria, can also be applied.

Another similar approach is described by Mallows (1973, 1995), Hocking (1976), Burman (1996), Der and Everitt (2006), and numerous others. Mallows's C_p statistic is estimated as the following,

$$C_p = \frac{\sum_{i=1}^{n}\left(Y_i - \hat{Y}_i\right)^2}{s^2} + 2p - n \tag{8.5}$$

where s^2 = the residual (error) mean square for the model including all available explanatory variables and an intercept. According to Hocking (1976, p. 18), "Cp is an estimate of the standardized total mean squared error of estimation for the current data, X." A set of explanatory variables are selected for possible inclusion and models built with increasing numbers of these variables incorporated. Mallows's Cp statistics are computed for the 2p − 1 possible models and tabulated beginning with models with the highest r^2 and ending with those with the lowest r^2. The best or most parsimonious one, two, three, and more variable models are identified as those with the lowest Cp statistics. Commonly, Mallows's Cp statistics for all 2p − 1 models are plotted against the number of estimated parameters in each model. The line for Cp = p is included in this plot because Cp values close to this line are those of the most parsimonious models.

The following SAS code implements AIC and Mallows's C_p statistic-based model selection for the bacterial bioluminescence inhibition by 20 metal ions (Appendix 8.1) using 6 candidate explanatory variables. The explanatory variables as defined in Newman et al. (1998) are the following: SOFTCON (Kinraide's σ_{con} softness index), ION (ionic index or the square of the ion charge divided by the Pauling ionic radius, Z^2/r), COVAL (covalence index, χ^2r, the square of the electronegativity times the radius), HYD (|log K_{OH}| where K_{OH} is the first hydrolysis constant of the metal ion), DELE (ΔE_0, the difference in electrochemical potential of the ion and its first stable reduced state), and ANIP (atomic number divided by ΔIP, the difference in ionization potentials for the ion oxidation numbers OX and OX − 1). Both of the PROC GLMSELECT procedures in the SAS code use forward model selection, that is, they begin with a model only fitting an intercept and then progressively add explanatory variables that most improve the model. Selection in this example is based on the AIC (CHOOSE = AIC) and C_p (SELECT = CP), although simply replacing these specifications at the end of the MODEL line with SELECTION = FORWARD(SELECT = ADJRSQ STOP = SL SLE = 0.2) would permit selection based on an $r^2_{adjusted}$ instead.

```
PROC GLMSELECT;
  MODEL TOTLEC = SOFTCON ION COVAL HYD DELE ANIP/SELECTION =
  FORWARD(SELECT = SL CHOOSE = AIC SLE = 0.2);
RUN;
PROC GLMSELECT;
  MODEL TOTLEC = SOFTCON ION COVAL HYD DELE ANIP/SELECTION =
  FORWARD(SELECT = CP);
RUN;
PROC REG;
  MODEL TOTLEC = SOFTCON ION COVAL HYD DELE ANIP/SELECTION = CP;
  PLOT CP.*NP./CMALLOWS = BLACK;
RUN;
```

The first PROC GLMSELECT in the code uses the AIC statistic to select the combinations of these six variables that produce the most parsimonious model. The model with the lowest AIC was that combining the softness, covalence, and ionic indices as shown in the inset table of Figure 8.2. Adding any of the other variables to the

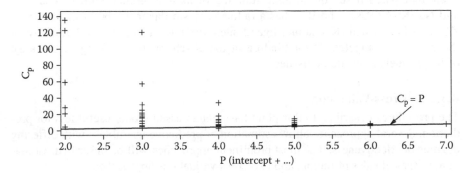

Model	AIC	Cp
Intercept	53.1	
plus σ_{con}	14.9	4.4
plus $\chi_m^2 r$	13.1	2.9
plus Z^2/r	11.3	1.9

FIGURE 8.2 Results from the SAS code that implements AIC and Mallows's C_p statistic-based model selection for the bacterial bioluminescence inhibition by 20 metal ions using 6 candidate explanatory variables (see text for details). The model with the lowest AIC was that including the softness, covalence, and ionic indices (inset table). Application of Mallows's C_p statistic also results in selection of the model containing the softness, covalence, and ionic indices.

models did not reduce the AIC any lower than 11.3. Note, however, that the selection used here involved significance levels for variables (SELECT = SL) and an associated p-value ≤ 0.2 was required (SLE = 0.2) for an explanatory variable to be considered in a model. The second PROC GLMSELECT in the code does the same except it selects models based on Mallows's C_p statistic. Again, the model containing the softness, covalence, and ionic indices was selected. This final model had an r^2 of 0.91 and $r^2_{adjusted}$ of 0.89.

The PROC REG that specifies forward variable selection with the C_p statistic produces a C_p versus p plot (Figure 8.2), and generates a table of models and associated r^2 and C_p values. The best 1, 2, and 3 explanatory variable models are highlighted in Table 8.1. Note that similar $r^2_{adjusted}$-p plots could also have been produced but break points for C_p-p plots tend to be clearer than for $r^2_{adjusted}$-p plots (Hocking 1976).

8.3.1.1 Nonmonotonic Models

It is important to mention at this point in discussions that, based on subject knowledge, some metal ion data sets should not be expected to conform to a monotonic trend. Model development should involve attention to such exceptions. The deviation of K^+ in the metal-valinomycin stability constant versus $AN(\Delta IP)^{-1}$ relationship discussed in Chapter 1 (Figure 1.2) is one important example. A similar example involves acute Ba^{2+} toxicity to the nematode, *Caenorhabditis elegans* (Tatara et al. 1998). This divalent cation is much more toxic than predicted by the general model constructed with 18 mono-, di- and trivalent cations. This could have been anticipated based on an understanding of its interference with K^+ channels and Na^+/K^+-ATPase. The Ba^{2+} has a radius very similar to K^+ but a much higher Z^2r^{-1}. Its bonds with K^+ channel ligand sites are much more stable than those of the K^+. It outcompetes K^+ for binding at the K^+-channels, resulting in blockage of K^+ channels in excitable tissues.

8.3.1.2 Cross-Validation

The procedures described to this point have not assessed model usefulness for prediction. Several approaches allow estimation of predictive usefulness with differing degrees of effectiveness. The most prominent approaches will be described: validation, statistical rules of thumb, and two cross-validations approaches.

The best approach, validation, generates a completely new data set and uses the model generated with the earlier data set to make predictions for these new data. The model is validated if new predictions are close to their corresponding observed bioactivities. Another approach might involve using the new data set to estimate new model parameters and subsequent comparison of those estimates to those generated earlier by fitting the first data set. Comparable estimates suggest good prediction for the first model. Understandably, all data are often pooled into a larger data set after successful validation to generate a final model.

Other approaches can generate acceptable estimates of usefulness if the above approach is not feasible. At the other extreme from the above validation approach is application of a statistical rule of thumb such as the γ_m criterion estimated from

TABLE 8.1
Results of Mallows's Cp Analysis

Number of Parameters	C(p)	r²	Variables in Model (plus an Intercept)
3	**1.9423**	**0.9083**	**SOFTCON ION COVAL**
3	2.3011	0.9060	SOFTCON COVAL HYD
3	2.4306	0.9051	SOFTCON COVAL DELE
2	**2.9164**	**0.8888**	**SOFTCON COVAL**
4	3.0258	0.9143	SOFTCON COVAL HYD DELE
2	3.3478	0.8859	SOFTCON DELE
4	3.3727	0.9121	SOFTCON ION COVAL DELE
4	3.6318	0.9104	SOFTCON ION COVAL ANIP
3	3.7654	0.8963	SOFTCON COVAL ANIP
4	3.7978	0.9093	SOFTCON COVAL HYD ANIP
4	3.9168	0.9085	SOFTCON ION COVAL HYD
4	4.3910	0.9054	SOFTCON COVAL DELE ANIP
1	**4.4219**	**0.8657**	**SOFTCON**
3	4.7552	0.8898	SOFTCON HYD DELE
2	4.8865	0.8758	SOFTCON HYD
3	4.9336	0.8886	SOFTCON DELE ANIP
5	5.0081	0.9145	SOFTCON ION COVAL HYD DELE
5	5.0134	0.9144	SOFTCON COVAL HYD DELE ANIP
3	5.3152	0.8861	SOFTCON ION DELE
5	5.3324	0.9123	SOFTCON ION COVAL DELE ANIP
4	5.3882	0.8988	SOFTCON ION HYD DELE
2	5.4163	0.8723	SOFTCON ION
5	5.5826	0.9107	SOFTCON ION COVAL HYD ANIP
4	5.6538	0.8971	ION COVAL HYD DELE
3	6.0443	0.8813	ION HYD DELE
3	6.0851	0.8811	COVAL HYD DELE
4	6.1357	0.8939	SOFTCON HYD DELE ANIP
2	6.3207	0.8664	SOFTCON ANIP
4	6.8304	0.8893	SOFTCON ION DELE ANIP
3	6.8336	0.8761	SOFTCON ION HYD
3	6.8852	0.8758	SOFTCON HYD ANIP
4	6.9399	0.8886	ION HYD DELE ANIP
5	6.9701	0.9016	SOFTCON ION HYD DELE ANIP
6	7.0000	0.9145	SOFTCON ION COVAL HYD DELE ANIP
3	7.4085	0.8724	SOFTCON ION ANIP

Note: Only the first 35 models of 2^p-1 or 63 possible models with 6 parameters (P) are tabulated.

F statistics (Draper and Smith 1988). The computed statistic is compared to some arbitrary threshold for model predictive usefulness. Alone, a statistically significant F statistic estimated for a regression model parameter provides only limited insight about how useful any predictions from a model might be; however, some F statistic value greater by a preestablished magnitude from an $F_{(df_m, df_r, 0.95)}$ has served as a tool for separating useful from nonuseful models. The general rule of thumb, (F statistic)/ $\left(F_{(df_m, df_r, 0.95)}\right) \geq 4$ to 5 is often applied for this purpose as detailed in Draper and Smith (1988).[*] For example, the estimated F statistic for the above model predicting bacterial bioluminescence inhibition based on the metal ions' softness (σ_{con}), ionic, and covalence indices was 52.83 and had an associated critical $F_{(3,16,0.95)}$ of 3.24. The resulting (F statistic)/($F_{(3,16,0.95)}$) = 52.83/3.24 = 16.31 is much greater than 5, suggesting that the model would be a useful one for prediction.

Two cross-validation methods provide a better approach than that just described but generally not as good as the validation method. The first involves splitting of the available data into two subsets and the second involves removal of one datum at a time from the data set prior to building models.

If large enough, a data set can be split into two subsets called the *training* and *validation* sets. This procedure simulates the validation technique by producing a data set not used to build the original model. The disadvantage of this approach is that all available data are not used to generate the model. As a general rule, the number of observations should be at least 6 to 10 times more than the number of explanatory variables in order to successfully apply this approach (Neter et al. 1990). Individual observations can be randomly split between the training and validation sets, but in some cases, a completely random assignment might not be the best approach. For example, it might be preferable to randomly pick observations from within regions along a gradient for some explanatory variable. This ensures that both the training and validation data sets will have observations representing all relevant regions along the gradient.

If the data set (n) is small, one observation can be removed at a time from the data set to produce a data set of size $n - 1$, a model is generated with the $n - 1$ observations, predictions made with the model for that one removed observation, and the difference between the observed value and predicted value calculated. The removed datum is then placed back into the data set and another datum removed and the above process repeated. This process is repeated for a data set to build n models. Each model has a different observation missing for which predictions are done. Analysis of the n differences between the observed and predicted values (prediction residuals) suggests how useful predictions will be from a model. Prediction residuals can be examined directly or some summary statistic might be generated from the prediction residuals. The following SAS code generates individual prediction residuals and also produces a summary statistic for the bacterial bioluminescence data set listed in Appendix 8.1. Figure 8.3 suggests good prediction (top panel) and no apparent trend in prediction residuals with predicted ln of the EC_{50} (bottom panel).

---

[*] The df_m = model degrees of freedom or the number of estimated parameters minus 1, df_r = the residuals degrees of freedom, and 0.95 = 1-α.

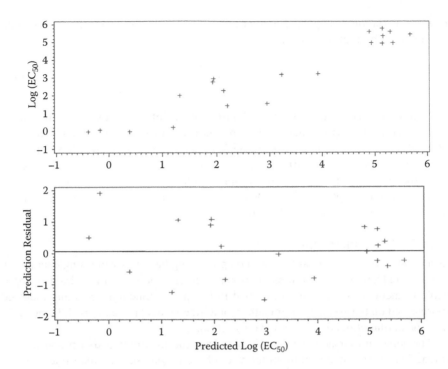

FIGURE 8.3 Good prediction is suggested in this plot of observed Ln (EC_{50}) versus the Ln (EC_{50}) predicted for each point that was omitted from the model (top panel). No trends in prediction quality over the range of predictions were evident in the bottom panel of prediction residuals versus the predicted ln of the EC_{50}.

```
PROC REG PRESS OUTEST = PSOFT;
  MODEL TOTLEC = SOFTCON/AIC BIC;
  OUTPUT OUT = LINEAR PREDICTED = PRED RESIDUAL = RES PRESS =
  PRES;
  RUN;
PROC PRINT;
  VAR TOTLEC PRED RES PRES;
RUN;
PROC PRINT DATA = PSOFT;
RUN;
```

The OUTPUT statement specifies that each prediction residual (PRESS = PRES) be listed in the output. The PROC REG specifies that the following prediction residual sum of squares statistic also be generated,

$$PRESS = \sum_{i=1}^{n} (y_i - \dot{y}_i)^2 . \qquad (8.6)$$

This PRESS statistic and the model total sum of squares (SS_T) can be combined to produce a statistic similar to the coefficient of determination (Equation [8.1]) except now the variation in expected predictions is quantified for a model that was built without the observation of interest.

$$r^2_{prediction} = 1 - \frac{PRESS}{SS_T} . \qquad (8.7)$$

Continuing with the bacterial bioluminescence inhibition example, the model including the softness, covalence, and ionic indices had a *PRESS* of 14.260 and a *SS_T* of 85.567. The $r^2_{prediction}$ is $1 - (14.269/85.567)$ or 0.83. This is slightly lower than the r^2 (0.91) or even the $r^2_{adjusted}$ (0.89) that were estimated earlier. Given the ions used to build the model are a representative sample of ions being modeled, the $r^2_{prediction}$ best reflects the amount of variation to be expected in bioactivity predictions for a metal ion not used to build the QICAR model.

8.3.1.3 Metal Interactions

Discussions to this point have focused on predicting bioactivity of a single metal ion in isolation from others; yet, metal ions are often present as mixtures. Models applicable to metal mixtures were described in Chapter 1 (Section 1.3.3) and included those based on the assumptions of either joint independent (Equations [1.3] and [1.4]) or joint similar (Equations [1.6] to [1.9]) action.

The joint similar action model is based on the assumption that probit models (bioactivity versus concentration) for a set of metal ions sharing the same mode of action (and toxicokinetics) will have a common slope (Equations [1.6] and [1.7]). So one simple metric gauging conformity to or deviation from the assumption of joint similar action is the absolute difference in the estimated slopes for two metal ions. To produce Figure 8.1 (bottom panel), maximum likelihood fitting of single metal ion concentration versus proportional inhibition of bacterial bioluminescence to a probit model was done using the SAS package procedure, PROC PROBIT. (Examples of such an application of PROC PROBIT are given for the combined influence of La^{3+} and Ce^{3+} concentration on bacterial bioluminescence in Appendix 8.2.) This was done for each metal separately and the absolute value of the difference in model slopes used to produce the figure. Note that a full factorial experimental design involving a matrix of two metal mixtures is not required. However, as the figure should also make clear, the more involved experimental design required for the approach based on independent action produced clearer metal interaction QICARs in this case.

An estimated interaction coefficient, ρ, was used in Chapter 1 to quantify potential interactions in binary mixtures of metal ions based on the independent joint action model (Figure 1.8, top panel). The associated data set was generated with a matrix of binary metal mixtures. The SAS code in Appendix 8.2 shows an example La^{3+} and Ce^{3+} data set with five La^{3+} concentrations (including 0) combined with five Ce^{3+} concentrations (including 0). The top row of data includes those for different concentrations of Ce^{3+} and no added La^{3+}. The leftmost column of data includes those for different concentrations of La^{3+} and no added Ce^{3+}. All other data reflect mixtures of the two metal ions at different concentrations. The first row and column of data can be used to generate the probit models for the bioactivity of each metal alone and then the slopes of the two models to be compared as described above under the assumption of similar mode of action. The confidence intervals for slopes of the La^{3+} (2.50 with a 95% confidence interval of 1.26 to 3.73) and Ce^{3+} (3.08 with a 95% confidence interval of 2.23 to 3.94) models suggest no obvious deviation from the assumption of similar action. All of the other data could be used to estimate the interaction coefficient ρ, under the assumption of independent action of the paired metal ions. The resulting interaction coefficient estimate of 1.50 (95% confidence interval: 0.91 to 2.10) provides little evidence for or against this assumption. (The interaction coefficient would be 1 if the metal ions had completely independent action.) Fortunately, the evidence was much clearer in the data sets used to generate Figure 1.8.

8.4   CONCLUSION

The general steps of QSAR production were described at the beginning of this chapter: (1) define the mode of action, (2) define the relevant toxicants sharing that mode of action, (3) define the variables with the most potential for quantifying differences among toxicants, and (4) generate a tool for selecting and relating the explanatory variable(s) to the effect of interest. Most attention was given in this chapter to the last step, although relevant aspects of the other steps were discussed. Approaches to selecting metal ions were identified, including spreading choices along the ranges of candidate explanatory variables, and in the case of a general metal ion QICAR, the periodic table corner calibration method. The coefficient of determination combined with regression residual plots was described as a useful but insufficient means of selecting the best model for making predictions. For candidate models with differing numbers of explanatory variables, MAICE and Mallows's C_p approaches were advocated for picking the model with the most information for making predictions per estimated parameter. Several techniques were described for quantifying how good predictions will be for metal ions not included during model fitting, including the γ_m criterion based on F statistics, model validation with a completely new data set, and cross-validation. The γ_m criterion based on F statistics is useful but involves an arbitrary threshold. Model validation with a completely new data set is ideal but might require more resources than are available to the researcher. Cross-validation involving splitting of a data set into training and validation data subsets simulates validation with new data but requires a relatively large data set. Often, the potential influence of the resulting data set sizes on model generation is assessed by conducting cross-validation twice. Data subset A is the training data and subset B the validation data during the first cross-validation, and then subset A becomes the validation data and subset B becomes the training data during a second cross-validation. How similar the results are for the two cross-validations suggests the influence of data splitting on the cross-validation process. The final cross-validation approach can be applied with smaller data sets. A set of n models are generated by omitting one observation from each model and then comparing predicted values for the omitted observation to the observed values for the observation. Prediction residuals or summary statistics are then generated. Summary statistics might include the PRESS (prediction sum of squares) or $r^2_{prediction}$.

In addition to the methods described above, those useful for quantifying metal ion mixture bioactivities were also discussed. The general models described in Chapter 1 are implemented with specific computer code to illustrate the two potential approaches. The first is based on similar joint action, assumes identical slopes for similar acting metal ions, and measures deviations from identical slopes in order to quantify departure from the assumption of similar action. This involves only probit models for the individual metal ions alone. The second approach, which is based on joint independent action, requires an experimental design in which different concentrations of each metal ion are in mixture with those of the second metal ion. An interaction coefficient, ρ, quantifies deviations from the assumption of independent action.

REFERENCES

Burman, P. 1996. Model fitting via testing. *Stat. Sinica* 6:589–601.

Der, G., and B.S. Everitt. 2006. *Statistical Analysis of Medical Data Using SAS*. Boca Raton, FL: Chapman and Hall/CRC.

Draper, N.R., and H. Smith. 1998. *Applied Regression Analysis, 3rd Edition*. New York: John Wiley & Sons, Inc.

Finney, D.J. 1942. The analysis of toxicity tests on mixtures of poisons. *Ann. Appl. Biol.* 29:82–94.

Finney, D.J. 1947. *Probit Analysis*. Cambridge: Cambridge University Press.

Hocking, R.R. 1976. The analysis and selection of variables in linear regression. *Biometrics* 32:1–49.

Kinraide, T.B. 2009. Improved scales for metal ion softness and toxicity. *Environ. Toxicol. Chem.* 28:525–533.

Mallows, C.L. 1973. Some comments on C_p. *Technometrics* 15:661–675.

Mallows, C.L. 1995. More comments on C_p. *Technometrics* 37:362–372.

McCloskey, J.T., M.C. Newman, and S.B. Clark. 1996. Predicting the relative toxicity of metal ions using ion characteristics: Microtox® bioluminescence assay. *Environ. Toxicol. Chem.* 15:1730–1737.

McKinney, J.D., A. Richard, C. Waller, M.C. Newman and F. Gerberick. 2000. The practice of structure activity relationships (SAR) in toxicology. *Toxicol. Sci.* 56:8–17.

Neter, J., W. Wasserman, and M.H. Kutner. 1990. *Applied Linear Statistical Models*. Homewood, IL: Richard D. Irwin, Inc.

Newman, M.C. 1995. *Quantitative Methods in Aquatic Ecotoxicology*. Boca Raton, FL: Lewis Publishers/CRC Press.

Newman, M.C., J.T. McCloskey, and C.P. Tatara. 1998. Using metal–ligand binding characteristics to predict metal toxicity: Quantitative ion character-activity relationships (QICARs). *Environ. Health Persp.* 106:1419–1425.

Tatara, C.P., M.C. Newman, J.T. McCloskey, and P.L. Williams. 1998. Use of ion characteristics to predict relative toxicity of mono-, di-, and trivalent metal ions. *Caenorhabditis elegans* LC50. *Aquat. Toxicol.* 42:255–269.

Walker, J.D., J.C. Dearden, T.W. Schultz, J. Jaworska, and M.H.I. Comber. 2003. QSARs for new practitioners. In *QSARs for Pollution Prevention, Toxicity Screening, Risk Assessment, and Web Applications*, ed. J.D. Walker, 3–18, Pensacola, FL: SETAC Press.

Wolterbeck, H.T. and T.G. Verburg. 2001. Predicting metal toxicity revisited: General properties vs. specific effects. *Sci. Total Environ.* 279:87–115.

APPENDIX 8.1:   SAS BACTERIAL BIOLUMINESCENCE EC$_{50}$ QICAR DATA SET

```
/* This code models ion characteristics against all metals */
/* from Microtox toxicity data - Ba included - McCloskey   */
/* et al. 1996 COVAL is covalence index which reflects the */
/* tendency to form covalent bonds with soft ligands such  */
/* as sulfur. It is the eletronegativity squared times the */
/* radius. ION is Z squared/radius. It is the polarizing   */
/* power or the energy of the metal ion during             */
/* electrostatic interaction with a ligand, SOFT is sigma  */
/* sub p or the softness index. It reflects the tendency   */
/* for the outer electron shell to deform (polarizability) */
/* and the ion's tendency to share electrons with ligands. */
/* ANIP reflects the ionization potential (IP) and inertia */
/* or size (AN). LGANZIP is the log of ANIP that           */
/* Kaiser (1980) preferred to ANIP. DELE this the absolute */
/* difference between the electrochemical potential of the */
/* ion and its first stable reduced state which is a       */
/* measure of the ion's ability to change electronic state.*/
/* HYD is the absolute value of the log of the first       */
/* hydrolysis constant which reflects the ion's affinity   */
/* to intermediate ligands such as oxygen donor atoms. The */
/* TOTLEC is the log (base 10) of the EC50 at 15 minutes   */
/* exposure and expressed as total dissolved metal, not the*/
/* free ion. This also includes Hg for which the chloride  */
/* species are not considered. Note that SOFTCON was added */
/* as a potentially better softness index. It is the       */
/* computed softness index Sigma Con Comp from             */
/* Kinraide 2009 Env. Tox. Chem. 28:525-533, Table 2.      */

OPTIONS PS = 58;
DATA REVIEW;
    INPUT METAL $ COVAL ION SOFT ANIP LGANIP DELE HYD TOTLEC
    SOFTCON @@;
    CARDS;
HG1+  4.08   3.92  0.065   9.62  0.983  0.91   3.40  -0.037   1.16
CA2+  1.00   4.00  0.181   3.47  0.540  2.76  12.7    4.976  -0.99
CD2+  2.71   4.21  0.081   6.07  0.783  0.40  10.1    1.424   0.17
CU2+  2.64   5.48  0.104   2.31  0.364  0.16   8.00   0.208   0.65
MG2+  1.24   5.56  0.167   1.62  0.210  2.38  11.6    4.941  -1.02
MN2+  1.99   4.82  0.125   3.05  0.484  1.03  10.6    3.196  -0.20
NI2+  2.52   5.80  0.126   2.66  0.425  0.23   9.90   2.753   0.29
PB2+  6.41   3.39  0.131  10.8   1.033  0.13   7.70   0.061   0.46
ZN2+  2.04   5.33  0.115   3.50  0.544  0.76   9.00   1.547  -0.09
CO2|  2.65   5.33  0.130   2.94  0.468  0.28   9.70   2.942   0.27
CR3+  1.71  14.5   0.107   1.66  0.220  0.41   4.00   2.265   0.02
FE3+  2.18  13.9   0.103   1.80  0.255  0.77   2.20   2.009   0.34
CS1+  1.06   0.59  0.218  14.1   1.149  2.92  14.9    5.606  -0.63
K1+   0.93   0.72  0.232   4.38  0.641  2.92  14.5    5.796  -0.73
SR2+  1.02   3.54  0.174   7.12  0.852  2.89  13.2    5.372  -0.88
```

```
LI1+   0.71   1.35   0.247   0.56  -.252   3.05   13.6    5.469   -0.97
NA1+   0.88   0.98   0.211   2.14   0.330  2.71   14.2    5.603   -0.80
BA2+   1.08   2.94   0.183  11.7    1.068  2.90   13.4    4.980   -0.76
LA3+   1.27   8.57   0.171   7.36   0.867  2.37    8.50   3.229   -0.53
AG1+   4.28   0.87   0.074   6.21   0.793  0.80   12.0   -0.034    0.84
;
```

APPENDIX 8.2:   SAS BACTERIAL BIOLUMINESCENCE—BINARY METAL MIXTURE EXAMPLE

```
/* LA AND CE ARE THE CONCENTRATIONS OF LANTHANUM AND CERUM. */
/* PAPB IS THE MEASURED BIOLUMINESCENCE AFTER 15 MINUTES OF
   EXPOSURE. */
DATA LACE;
INPUT LA CE PAPB @@;
PAPB = 100*((PAPB-.372)/(1-.372)); NORMZ = 100;
CARDS;
      0 0.372      0 3.125 .359      0 6.25 .385      0 12.50 .481      0 25.00 .662
  3.125 0.333 3.125 3.125 .370 3.125 6.25 .447 3.125 12.50 .533 3.125 25.00 .684
  6.250 0.368 6.250 3.125 .419 6.250 6.25 .449 6.250 12.50 .568 6.250 25.00 .747
  12.50 0.500 12.50 3.125 .548 12.50 6.25 .569 12.50 12.50 .629 12.50 25.00 .761
  25.00 0.667 25.00 3.125 .725 25.00 6.25 .708 25.00 12.50 .757 25.00 25.00 .821
;
DATA LAN; SET LACE; IF CE = 0; RUN;
PROC PROBIT LOG10 INVERSECL LACKFIT DATA = LAN; /* PROC PROBIT A */
   MODEL PAPB/NORMZ = LA/D = NORMAL ITPRINT;
   OUTPUT OUT = PLAN P = PPROB;
RUN;
DATA LAN2; SET PLAN; PAPB = PAPB/100; RUN;
DATA CEN; SET LACE; IF LA = 0; RUN;
PROC PROBIT LOG10 INVERSECL LACKFIT DATA = CEN; /* PROC PROBIT B */
   MODEL PAPB/NORMZ = CE/D = NORMAL ITPRINT;
   OUTPUT OUT = PCEN P = PPROB;
RUN;
DATA NEW;
   SET LACE;
   IF LA NE 0; IF CE NE 0;
   PAPB = PAPB/100;
   LCE = LOG10(CE); LLA = LOG10(LA);
   INTERLA = -3.5687+2.4985*LLA; /* Resulting Model from PROC
                                 PROBIT A */
   PLA = PROBNORM(INTERLA);
   INTERCE = -4.3893+3.0872*LCE; /* Resulting Model from PROC
                                 PROBIT B */
   PCE = PROBNORM(INTERCE);
   EXPECT = PLA+PCE;
   RUN;
PROC GLM DATA = NEW;
   MODEL PAPB = PLA PCE PLA*PCE/CLPARM;
RUN;
```

Index